The CD that accompanied this book has been replaced by a web site that can be found at the following address: http://www.phptr.com

Note that all references to the CD in the book now pertain to the web site.

Advanced Topics in LabWindows/CVI®

ISBN 0-13-089229-7
9 780130 892294
90000

NATIONAL INSTRUMENTS | VIRTUAL INSTRUMENTATION SERIES

Jeffrey Y. Beyon
- Hands-On Exercise Manual for LabVIEW Programming, Data Acquisition, and Analysis

Jeffrey Y. Beyon
- LabVIEW Programming, Data Acquisition, and Analysis

Mahesh L. Chugani, Abhay R. Samant, Michael Cerra
- LabVIEW Signal Processing

Rahman Jamal • Herbert Pichlik
- LabVIEW Applications and Solutions

Shahid F. Khalid
- Advanced Topics in LabWindows/CVI

Shahid F. Khalid
- LabWindows/CVI Programming for Beginners

Hall T. Martin • Meg L. Martin
- LabVIEW for Automotive, Telecommunications, Semiconductor, Biomedical, and Other Applications

Bruce Mihura
- LabVIEW for Data Acquisition

Jon B. Olansen • Eric Rosow
- Virtual Bio-Instrumentation: Biomedical, Clinical, and Healthcare Applications in LabVIEW

Barry Paton
- Sensors, Transducers, and LabVIEW

Jeffrey Travis
- LabVIEW for Everyone, second edition

Jeffrey Travis
- Internet Applications in LabVIEW

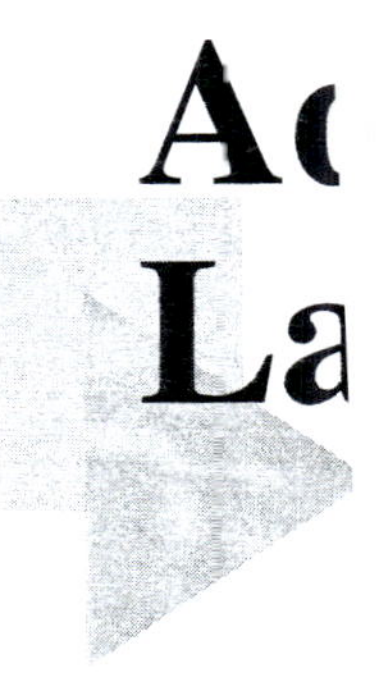

Prentice Hall PTR
Upper Saddle River, NJ 07458
www.phptr.com

Library of Congress Cataloging-in-Publication Data

Khalid, Shahid F.
Advanced topics in LabWindows/CVI / Shahid F. Khalid.
p. cm. -- (National Instruments virtual instrumentation series)
Includes bibliographical references and index.
ISBN 0-13-089229-7
1. Computer programming 2. LabVIEW. 3. Engineering instruments--Data processing.
I. Title. II. Series.

QA76.73.C15 K482 2001
006--dc21

2001036970

Editorial/Production Supervision: Joan L. McNamara
Acquisitions Editor: Bernard Goodwin
Editorial Assistant: Michelle Vincenti
Marketing Manager: Dan DePasquale
Manufacturing Manager: Alexis R. Heydt-Long
Cover Design: Nina Scuderi
Cover Design Direction: Jerry Votta
Series Design: Gail Cocker-Bogusz
Composition/Page Make-up: Ronnie K. Bucci
Project Coordinator: Anne R. Garcia

Prentice-Hall, Inc.
Upper Saddle River, NJ 07458

Prentice Hall books are widely used by corporations and government agencies for training, marketing, and resale.

The publisher offers discounts on this book when ordered in bulk quantities.
For more information, contact:
Corporate Sales Department
Prentice Hall PTR
One Lake Street
Upper Saddle River, NJ 07458
Phone: 800-382-3419; FAX: 201-236-7141
Email (Internet): corpsales@prenhall.com

Printed in the United States of America

10 9 8 7 6 5 4 3 2 1

ISBN 0-13-089229-7

Pearson Education LTD.
Pearson Education Australia PTY, Limited
Pearson Education Singapore, Pte. Ltd
Pearson Education North Asia Ltd
Pearson Education Canada, Ltd.
Pearson Educación de Mexico, S.A. de C.V.
Pearson Education—Japan
Pearson Education Malaysia, Pte. Ltd
Pearson Education, Upper Saddle River, New Jersey

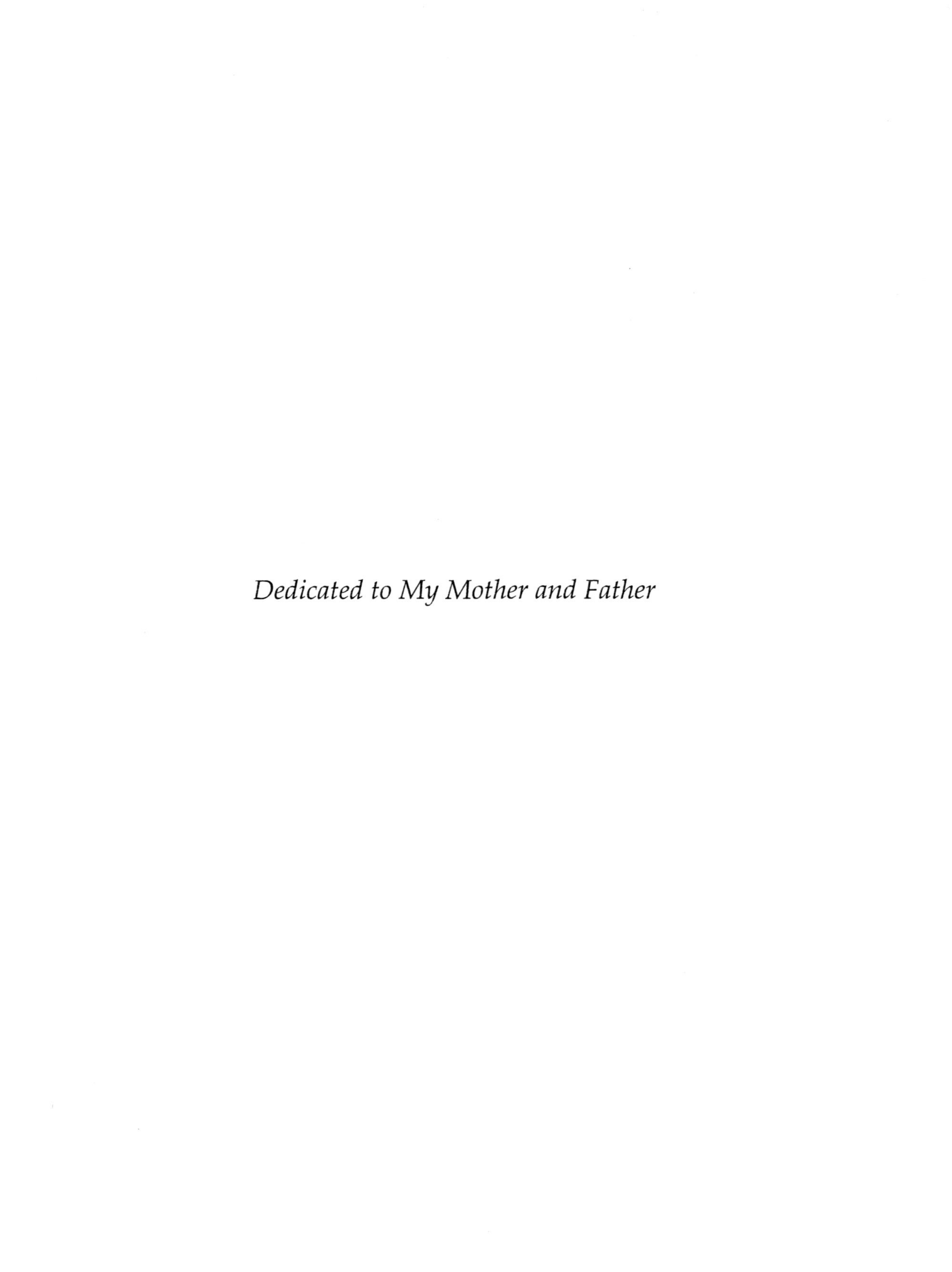

Dedicated to My Mother and Father

Contents

Contents vii
Illustrations xv
Tables xxi
Preface xxv
What Is *LabWindows/CVI*? xxv
Objectives of This Book. xxviii
What You Need to Run *CVI* xxx
Conventions Used in This Book xxxi
Acknowledgments xxxiii
Foreword xxxv

1

Programmatically Creating the Graphical User Interface 1

Introduction 2
Analyzing the Source Code 3
main Function 3
CreateGUI Function 5
RingCB Function 10
StartGUICB Function 10
SamplesControlCB Function 19
DoneGUICB Function 19
ClearGraphCB Function 21
PlotUniformCB Function 22
ExitGUICB Function 22
Summary 22
Library Function Prototypes and Definitions 24
DeleteGraphPlot Function 24
DiscardPanel Function 25
NewCtrl Function 26
NewPanel Function 26

PlotY Function 28
RefreshGraph Function 30
SetPanelAttribute Function 30

2

Plotting on Graph Controls 33

Graph Attributes and Cursors **34**
Manual Zooming and Panning **46**
Creating Graph Legends **46**
Plotting Geometric Patterns on Graph Control **50**
Summary **59**
Library Function Prototypes and Definitions **61**
CreateMetaFont Function 61
GetAxisScalingMode Function 62
GetGraphCursor Function 63
GetGraphCursorIndex Function 64
LGCreateLegendControl Function 64
LGInsertLegendItemForPlot Function 66
LGSetLegendCtrlAttribute Function 67
PlotArc Function 67
PlotLine Function 70
PlotOval Function 70
PlotRectangle Function 71
PlotText Function 72
SetAxisScalingMode Function 73

3

Using DataSocket 75

Introduction **76**
Communicating Using DataSocket **76**
DataSocket Data Files **77**
Creating a DataSocket Application **78**
Analyzing the Writer Code 80
Analyzing the Reader Code 90
DataSocket Applications **97**
Accessing the DataSocket Server **99**
DataSocket Server Manager Configurations **99**
Server Settings 100
Permissions Groups 101

Predefined Data Items ... 102
Summary ... **104**
Library Function Prototypes and Definitions ... **105**
DS_ControlLocalServer Function ... 105
DS_GetAttrValue Function ... 105
DS_GetDataType Function ... 108
DS_GetDataValue Function ... 109
DS_GetLastMessage Function ... 109
DS_GetLibraryErrorString Function ... 111
DS_GetStatus Function ... 112
DS_Open Function ... 113
DS_SetAttrValue Function ... 114
DS_SetDataValue Function ... 116
DS_Update Function ... 116
MakeDir Function ... 118
SetBreakOnLibraryErrors Function ... 118
SetDir Function ... 120

4

Table Control 121

Introduction ... **122**
Table Control Basics ... **123**
Table Control States ... 123
Moving Around in the Table Control ... 123
Resizing Rows and Columns ... 124
Using the System Clipboard ... 124
Table Control Events ... 125
Browsing the Table Control Dialog Windows ... **125**
Table Control Project ... **134**
Examining the Project Code ... **136**
Header and *main* Function ... 136
Load Data Function ... 136
Selecting Columns to Sort ... 142
Sort Ascending/Descending ... 143
Search Function ... 146
Highlighting/Pasting Rows to Clipboard ... 149
Summary ... **151**
Library Function Prototypes and Definitions ... **151**
ClipboardGetTableVals Function ... 151
ClipboardPutTableVals Function ... 151
DeleteTableRows Function ... 152

FillTableCellRange Function . . . 153
FileToArray Function . . . 153
GetActiveTableCell Function . . . 156
GetBitmapFromFile Function . . . 156
GetNumTableRows Function . . . 156
GetTableCellFromVal Function . . . 157
GetTableCellVal Function . . . 159
HideBuiltInCtrlMenuItem Function . . . 159
InsertTableColumns Function . . . 159
InsertTableRows Function . . . 159
MakePoint Function . . . 161
MakeRect Function . . . 163
NewCtrlMenuItem Function . . . 164
SetActiveTableCell Function . . . 166
SetTableCellAttribute Function . . . 166
SetTableCellVal Function . . . 166
SetTableCellRangeVals Function . . . 168
SetTableCellRangeAttribute Function . . . 169
SetTableColumnAttribute Function . . . 169
SetTableRowAttribute Function . . . 170
ShowBuiltInCtrlMenuItem Function . . . 170
SortTableCells Function . . . 172

5

VXI Communication Using VISA 175

Introduction . . . 176
Short History . . . 176
VXI Chassis, Modules, and Connectors . . . 177
Controlling the VXI System . . . 178
VXI Address Space and Configuration Registers . . . 181
VXI Device Classes . . . 186
Communicating with Message-Based Devices . . . 187
Resource Manager . . . 189
Basics of Programming with VISA . . . 191
VISA Project . . . 197
Header and *main* Function . . . 198
Finding System Resources . . . 200
Setting Up Communication with the Function Generator . . . 202
Configuring the Function Generator . . . 205
Summary . . . 209

Library Function Prototypes and Definitions 209
AssertSysReset Function 210
viAssertUtilSignal Function 210
viClose Function 211
viFindNext Function 212
viFindRsrc Function 212
viIn16 Function 212
viOpen Function 212
viOpenDefaultRM Function 215
viOut16 Function 215
viRead Function 215
viSetAttribute Function 216
viStatusDesc Function 217
viWrite Function 218

6

Data Acquisition 219

Introduction 220
Data Acquisition Board Architecture 222
Signal Conditioning 223
Analog Input/Output Parameters 225
Range, Gain, and Code Width 226
DAQ Designer Tool 227
Installing and Setting Up the DAQ Board 228
Using the DAQ Channel Wizard 238
Hardware Configurations 246
Using DAQ Library Functions 248
Analog Input 248
Analog Output 259
Digital Input/Output 261
Counter Fundamentals 263
Counter Applications 265
Event Counting and Timing 266
Pulse Generation 268
Pulse Measurement 272
Frequency Measurement 273
Summary 274

Library Function Prototypes and Definitions 275
AOClearWaveforms Function 275
AOUpdateChannel Function 275
AOUpdateChannels Function 276
AOGenerateWaveforms Function 276
ContinuousPulseGenConfig Function 278
CounterEventOrTimeConfig Function 278
CounterMeasureFrequency Function 281
CounterRead Function 281
CounterStart Function 281
CounterStop Function 281
DelayedPulseGenConfig Function 284
DIG_Line_Config Function 284
DIG_Prt_Config Function 284
GroupByChannel Function 286
nidaqAICreateTask Function 288
nidaqAIDestroyTask Function 288
nidaqAIRead Function 290
nidaqAIScanOp Function 290
nidaqAISinglePointOp Function 291
nidaqAISingleScanOp Function 292
nidaqAIStart Function 292
nidaqAIStop Function 293
nidaqGetErrorString Function 294
PulseWidthOrPeriodMeasConfig Function 294
ReadFromDigitalLine Function 296
ReadFromDigitalPort Function 297
WriteToDigitalLine Function 298
WriteToDigitalPort Function 299

7

Creating and Using Function Panels 301

Purpose of a Function Panel 302
Creating a Function Tree 304
Creating a Function Panel 308
Testing the Function Panel Functions 319
Function Panel Controls 321
Numeric Control 321
Slide Control 323
Binary Control 324

Ring Control ... 326
Global Variable ... 327
Message Control ... 328
Summary ... 330

8

Creating Instrument Drivers 331

Introduction ... 332
Creating an Instrument Driver ... 334
Generating Driver Files Review ... 347
Function Panel File ... 347
Initialize Functions ... 347
Configuration Functions ... 348
Measure Output Functions ... 349
Action/Status Functions ... 349
Utility Functions ... 349
Close Function ... 349
Source File ... 349
Include File ... 350
.sub File ... 350
Using the Attribute Editor ... 350
Attribute Editor Controls ... 353
Editing High-Level Instrument Driver Functions ... 356
Deleting High-Level Instrument Driver Functions ... 359
Adding High-Level Instrument Driver Functions ... 359
Creating Instrument Driver Documentation ... 360
Creating the Instrument Driver Text File ... 360
Creating the Instrument Driver Windows Help ... 362
Testing the Instrument Driver ... 362
Summary ... 363

9

OpenGL 365

Introduction ... 366
OpenGL Project ... 367
Source Code Analysis ... 370
Header and *main* Function ... 370
Load Data File ... 372

Setting OpenGL Attributes 373
Plotting Data 376
Creating a Color Map 381
Creating a Color Scale 382
Printing the OpenGL Plot 383
OpenGL Properties Panel 385
Summary 385
Library Function Prototypes and Definitions 386
OGLConvertCtrl Function 386
OGLDeletePlot Function 387
OGLDiscardCtrl Function 388
OGLGetCtrlAttribute Function 388
OGLGetErrorString Function 388
OGLPlot3DUniform Function 389
OGLPropertiesPopup Function 391
OGLRefreshGraph Function 391
OGLSetCtrlAttribute Function 392
OGLSetPlotAttribute Function 393
OGLSetPlotColorScheme Function 393
PlotIntensity Function 393
PrintPanel Function 397
SetWaitCursor Function 398

Bibliography 399
Index 401
The Author 423

Illustrations

Figure 1–1 `project1-1` Window List . . . 2
Figure 1–2 `Project1-1` GUI . . . 3
Figure 1–3 `Project1-1` Header and *main* Function . . . 4
Figure 1–4 *CreateGUI* Function Listing . . . 5
Figure 1–5 *SetPanelAttribute* Function Panel . . . 8
Figure 1–6 *RingCB* Function Listing . . . 11
Figure 1–7 **Select Function** Control Pull-Down List . . . 11
Figure 1–8 *StartGUICB* Function Listing . . . 12
Figure 1–9 *CreateDisplayDataPanel* Function Listing . . . 16
Figure 1–10 **SELECT DATA** Child Panel . . . 18
Figure 1–11 *SamplesControlCB* Function Listing . . . 19
Figure 1–12 *DoneGUICB* Function Listing . . . 20
Figure 1–13 **Sine Wave** Sample Run . . . 21
Figure 1–14 *ClearGraphCB* Function Listing . . . 22
Figure 1–15 *PlotUniformCB* Function Listing . . . 23
Figure 1–16 *ExitGUICB* Function Listing . . . 23

Figure 2–1 `project2-1` GUI . . . 34
Figure 2–2 **Edit Cursors** Dialog Box . . . 36
Figure 2–3 **Cross Hair Style:** Pull-Down Menu . . . 36
Figure 2–4 **Edit Graph** Dialog Box . . . 37
Figure 2–5 **Edit Axis Settings** Dialog Box . . . 38
Figure 2–6 `project2-1` Header and *main* Function Listing . . . 40
Figure 2–7 *PlotData* Function Listing . . . 41
Figure 2–8 *UpdateMarkerCB* Function Listing . . . 43
Figure 2–9 *DisplayCursorPosition* Function Listing . . . 43
Figure 2–10 *SelectZoomCB* Function Listing . . . 44
Figure 2–11 **Zoom Control Label/Value Pairs** . . . 45

Figure 2–12 **Legend Control** Function Panel . . . 47
Figure 2–13 `project2-2` GUI . . . 48
Figure 2–14 *AddLegendCB* Listing . . . 49
Figure 2–15 `project2-3` **Graph Plots** GUI . . . 51
Figure 2–16 **Select Plot** Ring Control . . . 51
Figure 2–17 **Select Color** Control . . . 51
Figure 2–18 Line Plotting and Text Sample . . . 52
Figure 2–19 **TEXT ENTRY** and **Font Attributes** GUI . . . 53
Figure 2–20 Existing Metafont Selections . . . 53
Figure 2–21 *main* Function Source Code Listing . . . 54
Figure 2–22 *UpdateMarkerCB* Source Code Listing . . . 55
Figure 2–23 **Arc Plot Angles** GUI . . . 57
Figure 2–24 *AnglesSelectedCB* Source Code Listing . . . 57
Figure 2–25 Sample of Various Plots . . . 58
Figure 2–26 **TEXT ENTRY** GUI . . . 59
Figure 2–27 *TextOKCB* Source Code Listing . . . 60

Figure 3–1 **WRITE DATA** GUI with **DataSocket Server** Launched . . . 79
Figure 3–2 **READ DATA** GUI . . . 80
Figure 3–3 `project3` Sample Run Showing Data Path . . . 81
Figure 3–4 `project3Send` Header and *main* Function . . . 82
Figure 3–5 `project3Send` *ConnectCB* Callback Function . . . 83
Figure 3–6 *UpdateDSCallback* Function . . . 85
Figure 3–7 *TimerCallback* Function . . . 86
Figure 3–8 *DisconnectCB* Function . . . 88
Figure 3–9 *WriteToFileCB* Function . . . 89
Figure 3–10 `project3Receive` Header and *main* Function . . . 90
Figure 3–11 *ConnectReadCB* Function . . . 92
Figure 3–12 *DSCallback* Function . . . 93
Figure 3–13 *ReadFileCB* Function . . . 96
Figure 3–14 *UpdateCB* Function . . . 97
Figure 3–15 DataSocket Server Path . . . 100
Figure 3–16 DataSocket Server . . . 100
Figure 3–17 **DataSocket Server Manager** Configuration . . . 101
Figure 3–18 Setting Up Permission Groups for **DefaultReaders** . . . 102
Figure 3–19 `MyGroup` Permission Group . . . 103
Figure 3–20 **NewItem** Creation Dialog Window . . . 103

Figure 4–1 Sample Table Control . . . 122
Figure 4–2 **Edit Table** Control . . . 126

Figure 4–3 **Table Mode:** Dialog Box 126
Figure 4–4 **Row** Controls 127
Figure 4–5 **Edit Row** Window 128
Figure 4–6 **Edit Default Cell Values (Row)** Window 129
Figure 4–7 **Column** Controls 131
Figure 4–8 **Edit Column** Window 131
Figure 4–9 **Edit Default Cell Values (Column)** Window 132
Figure 4–10 **Edit Default Cell Values** Window 133
Figure 4–11 **Size/Scroll Options** Window 134
Figure 4–12 `project4` Table Control GUI 135
Figure 4–13 `project4` Header and *main* Function 137
Figure 4–14 Pop-up Menu for Table Control 138
Figure 4–15 *LoadDataCB* Function Listing 139
Figure 4–16 **Select Sort Column** Ring Control 142
Figure 4–17 *SortColumnSelectCB* Function Listing 143
Figure 4–18 **Sort Results as:** Ring Control 143
Figure 4–19 *SortResultsCB* Function Listing 144
Figure 4–20 Sorting Using the Built-in Menu Table Control 145
Figure 4–21 *SearchCB* Function Listing 147
Figure 4–22 *HighlightRow* Function Listing 149
Figure 4–23 **SEARCHED DATA** GUI 150
Figure 4–24 Example Cell Range 164

Figure 5–1 13-Slot VXI Chassis with Plug-in Modules 178
Figure 5–2 VXI Module Sizes and Connectors 179
Figure 5–3 PC Using MXI to Control Two-VXIbus System 180
Figure 5–4 VXI-MXI Used for Connecting Multiple VXI Mainframes 180
Figure 5–5 VXI A16 Space Address Mapping 182
Figure 5–6 ID/Logical Address Register Bit Allocation 183
Figure 5–7 Device Type Register Bit Allocation 184
Figure 5–8 Status/Control Register Bit Allocation 185
Figure 5–9 A24/A32 Offset Register Bit Allocation 186
Figure 5–10 Arbitrary Waveform Generator GUI 198
Figure 5–11 `project5` Header and *main* Function Listing 199
Figure 5–12 *FindResourcesCB* Function Listing 201
Figure 5–13 Expression List in *viFindRsrc* Function 202
Figure 5–14 *SetupCommunication* Function Listing 203
Figure 5–15 Attributes in *viSetAttribute* Function Panel 204
Figure 5–16 *ConfigAWGCB* Function Listing 205
Figure 5–17 *ExitAWGCB* Function Listing 209

Figure 6–1 Plug-in DAQ Board Configuration 220
Figure 6–2 PCMCIA DAQ Board Configuration 221
Figure 6–3 Remote DAQ Board Configuration 221
Figure 6–4 Multifunction DAQ Board Block Diagram 222
Figure 6–5 Resolution Example Using 3- and 16-bit ADC 226
Figure 6–6 DAQ Designer Query Screen 228
Figure 6–7 NI-DAQ Installation 229
Figure 6–8 **Measurement & Automation Explorer (MAX)** Window 230
Figure 6–9 **MAX** Window with Software Installed 231
Figure 6–10 Devices and Interfaces Configuration: **System** Tab 231
Figure 6–11 Devices and Interfaces Configuration: **AI** Tab 232
Figure 6–12 Devices and Interfaces Configuration: **AO** Tab 233
Figure 6–13 Devices and Interfaces Configuration: **Accessory** Tab 234
Figure 6–14 Run Test Panel: **Analog Input** Window 234
Figure 6–15 Run Test Panel: **Analog Output** Window 235
Figure 6–16 Run Test Panel: **Counter I/O** Window 236
Figure 6–17 Run Test Panel: **Digital I/O** Window 236
Figure 6–18 **MAX** Window: `IVI Instruments` 237
Figure 6–19 DAQ Channel Wizard: **Create New** Window 238
Figure 6–20 DAQ Channel Wizard: **Create New Channel** Window 239
Figure 6–21 DAQ Channel Wizard: **Channel Name and Description** Window 240
Figure 6–22 DAQ Channel Wizard: Type of **Sensor** or **Measurement** Window 240
Figure 6–23 DAQ Channel Wizard: **Signal Scaling Choices** Window 241
Figure 6–24 DAQ Channel Wizard: **Create New Custom Scale** Window 242
Figure 6–25 DAQ Channel Wizard: **Scale Coefficient** Window 242
Figure 6–26 DAQ Channel Wizard: **Verify Scale** Window 243
Figure 6–27 DAQ Channel Wizard: Hardware Settings 244
Figure 6–28 **Data Neighborhood:** Created Channels 244
Figure 6–29 Channel Configuration Window 245
Figure 6–30 DAQ Channel Wizard: Test Panel 245
Figure 6–31 DAQ Channel Wizard: **Scales** Window 246
Figure 6–32 NI-DAQ Setup Path 247
Figure 6–33 Analog Input Example GUI for `project6-1` 249
Figure 6–34 Source Code Listing for `project6-1` 250
Figure 6–35 `project6-1` Sample Run 256
Figure 6–36 `project6-2` Sample Run 257
Figure 6–37 `project6-2` Code Segment 258
Figure 6–38 Typical Counter 264
Figure 6–39 Counter Signal Characteristics 265
Figure 6–40 *CounterEventOrTimeConfig* Code Fragment 266
Figure 6–41 Pulse Duty Cycle 269

Figure 6–42 Pulse Polarity . . . 269
Figure 6–43 *ContinuousPulseGenConfig* Code Fragment . . . 270
Figure 6–44 *DelayedPulseGenConfig* Code Fragment . . . 271

Figure 7–1 Sample Function Panel . . . 302
Figure 7–2 Sample **Select Function Panel** . . . 304
Figure 7–3 Blank **Function Tree Editor** Window . . . 305
Figure 7–4 **Create Instrument Node** Window . . . 306
Figure 7–5 `MyFunctions` Function Tree Window . . . 306
Figure 7–6 **Create Function Panel Window Node** Window . . . 307
Figure 7–7 Complete Function Tree Window . . . 308
Figure 7–8 Blank Function Panel . . . 309
Figure 7–9 **Create Input Control** Dialog Box . . . 310
Figure 7–10 Function Panel with Input Control Boxes . . . 311
Figure 7–11 **Create Output Control** Dialog Box . . . 312
Figure 7–12 **Create Return Value Control** Dialog Box . . . 313
Figure 7–13 *my_SolveQuadratic* Function Panel . . . 314
Figure 7–14 *my_Factor* Function Panel . . . 314
Figure 7–15 *MyFunctions* Help Window . . . 315
Figure 7–16 *MyFunctions* Class Help Window . . . 316
Figure 7–17 *MyFunctions* Header File . . . 317
Figure 7–18 *MyFunctions* Source Code File (continued) . . . 318
Figure 7–19 Testing *my_SolveQuadratic* Function Panel Using IW . . . 320
Figure 7–20 Testing *my_Factor* Function Panel Using IW . . . 321
Figure 7–21 Numeric Control . . . 322
Figure 7–22 **Create Numeric Control** Dialog Box . . . 323
Figure 7–23 **Edit Value Set** Dialog Box . . . 323
Figure 7–24 Slide Control . . . 324
Figure 7–25 **Edit Slide Control** Dialog Box . . . 324
Figure 7–26 Slide Control **Label/Value Pairs...** Box . . . 325
Figure 7–27 Binary Control . . . 325
Figure 7–28 **Create Binary Control** Dialog Box . . . 325
Figure 7–29 **Edit On/Off Settings** Dialog Box . . . 326
Figure 7–30 Ring Control . . . 327
Figure 7–31 **Create Ring Control** Dialog Box . . . 327
Figure 7–32 Global Variable Control . . . 328
Figure 7–33 **Edit Global Variable Control** Dialog Box . . . 328
Figure 7–34 Message Control . . . 329
Figure 7–35 **Create Message Control** Dialog Box . . . 329
Figure 7–36 **Change Input Control Type** Selection Box . . . 330

Figure 8–1 Instrument Driver Development Wizard: **Welcome Window** 335
Figure 8–2 Instrument Driver Development Wizard: **Select an Instrument Driver** 336
Figure 8–3 Instrument Driver Development Wizard: **General Information** Window 337
Figure 8–4 Instrument Driver Development Wizard: **General Command Strings** Window 338
Figure 8–5 Instrument Driver Development Wizard: **Standard Operations** Window. 339
Figure 8–6 Instrument Driver Development Wizard: **ID Query** Window. 340
Figure 8–7 Instrument Driver Development Wizard: **Reset Command** Window. 341
Figure 8–8 Instrument Driver Development Wizard: **Self-Test** Window. 341
Figure 8–9 Instrument Driver Development Wizard: **Error Query** Window 342
Figure 8–10 Instrument Driver Development Wizard: **Revision** Window 343
Figure 8–11 Instrument Driver Development Wizard: **Test** Window. 344
Figure 8–12 Instrument Driver Development Wizard: **Test Results** Window. 344
Figure 8–13 Instrument Driver Development Wizard: **Finish** Window 345
Figure 8–14 Power Supply Function Panel Tree 348
Figure 8–15 **Select Attribute Constant** Dialog Window 351
Figure 8–16 Instrument Driver Development Wizard: **Edit Driver Attributes** Window. 352
Figure 8–17 Attribute Editor: **Edit Attribute** Window. 353
Figure 8–18 Attribute Editor: **Edit Group** Window 354
Figure 8–19 Function Tree Editor Context Menu 356
Figure 8–20 *E3631A_error_message* Driver Created Source Code 357
Figure 8–21 *E3631A_error_message* Modified Source Code 358
Figure 8–22 **Generate Documentation** Dialog Window 361
Figure 8–23 **Generate Windows Help** Window 362

Figure 9–1 `project9` GUI 368
Figure 9–2 `project9` Data Plot 369
Figure 9–3 `project9` Using **OpenGL Controls** Help Panel 369
Figure 9–4 `project9` Header and *main* Function Listing 370
Figure 9–5 *LoadPlotCB* Source Listing 373
Figure 9–6 *SetOGLAttributes* Source Listing 374
Figure 9–7 *PlotOGLData* Source Listing 377
Figure 9–8 *CreateColorMap* Source Listing. 381
Figure 9–9 *CreateColorScale* Source Listing 382
Figure 9–10 *PrintCB* Source Listing. 384
Figure 9–11 OpenGL Control **Properties** Panel 386

Tables

Table 1–1 *DeleteGraphPlot* Function 24
Table 1–2 *DiscardPanel* Function 25
Table 1–3 *NewCtrl* Function 26
Table 1–4 *NewPanel* Function 27
Table 1–5 *PlotY* Function 28
Table 1–6 *RefreshGraph* Function 30
Table 1–7 *SetPanelAttribute* Function 31

Table 2–1 *CreateMetaFont* Function 61
Table 2–2 *GetAxisScalingMode* Function 62
Table 2–3 *GetGraphCursor* Function 63
Table 2–4 *GetGraphCursorIndex* Function 64
Table 2–5 *LGCreateLegendControl* Function 65
Table 2–6 *LGInsertLegendItemForPlot* Function 66
Table 2–7 *LGSetLegendCtrlAttribute* Function 67
Table 2–8 Legend Control Attributes 68
Table 2–9 *PlotArc* Function 69
Table 2–10 *PlotLine* Function 70
Table 2–11 *PlotOval* Function 71
Table 2–12 *PlotRectangle* Function 72
Table 2–13 *PlotText* Function 73
Table 2–14 *SetAxisScalingMode* Function 74

Table 3–1 *DS_ControlLocalServer* Function 106
Table 3–2 *DS_GetAttrValue* Function 106
Table 3–3 *DS_GetDataType* Function 108
Table 3–3 *DS_GetDataType* Function 109
Table 3–4 *DS_GetDataValue* Function 110

Table 3–5 *DS_GetLastMessage* Function 111
Table 3–6 *DS_GetLibraryErrorString* Function 112
Table 3–7 *DS_GetStatus* Function 112
Table 3–8 *DS_Open* Function 113
Table 3–9 *DS_SetAttrValue* Function 115
Table 3–10 *DS_SetDataValue* Function 117
Table 3–11 *DS_Update* Function 118
Table 3–12 *MakeDir* Function 119
Table 3–13 *SetBreakOnLibraryErrors* Function 119
Table 3–14 *SetDir* Function 120

Table 4–1 *ClipboardGetTableVals* Function 152
Table 4–2 *ClipboardPutTableVals* Function 152
Table 4–3 *DeleteTableRows* Function 153
Table 4–4 *FillTableCellRange* Function 154
Table 4–5 *FileToArray* Function 154
Table 4–6 *GetActiveTableCell* Function 156
Table 4–7 *GetBitmapFromFile* Function 157
Table 4–8 *GetNumTableRows* Function 157
Table 4–9 *GetTableCellFromVal* Function 158
Table 4–10 *GetTableCellVal* Function 160
Table 4–11 *HideBuiltInCtrlMenuItem* Function 160
Table 4–12 *InsertTableColumns* Function 161
Table 4–13 *InsertTableRows* Function 162
Table 4–14 *MakePoint* Function 162
Table 4–15 *MakeRect* Function 163
Table 4–16 *NewCtrlMenuItem* Function 165
Table 4–17 *SetActiveTableCell* Function 166
Table 4–18 *SetTableCellAttribute* Function 167
Table 4–19 *SetTableCellVal* Function 167
Table 4–20 *SetTableCellRangeVals* Function 168
Table 4–21 *SetTableCellRangeAttribute* Function 169
Table 4–22 *SetTableColumnAttribute* Function 170
Table 4–23 *SetTableRowAttribute* Function 171
Table 4–24 *ShowBuiltInCtrlMenuItem* Function 171
Table 4–25 *SortTableCells* Function 172

Table 5–1 Address Space Bits Decoding 184
Table 5–2 Device Class Bits Decoding (ID/Logical Address Register) 186
Table 5–3 Instrument Descriptor Syntax 193
Table 5–4 *AssertSysReset* Function 210

Table 5–5 *viAssertUtilSignal* Function 211
Table 5–6 *viClose* Function 211
Table 5–7 *viFindNext* Function 213
Table 5–8 *viFindRsrc* Function 213
Table 5–9 *viIn16* Function 214
Table 5–10 *viOpen* Function 214
Table 5–11 *viOpenDefaultRM* Function 216
Table 5–12 *viOut16* Function 216
Table 5–13 *viRead* Function 217
Table 5–14 *viSetAttribute* Function 217
Table 5–15 *viStatusDesc* Function 218
Table 5–16 *viWrite* Function 218

Table 6–1 *AOClearWaveforms* Function 275
Table 6–2 *AOUpdateChannel* Function 276
Table 6–3 *AOUpdateChannels* Function 277
Table 6–4 *AOGenerateWaveforms* Function 277
Table 6–5 *ContinuousPulseGenConfig* Function 279
Table 6–6 *CounterEventOrTimeConfig* Function 279
Table 6–7 *CounterMeasureFrequency* Function 282
Table 6–8 *CounterRead* Function 283
Table 6–9 *CounterStart* Function 283
Table 6–10 *CounterStop* Function 283
Table 6–11 *DelayedPulseGenConfig* Function 285
Table 6–12 *DIG_Line_Config* Function 286
Table 6–13 *DIG_Prt_Config* Function 287
Table 6–14 *GroupByChannel* Function 288
Table 6–15 *nidaqAICreateTask* Function 289
Table 6–16 *nidaqAIDestroyTask* Function 289
Table 6–17 *nidaqAIRead* Function 290
Table 6–18 *nidaqAIScanOp* Function 291
Table 6–19 *nidaqAISinglePointOp* Function 292
Table 6–20 *nidaqAISingleScanOp* Function 293
Table 6–21 *nidaqAIStart* Function 293
Table 6–22 *nidaqAIStop* Function 294
Table 6–23 *nidaqGetErrorString* Function 295
Table 6–24 *PulseWidthOrPeriodMeasConfig* Function 295
Table 6–25 *ReadFromDigitalLine* Function 296
Table 6–26 *ReadFromDigitalPort* Function 297
Table 6–27 *WriteToDigitalLine* Function 298
Table 6–28 *WriteToDigitalPort* Function 299

Table 9–1 *OGLConvertCtrl* Function 387
Table 9–2 *OGLDeletePlot* Function. 387
Table 9–3 *OGLDiscardCtrl* Function 388
Table 9–4 *OGLGetCtrlAttribute* Function. 389
Table 9–5 *OGLGetErrorString* Function. 389
Table 9–6 *OGLPlot3DUniform* Function 390
Table 9–7 *OGLPropertiesPopup* Function 391
Table 9–8 *OGLRefreshGraph* Function 392
Table 9–9 *OGLSetCtrlAttribute* Function. 392
Table 9–10 *OGLSetPlotAttribute* Function 394
Table 9–11 *OGLSetPlotColorScheme* Function 394
Table 9–12 *PlotIntensity* Function. 395
Table 9–13 *PrintPanel* Function 397
Table 9–14 *SetWaitCursor* Function. 398

Preface

What Is *LabWindows/CVI*?

LabWindows/CVI is a fully integrated interactive development environment with easy-to-use development tools that allow you quickly to create, configure, and display measurements on an easy-to-create graphical user interface (GUI). Engineers and scientists use *LabWindows/CVI* to create virtual instruments to acquire, analyze, and display data. *Virtual instrumentation* refers to the combination of hardware and software elements that provide you complete flexibility of designing and controlling the elements of stand-alone or embedded instruments from your computer system.

LabWindows/CVI was developed by National Instruments Corporation and uses all the features of the American National Standards Institute (ANSI) C programming language to develop your source code and create your applications expeditiously, using the power of its built-in tools and libraries. The *LabWindows/CVI* environment provides for a drag-and-drop editor for building GUIs, generating code automatically from the user interface to create the "skeleton" of an application source code, and includes a multitude of runtime libraries for instrumentation control and analysis. Optionally, when you obtain the data acquisition hardware from National Instruments, you also get the NI-DAQ (data acquisition) libraries to communicate and perform the necessary data acquisition functions. *LabWindows/CVI* has built-in libraries to communicate with hardware interfaces such as: General Purpose Interface Bus (GPIB), VME (Versa-Modular Eurocard) eXtensions for Instrumentations (VXI), PCI (Peripheral Component Interconnect) eXtensions for Instrumentation (PXI), serial interface communications, and data acquisition, to name a few. For simplicity, *LabWindows/CVI* is referred to throughout as *CVI.*

CVI is included as part of National Instruments' *Measurement Studio* package, which enables you to build virtual instrumentation applications in your favorite programming environments: Microsoft Visual C++, Visual Basic, or *CVI.* Measurement Studio adds measurement tools to the Microsoft visual development programs that are necessary for you to create measurement-ready development environments. In this book we do not discuss the other products included in the Measurement Studio but focus exclusively on *CVI.*

The measurement tools for Microsoft Visual C++ include a Measurement Studio AppWizard that extends the Microsoft Foundation Class (MFC) AppWizard to augment the application template with measurement and automation capabilities. By using the Measurement Studio AppWizard you can choose between creating a new measurement application based on MFC or importing an existing *CVI* application. Additionally, the tools consist of a set of class libraries that integrate completely with the Visual C++ environment, including the Class Wizard and IntelliSense.

The Measurement Studio tools for Microsoft Visual Basic provide ActiveX controls for visualization, data acquisition, and analysis. Using the visualization controls in Measurement Studio, you can configure real-time two- and three-dimensional graphs, knobs, meters, gauges, dials, tanks, thermometers, binary switches, and LEDs (light-emitting diodes) to create the front panels of instruments. Using the Measurement Studio ActiveX data acquisition controls allows you to perform analog, digital, and timing I/O operations on National Instruments Data Acquisition (DAQ) boards. In Measurement Studio, the instrument control components display property pages to configure communications with your instrument via GPIB, Virtual Instrumentation Software Architecture (VISA), or serial interfaces. You can also share live measurement data between applications via the Internet using DataSocket.

National Instruments developed the DataSocket technology (included with *CVI* versions 5.5 and later) to simplify the broadcast, exchange, and control of measurement data between a measurement system and an unlimited number of subscribing applications, or Web pages on the Internet. Using DataSocket, you can distribute live data across the Internet to Java applets, *LabVIEW*, and Visual Basic applications using the standard URL (uniform resource locator) address format.

CVI provides library functions to perform time/frequency analysis, curve fitting, digital filters, integration and differentiation functions, statistical functions, linear equation solutions, signal processing, and many more, that comprise the Advanced Analysis Library.

Using *CVI*'s Utility Library functions, you can easily create a multithread-safe application that gives you complete control for creating, debugging, and deploying multithreading applications.

You can create distributed computing applications using the power of ActiveX, Transmission Control Protocol/Internet Protocol (TCP/IP), and Dynamic Data Exchange (DDE) libraries included with *CVI*. Among other things, ActiveX-enabled applications on one machine can control an ActiveX application on another machine across the network.

CVI allows you to create Interchangeable Virtual Instruments (IVI) instrument drivers for various instrument classes using the Instrument Driver Development Wizard. *CVI* also contains a User Interface (UI) Localization Utility to translate and display the text on the GUI in the selected language of your choice.

CVI interfaces with configuration management systems such as Microsoft Visual Source Safe, Perforce, Continuous, and ClearCase, allowing you to manage individual and team projects easier. This is useful, as it protects code from inadvertently being changed when shared across various development teams, keeping track of code changes, archiving, and controlling the software versions released.

Microsoft Word Reports and NI-Reports instrument drivers, included with *CVI*, allow you to save and print professional-looking text reports. Using these instrument drivers, you can set margins, add headers and footers, and set font style and size programmatically to create a report format to your liking.

CVI supports all OpenGL functionalities. The OpenGL instrument driver can be used to display and view data using different perspectives and fields of view in two or three dimensions.

The *CVI* environment supports open software architecture, enabling you to reuse existing programs within its environment. If you are programming in C using your preferred environment, *CVI* complements your present efforts and streamlines your future development. You can incorporate standard ANSI C source code, object files, and dynamic link libraries (DLLs) within *CVI*. You have the flexibility to use the projects created in *CVI* in any of the following external C/C++ compilers with which you may be more familiar: Microsoft Visual C++, Borland C++, C++ Builder, Symantec C, or WATCOM C/C++. Note that starting with *CVI* version 6.0, the Symantec and Watcom compilers will not be supported.

Objectives of This Book

The success of my first book, *LabWindows/CVI Programming for Beginners,* was responsible for the creation of this book. After I had written the "beginner's" book, I was approached by many who wanted to learn more about the advanced features of *CVI* that were not included in the first book. I found that there was a need in the industry and academia to introduce to the beginner and the novice alike advanced topics of *CVI.* There remain many topics in *CVI* that cannot possibly be included in this book due to the vastness of *CVI*'s programming environment. There are topics in *CVI* that are so extensive that one can write a complete book on each topic. In this book an attempt is made to cover what I feel are some of the more useful features of *CVI* that would be beneficial to a wide range of audiences.

As the name of this book suggests, *Advanced Topics in LabWindows/CVI* explains some of the advanced topics in *CVI.* You should be familiar with the basics before you use this book. The basic concepts that were introduced in *LabWindows/CVI Programming for Beginners* are not covered here. You should either refer to the beginner's book or to the manuals (online or otherwise) that are available with *CVI.* You must be familiar with creating a graphical user interface and know how to add the various control objects, incorporate callback functions to the objects, and know how to compile, build, debug, and run a complete *CVI* project.

Projects with source code, user interface resource files, header files, and instrument drivers are included on the CD-ROM distributed with the book. These projects are ready for you to build and run. They have been thoroughly tested and executed numerous times to expose any possible bugs in the projects.

Each chapter contains an overview giving the intent of the chapter. Where applicable, *CVI* project(s) is/are created, explaining the chapter's features by means of example(s). Seeing an example gives you a better understanding of how to use that feature in your application. The purpose and use of the library functions encountered in the project's source code are explained. The function prototypes, with a brief explanation of the function arguments, are given in alphabetical order for your convenience in the section *Library Function Prototypes and Definitions* at the back of each chapter (where applicable). To obtain complete details on these library functions, it is recommended that you refer to On-line Help. At the end of each chapter, a summary section outlines the important features introduced in the chapter.

Note that if you are using a *CVI* version prior to 5.5, you will not be able to benefit fully from Chapters 3 and 4 since you will not be able to run the projects included in these chapters. These two topics were included in *CVI* starting with version 5.5. You can, however, read the remaining chapters, as they are not based on the concepts explained in these chapters. Each chapter is organized as a separate topic and does not require knowledge of preceding chapters. The exception to this rule is *Chapter 7, Creating and Using Function Panels,* which must be understood before reading *Chapter 8, Creating Instrument Drivers,* since function panels are used in creating instrument drivers.

Here is a short description of the chapters:

Chapter 1, "Programmatically Creating the Graphical User Interface," shows you how to create the graphical user interface entirely using the *CVI* library functions. You can create a panel and manipulate controls on the panel during run-time, change their attributes, and hide/display the control(s) as needed. These features enhance your application by making the GUI more versatile and providing you further flexibility.

Chapter 2, "Plotting on Graph Controls," discusses the many features of graph control. You are shown the graph control features of using the cursor controls, zooming and panning, and creating graph legends using the toolbox. This chapter shows you how to draw geometric objects such as lines, arcs, rectangles, and ovals on the graph control and to label these objects. You are shown how to add a header to your graph control using different font attributes.

Chapter 3, "Using DataSocket," explains the fundamentals of DataSocket, a National Instruments' proprietary technology. DataSocket transfers data between applications on the local computer, computers on the network, or to remote computers via the Internet. The various features of configuring the DataSocket Server Manager are also explained.

Chapter 4, "Table Control," shows you the features and capabilities of the table control to display, sort, select, and manipulate data in the cells that comprise the rows and columns of the table control, similar to a spreadsheet.

Chapter 5, "VXI Communication Using VISA," introduces you to the basic features of a VXI system and to establish communication with the VXI devices. You are introduced to the VXI hardware and to the address space mapping. The different VXI device classes and their configuration registers are explained. The functions performed by the VXI Resource Manager and the VISA Default Resource Manager are shown. You are shown how the VISA functions are used to set up communication and "talk" to devices.

Chapter 6, "Data Acquisition," explains the basic hardware and software features of data acquisition. After an overview of the data acquisition hardware, you are shown how to select the appropriate DAQ board using the DAQ Designer Tool and to install and set up the DAQ board. The DAQ Channel Wizard, used to configure and verify operations of DAQ channels, is explained, and use of the DAQ library functions for analog input/output, digital input/output, and counter applications is shown by means of code fragments.

Chapter 7, "Creating and Using Function Panels," steps you through the process of creating a function tree, including classes, and adding functions to the function tree. You are shown how to create a function panel using the different types of function panel controls and adding help text to the function panel and the controls. You will learn how to test the function panel using the Interactive Execution Window before including it in your code.

Chapter 8, "Creating Instrument Drivers," introduces you to the basic features of creating instrument drivers using Interchangeable Virtual Instruments. The Instrument Driver Development Wizard for creating an instrument driver is explained. Use of the Attribute Editor to add, delete, or modify the instrument driver attributes is shown. You are also shown how to add, edit, and delete the high-level instrument driver functions and modify the source code file appropriately and create documentation for the instrument driver.

Chapter 9, "OpenGL," introduces you to creating OpenGL applications using *CVI*'s OpenGL instrument driver. The basic concepts and terms of OpenGL are explained as applicable to the project. Using the project created in this chapter, you will learn to control, display, and manipulate the plotted data in three dimensions on the OpenGL control.

What You Need to Run *CVI*

CVI versions prior to 5.5 use the Windows 95/98/NT/3.1 operating system on personal computers (PCs) and Sun Solaris operating systems on Sun SPARC stations. *CVI* versions 5.5 and above run on PCs using Windows 2000/NT/Me/98/95 and no longer support UNIX operating systems or Windows 3.1. The discussion and examples in this book are limited to the Windows environment. (In this book, Windows refers to Windows 2000/NT/Me/98/95.)

To install and run *CVI* on Windows requires using at least a Pentium 133-MHz processor (Pentium 266 MHz or higher recommended), a minimum of 70 MB of free hard disk space for full installation, and at least 32 MB of RAM (64 MB recommended). To install versions prior to *CVI* 5.5 on a SPARC station, you will need 12 MB of free disk space, a minimum of 32 MB of disk swap space, and 23 MB of main memory.

To get really effective results in creating and displaying GUIs, it is recommended that you have at least a 17-inch Super VGA monitor and a high-resolution video card which supports a resolution of at least 800 by 600. The GUIs included on the CD-ROM with this book are designed using a resolution of 1024 by 768 pixels on the monitor. The GUIs have been tested to work on 17-inch through 21-inch monitors. However, if your display is different, you may have to resize the panel(s) in the User Interface Editor.

As in any Windows programming task, use of a Microsoft-compatible mouse or trackball is a great convenience and is recommended for use with *CVI*. In this book, only mouse interactions are used to access objects or to select menu commands. Keyboard commands are only used to show alternative capabilities in controlling and running projects.

It is assumed that you are familiar with the Windows environment on the PC or the operating environment of Sun workstations.

Conventions Used in This Book

For compatibility purposes I am using the same conventions in this book as in the *LabWindows/CVI Programming for Beginners*. These conventions are similar to the conventions used by National Instruments in their manuals. The following conventions are used in this book.

<>	The names of keyboard keys are enclosed between angle brackets (e.g., <Ctrl> means select the Control key, <F1> means select Function 1 key on the keyboard).
-	A hyphen between two or more keys within angle brackets means to press these keys simultaneously: for example, <Ctrl-Alt>, meaning hold down the Control key and press the Alternate key at the same time.

bold	Bold text denotes the names of menus, menu commands, arguments, dialog buttons, or user-created items.
>>	This symbol directs the user to follow nested menu commands, or dialog boxes, one after the other (e.g., **Library>>User Interface>> Panels>> Default-Panel** tells you to select **User Interface**, select **Panels,** and finally, **DefaultPanel** from the **Library** menu.
bold italic	Bold italic text denotes a caution, warning, or other kind of important message.
italic	Italicized text denotes program functions, emphasis, a cross-reference, or text for which you supply the appropriate word or value.
`monospace`	Monospace text is used to enter exact values from the keyboard, or any literal examples of programming, file names and extensions, and for folder (directory) names, programs, or functions.
`monospace bold`	Bold monospace text is used for messages that the computer prints to the screen automatically.

At various places in the text, I have referenced *CVI* manuals for you to use to obtain more information on certain topics. This information may refer either to paper manual(s) or to On-line Help. You have to determine where it is located depending on the *CVI* version you are using since many of the paper manual(s) of earlier *CVI* versions are available as On-line Help or merged into other manuals.

The user interface controls, menus, menu items, context menus, and various options and features to perform certain tasks may vary from that shown in the book, as they depend on the *CVI* version you are using. The menus in this book are based on *CVI* version 5.5. You should have little problem in relating these to different version(s) of *CVI.*

The *CVI* library functions introduced in this book are explained at the end of the chapter in which they are first introduced. However, the library functions that were introduced in the book *LabWindows/CVI Programming for Beginners* are not repeated in this book. The reader should consult this book or the manual(s) to learn more about these functions.

All the projects in this book are included on the CD-ROM accompanying this book. These projects are included in their respective chapter folder names under the `AdvProjects` folder. For example, the projects explained in Chapter 2 are included in the `Chapter2_Proj` folder located under the `AdvProjects` folder. To use these projects, it is best to copy these folders from the CD-ROM on to your hard disk and remove the read-only (**R**) and archive (**A**) attributes from all the files. You can change the file attributes by first selecting the file(s) and right-clicking with the mouse button. Select **Properties** from the pop-up menu that is displayed and remove the check marks from the **Read-only** and **Archive** boxes. These projects are created using *CVI* version 5.5 and are compatible with higher versions of *CVI*. Some of the projects may not run with *CVI* versions released prior to version 5.5.

Acknowledgments

To write a book requires the help of many people without whose help the book would have been deficient in many aspects. There are people who helped edit the book for correctness, completeness, and linguistic style. Others supplied me with the appropriate literature and necessary hardware to enable me to write the chapters and run and test the programs. I would like to acknowledge the following people, without whose help this book would not have been possible.

I would like to thank Ravi Marawar, Academic Program Manager at National Instruments, who introduced me to a group of *CVI* experts at National Instruments who reviewed the book and made extremely positive suggestions in improving this book. Ravi Marawar also supplied me with interface cards, test equipment, and course manuals that enabled me to write the various chapters.

Luis Gomes, LabWindows/CVI Group Manager at National Instruments, deserves special appreciation and thanks for doing such a thorough job of reviewing all the chapters and projects included in this book. He sacrificed many weekends and numerous hours working days and nights that were crucial in improving the quality of the book.

I greatly appreciate the willingness of Jeff Laney, Product Marketing Manager at National Instruments, to write a very informative foreword to the book.

Some of the engineers at National Instruments were assigned the task of reviewing the specific chapters that fall into their *CVI* specialty. I am thankful to Madras Mohanasundaram, Ludek Pekarek, and Chris Matthews for performing such a thorough review and making constructive suggestions, corrections, and improvements that enhanced the book greatly.

I am grateful to John Koontz, National Instrument's District Sales Manager of Southern California, for supplying me with the Data Acquisition hardware that enabled me to write Chapter 6. I am thankful to James Fithian, President, The Spacemark Corporation, for letting me use test equipment that enabled me to test the code included with this book.

My special thanks to the people who reviewed and made constructive suggestions on various parts of the book. I appreciate Yaakov Ben-Ami's guidance in designing the book and for his help in reviewing earlier drafts.

Special thanks go to my family for their understanding of why I could not spend time with them during the multitude of hours that I had to work on the book.

I specially acknowledge the efforts of Anne R. Garcia and Joan L. McNamara at the Prentice Hall PTR Production Department for the countless hours they spent bringing this book to its present form, with many editorial improvements.

Finally, thanks to the staff at Prentice Hall, whose help during various phases of the book is highly appreciated. I would like especially to thank Bernard Goodwin, the publisher, and his assistant, Michelle Vincenti, for making this book happen.

Foreword

Over the years National Instruments has focused on providing engineers with the tools they need to define their own solutions. By providing tools that greatly increase their productivity, National Instruments has become a leader in computer-based test and measurement systems. Starting with LabWindows for DOS in 1988, there was the idea of integrating commercially available PC hardware and software to build a complete system that was never obsolete. This is because the software powered the system, so the system could be changed to do something else, scaled to add more features, or updated to leverage new technological advancements.

When the computer evolved to provide a graphical interface for its operating systems, this validated LabWindows ideas about what an application should be. The GUI technology was adopted and LabWindows/CVI was created. LabWindows/CVI remained flexible in its approach by leveraging the standard development language of ANSI C, but provided libraries of functions that eased the development of measurement systems. By basing the development environment on C, LabWindows/CVI became the premier open standard for instrument driver creation and use. Almost every instrument available has a driver for LabWindows/CVI, and both the VXI plug and play and IVI driver standards started in the LabWindows/CVI environment.

As computer hardware and software advanced over the years, so did LabWindows/CVI. Many users have thanked us for making it easy to migrate their applications from DOS through Windows 2000, and users have also enjoyed the thought that the developers have put into making advanced features available to them in the same way as all other features. The interactive function libraries are the way that LabWindows/CVI simplifies development and remains the standard way in which all features are presented to the developer. The LabWindows/CVI development team has strived to add new features, such as three-dimensional capabilities, ActiveX automation, and simplified networking, while maintaining backward compatibility and ease of use. We also listen to our users when they ask us to add features that they find important, such as simplified multithreaded libraries and a complete multithreaded environment. National Instruments has also pushed LabWindows/CVI to become a broad development solution so that many engineers can use it to solve their problems. It has been used in applications from simple laboratory automation, to automobile parts testing, to telecommunications signal analysis.

With all the thought and work that was put into LabWindows/CVI, and its popularity among many engineers, a community of developers was soon created to help others become familiar with and be productive using LabWindows/CVI. These forums grew quickly inside some companies to the point where LabWindows/CVI was adopted as the standard for development of internal systems. Next, a worldwide forum was formed over the Internet with the LabWindows/CVI listserv, info-lw@listserv.tamu.edu. This allowed advanced users to help beginners all over the world. National Instruments has acknowledged LabWindows/CVI popularity and provides its own forum in the form of a developer network called Developer Zone, ni.com/zone.

Shahid's second book on advanced techniques is another form of the developer community. It is a way for developers to pass on their knowledge to those who have not found all the advanced features inside the LabWindows/CVI development environment. This book follows his beginner's book by covering most of the advanced features that have been added to LabWindows/CVI over the 13 years of development. This book provides substantial examples to help readers learn the material through use and understanding.

We hope that upon completing this book you will have all the knowledge and tools necessary with LabWindows/CVI to create revolutionary test and measurement systems for many years to come.

Jeff Laney
Product Marketing Manager
National Instruments

1

PROGRAMMATICALLY CREATING THE GRAPHICAL USER INTERFACE

Chapter Highlights

- Introduction
- Analyzing the Source Code
- Summary
- Library Function Prototypes and Definitions

Creating the graphical user interface (GUI) programmatically gives you the flexibility to add and replace controls during run-time, change their attributes, make the control invisible when not required, and display the control again when needed. Such features enhance your application by making your GUI more versatile.

In this chapter you will not be using the **User Interface Editor** but create all the controls on the GUI using the *CVI* library functions through the **Source Editor**. Creating a GUI programmatically has its advantages and disadvantages, which are mentioned in this chapter. Only the requirements of your project will decide which is a more favorable approach for you to use. A balanced approach would be to create most of the GUI using the **User Interface Editor** and perform control manipulations programmatically.

Introduction

The process of creating a graphical user interface (GUI) is usually accomplished through use of the *CVI* **User Interface Editor**. In this section you will learn how to create a GUI programmatically using the **Source Editor**. Creating a GUI using the **Source Editor** gives you the capability of changing the controls and their attributes dynamically at run-time. Sometimes you would like to create most of the GUI using the **User Interface Editor** and programmatically modify the controls during execution.

Here you will be shown how to programmatically create the panel and add controls to the panel using the *CVI* library functions. The panel and control attributes that you configure using the **Edit** dialog box will be demonstrated using the *CVI* library functions. You will also be shown how to hide a control and display another control in its place on the panel. These features add flexibility to the creation and manipulation of controls on your GUI.

Load `project1-1.prj` from the `AdvProject\Chapter1_Proj` folder located on the CD that came with this book. For best results, copy the project to your local hard disk and load the project from there. The **Project** window list shown in Figure 1–1 will be displayed. Notice that in this **Project** window list there is no user interface file(s) or associated header file(s). There is only one file in the project: `project1-1.c`. You cannot, therefore, use the *CodeBuilder* to create the "skeleton code."

The code for this project has to be entered manually in the **Source Editor**. The GUI that you will create from the source code is shown in Figure 1–2.

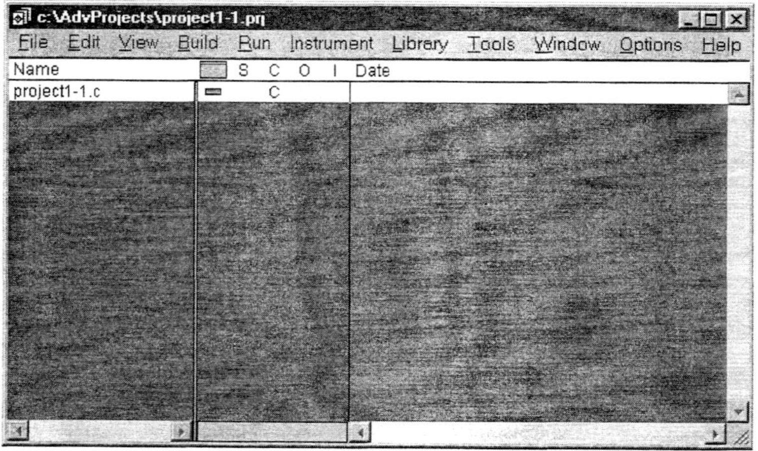

Figure 1–1
`project1-1` Window List

Note that functions listed in the **Select Function** control in Figure 1–2 are part of the *CVI* Advanced Analysis Library package that uses the header file `analysis.h`. The Advanced Analysis Library is part of the full *CVI* package. Unless you have installed this library package, these functions will not run. You can, however, substitute your own functions at the appropriate places in the code to execute this program if you desire. Please note that this program is not a demonstration of using functions, but demonstrates the concepts in creating and manipulating controls using the source code.

Analyzing the Source Code

In this section the code to create the GUI for this project is explained.

main Function

The program header files, function prototypes, global variables, and the *main* function listing are shown in Figure 1–3. The function prototype for the

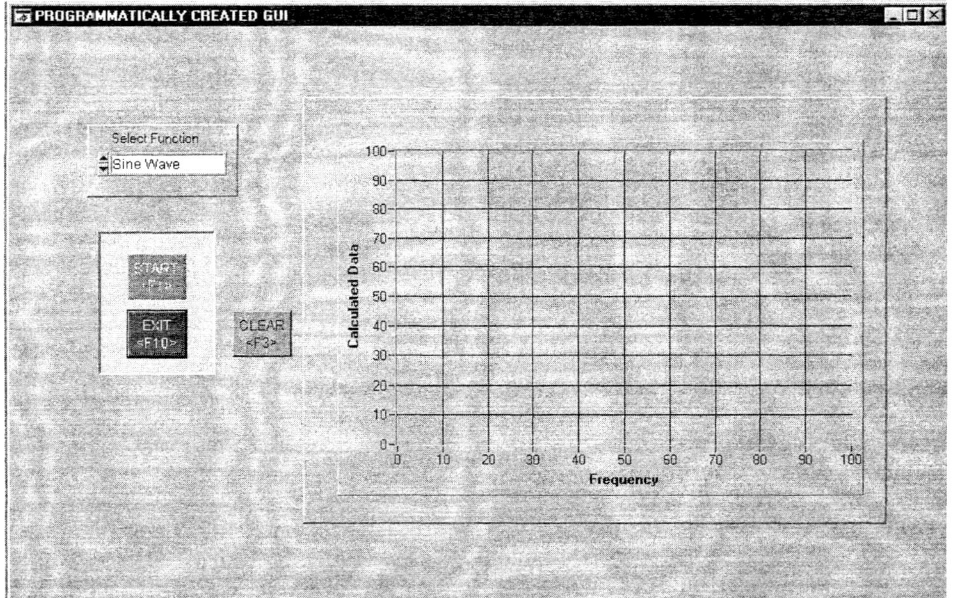

Figure 1–2
`Project1-1` GUI

user-created files and the callback functions are shown at lines 9–34. At line 54 the user-defined function *CreateGUI* is called to create the GUI shown in Figure 1–2. The GUI created by *CreateGUI* function is displayed by the library function *DisplayPanel* at line 55 and the panel discarded by the library function *DiscardPanel* when the panel is no longer needed.

```
1   #include <formatio.h>
2   #include <analysis.h>
3   #include <utility.h>
4   #include <userint.h>
5   #include <ansi_c.h>
6
7   /*****************Function Prototypes ************************/
8
9   void CreateGUI(void);         //Create the GUI
10  void CreateDisplayDataPanel(void); //Create Data panel
11
12  //Invoked when any item on Ring control is selected
13  int CVICALLBACK RingCB (int panel, int control, int event,
14                          void *callbackData, int eventData1, int eventData2);
15  //Invoked by "Start" command button
16  int CVICALLBACK StartGUICB (int panel, int control, int event,
17                          void *callbackData, int eventData1, int eventData2);
18  //Invoked by "Stop" command button
19  int CVICALLBACK StopGUICB (int panel, int control, int event,
20                          void *callbackData, int eventData1, int eventData2);
21  //Invoked by "Exit" command button
22  int CVICALLBACK ExitGUICB (int panel, int control, int event,
23                          void *callbackData, int eventData1, int eventData2);
24  //Invoked by "CLEAR" command button
25  int CVICALLBACK ClearGraphCB (int panel, int control, int event,
26                          void *callbackData, int eventData1, int eventData2);
27  //Plot Uniform random numbers
28  int CVICALLBACK PlotUniformCB (int panel, int control, int event,
29                          void *callbackData, int eventData1, int eventData2);
30  //Done entering data
31  int CVICALLBACK DoneGUICB (int panel, int control, int event,
32                          void *callbackData, int eventData1, int eventData2);
33  //Change Samples Callback
34  int CVICALLBACK SamplesControlCB(int panel, int control, int event,
35                          void *callbackData, int eventData1, int eventData2);
36
37  /************************************************************/
38  static int GUI_Handle, Child_Handle;
39
40  //User Interface control IDs
41  int GUI_START_ID, GRAPH_ID, TEXT_BOX_ID, CLEAR_PLOT_ID, PLOT_UNI_ID,
42  AMPLITUDE_ID, PHASE_ID, CYCLES_ID, SAMPLES_ID,TEXT_MSG, SELECTED_ITEM,
43  TABLE_MSG, RING_BOX;
44
```

Figure 1–3
Project1-1 Header and *main* Function *(continued)*

```
double amp, Y_axis[101], phase;
int seed, samples, PlotHandle, cycles, status;

//main
int main (int argc, char *argv[])
{
  if (InitCVIRTE (0, argv, 0) == 0)   //Used for DLLs or external compilers
          return -1;

  CreateGUI();   //Create the GUI
  DisplayPanel (GUI_Handle);

  RunUserInterface();
  DiscardPanel(GUI_Handle);

return 0;
} //main
```

Figure 1–3
`Project1-1` Header and *main* Function *(continued)*

CreateGUI Function

The code for the *CreateGUI* function is listed in Figure 1–4.

```
void CreateGUI(void)
{

int DEC_BOX_CMDS,DEC_BOX_GRAPH, DEC_BOX_FUNC,TEXT_MSG_FUNC,
GUI_EXIT_ID;

        //Create a new panel
        GUI_Handle = NewPanel (0, "PROGRAMMATICALLY CREATED GUI",
                               VAL_AUTO_CENTER, VAL_AUTO_CENTER, 500, 800);

        //Make the Panel movable
        SetPanelAttribute (GUI_Handle, ATTR_MOVABLE, 1);
        //Create Recessed decoration box for Commands
        DEC_BOX_CMDS = NewCtrl (GUI_Handle, CTRL_RECESSED_BOX, "", 180, 75);

        //Set dimensions of Decoration box
        SetCtrlAttribute (GUI_Handle, DEC_BOX_CMDS, ATTR_LEFT, 75);
        SetCtrlAttribute (GUI_Handle, DEC_BOX_CMDS, ATTR_HEIGHT, 125);
        SetCtrlAttribute (GUI_Handle, DEC_BOX_CMDS, ATTR_WIDTH, 100);
        //Set frame color of Decoration box
        SetCtrlAttribute (GUI_Handle, DEC_BOX_CMDS, ATTR_FRAME_COLOR,
                                                            VAL_OFFWHITE);

/************ Create a "START" square command button ***************/
        GUI_START_ID = NewCtrl (GUI_Handle, CTRL_SQUARE_COMMAND_BUTTON,
                                                      "START\n<F1>", 200, 100);

```

Figure 1–4
CreateGUI Function Listing *(continued)*

```
        //Set the width of the command button

        SetCtrlAttribute (GUI_Handle, GUI_START_ID, ATTR_WIDTH, 50);
        //Create a callback function for the start button
        InstallCtrlCallback (GUI_Handle, GUI_START_ID,  StartGUICB, 0);
        //Set the command button color
        SetCtrlAttribute (GUI_Handle, GUI_START_ID, ATTR_CMD_BUTTON_COLOR,
                                                        VAL_DK_GREEN);
        //Set the label color to WHITE
        SetCtrlAttribute (GUI_Handle, GUI_START_ID, ATTR_LABEL_COLOR,
                                                        VAL_WHITE);
        //Set Fonts to "Meta Fonts"
        SetCtrlAttribute (GUI_Handle, GUI_START_ID, ATTR_LABEL_FONT,
                                                        VAL_APP_META_FONT);
        //Assign "F1" as the shortcut key to this button
        SetCtrlAttribute (GUI_Handle, GUI_START_ID, ATTR_SHORTCUT_KEY,
                                                        VAL_F1_VKEY);

        //Disable the START command button
        SetCtrlAttribute (GUI_Handle, GUI_START_ID, ATTR_DIMMED, 1);

/********** Create a "EXIT" square command button ******************/
        GUI_EXIT_ID = NewCtrl (GUI_Handle, CTRL_SQUARE_COMMAND_BUTTON,
                                   "EXIT\n<F10>", 250, 100);
        //Set the width of the command button
        SetCtrlAttribute (GUI_Handle, GUI_EXIT_ID, ATTR_WIDTH, 50);

        //Create a callback function for the EXIT button
        InstallCtrlCallback (GUI_Handle, GUI_EXIT_ID,  ExitGUICB, 0);
        //Set the command button color
        SetCtrlAttribute (GUI_Handle, GUI_EXIT_ID, ATTR_CMD_BUTTON_COLOR, VAL_RED);

        //Set the label color to yellow
        SetCtrlAttribute (GUI_Handle, GUI_EXIT_ID, ATTR_LABEL_COLOR, VAL_YELLOW);

        //Set Fonts to "Meta Fonts"
        SetCtrlAttribute (GUI_Handle, GUI_EXIT_ID, ATTR_LABEL_FONT,
                                                        VAL_APP_META_FONT);
        //Assign "F10" as shortcut to this button
        SetCtrlAttribute (GUI_Handle, GUI_EXIT_ID, ATTR_SHORTCUT_KEY,
                                                        VAL_F10_VKEY);

/********** Create Decoration box for Functions ************/
        DEC_BOX_FUNC = NewCtrl (GUI_Handle, CTRL_RAISED_BOX, 0, 85, 75);

        //Set dimensions of Decoration box
        SetCtrlAttribute (GUI_Handle, DEC_BOX_FUNC, ATTR_LEFT, 65);
        SetCtrlAttribute (GUI_Handle, DEC_BOX_FUNC, ATTR_HEIGHT, 65);
        SetCtrlAttribute (GUI_Handle, DEC_BOX_FUNC, ATTR_WIDTH, 130);
        //Add Label to Function box and center it over the ring control
        TEXT_MSG_FUNC = NewCtrl (GUI_Handle, CTRL_TEXT_MSG, "Select Function", 90, 85);

/********* Create Ring Box ****************/
         //Label for the control could be inserted within quotes in NewCtrl function
        RING_BOX = NewCtrl (GUI_Handle, CTRL_RING, "", 110, 75);
        SetCtrlAttribute (GUI_Handle, RING_BOX, ATTR_WIDTH, 110);

```

Figure 1–4
CreateGUI Function Listing *(continued)*

```
        //Create a callback function for the Ring button
        InstallCtrlCallback (GUI_Handle, RING_BOX, RingCB, 0);
        //Add the list items
        InsertListItem (GUI_Handle, RING_BOX, 0, "Sine Wave",0);
        InsertListItem (GUI_Handle, RING_BOX, 1, "Uniform Random Numbers",1);
        InsertListItem (GUI_Handle, RING_BOX, 2, "White Noise",2);

        //Set the Fonts style for the ring control items
        SetCtrlAttribute (GUI_Handle, RING_BOX, ATTR_TEXT_COLOR, VAL_DK_BLUE);
        SetCtrlAttribute (GUI_Handle, RING_BOX, ATTR_TEXT_FONT,
                                                    VAL_APP_META_FONT);

/************ Create Raised decoration box for Graph ************/
        DEC_BOX_GRAPH = NewCtrl (GUI_Handle, CTRL_RAISED_BOX, "", 60, 250);

        //Set dimensions of Decoration box
        SetCtrlAttribute (GUI_Handle, DEC_BOX_GRAPH, ATTR_HEIGHT, 375);
        SetCtrlAttribute (GUI_Handle, DEC_BOX_GRAPH, ATTR_WIDTH, 500);
        //Set frame color of Decoration box
        SetCtrlAttribute (GUI_Handle, DEC_BOX_GRAPH, ATTR_FRAME_COLOR,
                                                    VAL_LT_GRAY);
/************ Create Graph control *****************/
        GRAPH_ID = NewCtrl (GUI_Handle, CTRL_GRAPH, "", 100, 280);
        SetCtrlAttribute (GUI_Handle, GRAPH_ID, ATTR_HEIGHT, 310);
        SetCtrlAttribute (GUI_Handle, GRAPH_ID, ATTR_WIDTH, 450);

        //Create Graph Labels
        SetCtrlAttribute (GUI_Handle, GRAPH_ID, ATTR_XNAME,"Frequency");
        SetCtrlAttribute (GUI_Handle, GRAPH_ID, ATTR_YNAME,"Calculated Data");
        SetCtrlAttribute (GUI_Handle, GRAPH_ID, ATTR_XYNAME_BOLD, 1);

        SetCtrlAttribute (GUI_Handle, GRAPH_ID, ATTR_XYNAME_COLOR, VAL_BLACK);

/************ Create a "CLEAR" square command button *************/
        CLEAR_PLOT_ID = NewCtrl (GUI_Handle, CTRL_SQUARE_COMMAND_BUTTON,
                                                    "CLEAR\n<F3>",  250, 190);
        //Set the width of the command button
        SetCtrlAttribute (GUI_Handle, CLEAR_PLOT_ID, ATTR_WIDTH, 40);

        //Create a callback function for the CLEAR  button
        InstallCtrlCallback (GUI_Handle, CLEAR_PLOT_ID,  ClearGraphCB, 0);
        //Set the command button color
        SetCtrlAttribute (GUI_Handle, CLEAR_PLOT_ID, ATTR_CMD_BUTTON_COLOR,
                                                    VAL_GREEN);
        //Set the label color to black
        SetCtrlAttribute (GUI_Handle, CLEAR_PLOT_ID, ATTR_LABEL_COLOR,
                                                    VAL_BLACK);
        //Set Fonts to "Meta Fonts"
        SetCtrlAttribute (GUI_Handle, CLEAR_PLOT_ID, ATTR_LABEL_FONT,
                                                    VAL_APP_META_FONT);
        //Assign "F3" as shortcut to this button
        SetCtrlAttribute (GUI_Handle, CLEAR_PLOT_ID, ATTR_SHORTCUT_KEY,
                                                    VAL_F3_VKEY);
        //Disable the CLEAR command button
        SetCtrlAttribute (GUI_Handle, CLEAR_PLOT_ID, ATTR_DIMMED, 1);
        //Close the panel by clicking on the "X" on the panel box
        SetPanelAttribute (GUI_Handle, ATTR_CLOSE_CTRL, GUI_EXIT_ID);

}//CreateGUI
```

Figure 1–4
CreateGUI Function Listing *(continued)*

Whether you create the GUI using the **User Interface Editor** or from the **Source Editor**, you always start by creating the panel on which you will place the controls. The panel for this GUI is created at line 8 using the library function *NewPanel.*

The *NewPanel* function creates either a new top-level panel (called the *parent panel*) or a *child panel* within a parent panel. Using the *NewPanel* function, you can specify the panel title for this panel and set the panel coordinates, height, and width. This function returns a panel handle that is assigned uniquely to this panel and is used when referencing the panel.

The *SetPanelAttribute* library function on line 12 is used to set the panel attributes: for example, to make the panel movable during program execution. There are other panel attributes that you can set using the *SetPanelAttribute* function. To understand these attributes, highlight *SetPanelAttribute* in your code and select **View>>Recall Function Panel** (or <Ctrl-P> for short). From

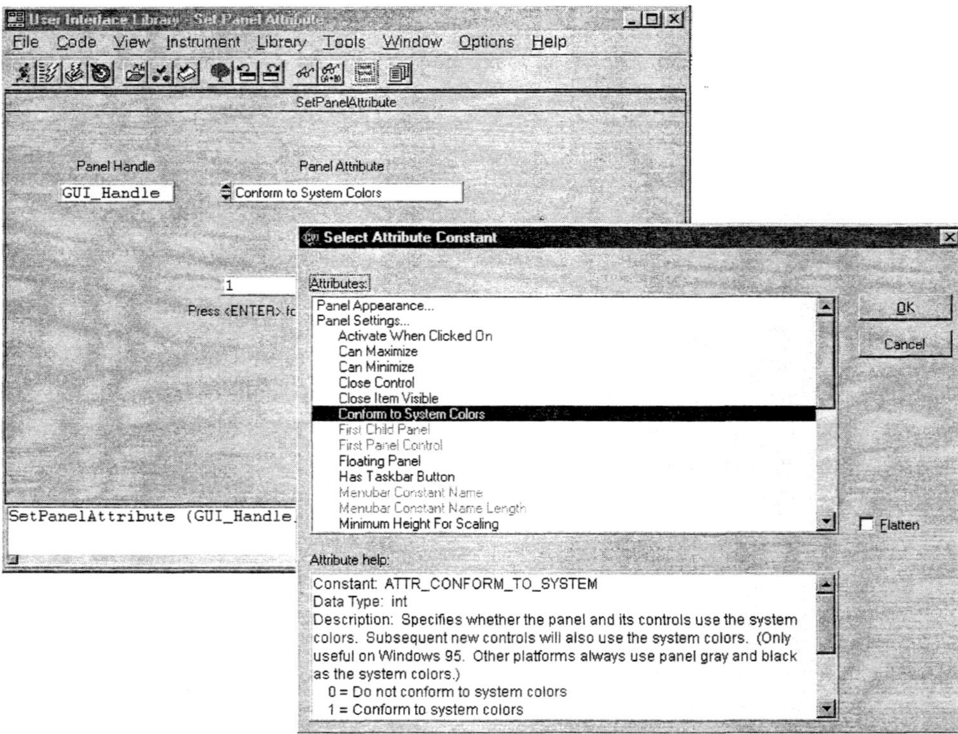

Figure 1–5
SetPanelAttribute Function Panel

the displayed function panel, click on the **Panel Attribute** box and Figure 1–5 is displayed, showing the various panel attributes. You can select other panel attributes from this function panel.

Line 14 creates a new control using the *NewCtrl* library function. Here a recessed decoration box is created. The *NewCtrl* function can be used to create any control on the panel that is normally developed using the **Create** menu command from the **User Interface Editor**. Again, using the function panel gives you the numerous choices of the available controls and various associated attributes. The return value of this function (**DEC_BOX_CMDS** in this case) is the control identifier used to reference this control from other functions.

The *SetCtrlAttribute* library function is used to set the attributes of the control. When you create a GUI using the **User Interface Editor**, you select the attributes of the control from the controls attribute dialog box. At lines 17–21 the attributes of the decoration box are set using these functions. The attributes configured here are the horizontal position, height, width, and color of the decoration box. Notice that the control identifier **DEC_BOX_CMDS** for the decoration box created at line 14 is used in all *SetCtrlAttribute* function calls to reference this decoration box.

The **START** square command button is created at line 25 at vertical and horizontal coordinates 200 and 100, respectively, using the *NewCtrl* library function. The label assigned to this command button is "START," with "<F1>" on the next line on the command button below "START." Line 30 assigns the width of the command button, and lines 34–40 set the color, label, and font style for this command button. At line 40, "metafonts" is used for the font type. Metafonts are scalable fonts that are independent of the monitor screen resolution. Line 43 assigns the shortcut key **F1** to the **START** command button.

The *InstallCtrlCallback* library function at line 32 assigns the user-defined callback function *StartGUICB* to the **START** command button. Whenever you click on the **START** command button, the *StartGUICB* function is called. This is similar to when you assign the callback function name by entering it in the **Callback Function** box of the control using the **User Interface Editor**.

To disable (dim or gray-out) the control, assign a value of 1 to the value of `ATTR_DIMMED` in the *SetCtrlAttribute* function, as shown on line 47. Similarly, lines 50–67 create and set the attributes for the **EXIT** command button. The callback function *ExitGUICB* is assigned to this control.

Lines 71–76 create another decoration box, where you will add a ring control. Lines 82 and 83 create a new ring control at the specified location with the assigned width. A callback function *RingCB* is assigned to this control at line 86. On lines 88–90 three list items are added to the ring using the library function *InsertListItem*. Note that the library functions for the ring control are the same as for list box control. List box functions, though, have more capabilities than do ring control functions. The three functions that are listed in the **Select Function** ring control, namely, **Sine Pattern**, **Uniform Random Numbers**, and **White Noise**, will be plotted on the graph control shown later in this section. These functions are part of the *CVI* Advanced Analysis Library package that is part of the full *CVI* package, as mentioned earlier. If you do not have the full *CVI* package, you can substitute your own functions here.

A decoration box to place the graph control is created at line 98 and its attributes set at lines 101–104. The graph control is created using the *NewCtrl* function at line 107. The labels for the X and Y axes are created using the *SetCtrlAttribute* function at lines 112–114.

In this project three different graphs will be plotted on the same graph control. If the graph is not cleared, these functions will be plotted over each other. Creating the **CLEAR** command button will give you the option to clear the graph before plotting another graph. The **CLEAR** command button is created at line 119 and its attributes assigned at lines 122–139.

At line 141 the *SetPanelAttribute* library function causes the panel to close and quits the project when you click on the "X" box on the top right corner of the panel window. This is accomplished by setting the panel attribute to `ATTR_CLOSE_CTRL` and pointing to the control ID of the command that will receive the *commit event* when the "X" box is selected. The control ID for this command is entered as the last argument in the *SetPanelAttribute* function. In this project the **EXIT** command button is assigned to closing the panel and exiting the project. Therefore, the control ID of the **EXIT** command button (**GUI_EXIT_ID**) is entered as the last argument in the *SetPanelAttribute* function.

RingCB Function

When you click on the **Select Function** ring control to select a function for which you want to display the data on the text box or plot on the graph control, the *RingCB* function is called. The listing for the *RingCB* function is shown in Figure 1–6. Here the **START** command button is enabled. To display the data or plot the graph for the function selected, select **START** from the GUI to invoke the callback function *StartGUICB.*

StartGUICB Function

Recall that when you created the **START** command button above, you had disabled it using the *SetCtrlAttribute* function. The reason is that the user must first make a selection from among one of the functions in the ring control before

```
//Callback function assigned to "Ring" control
int CVICALLBACK RingCB (int panel, int control, int event,
          void *callbackData, int eventData1, int eventData2)
{
     switch (event)
          {
          case EVENT_COMMIT:
               //Enable the START command button
               SetCtrlAttribute (GUI_Handle, GUI_START_ID, ATTR_DIMMED, 0);
          break;
          }
     return 0;
}//RingCB
```

Figure 1–6
RingCB Function Listing

he/she can plot on a graph control. In this project you have a choice from among the **Sine Pattern**, **Uniform Random Numbers**, **White Noise** functions, as shown in Figure 1–7. When **Uniform Random Numbers** is selected from the **Select Function** ring control, data is displayed on the text box. To plot this data on the graph control, select the **PLOT** command button that replaces the

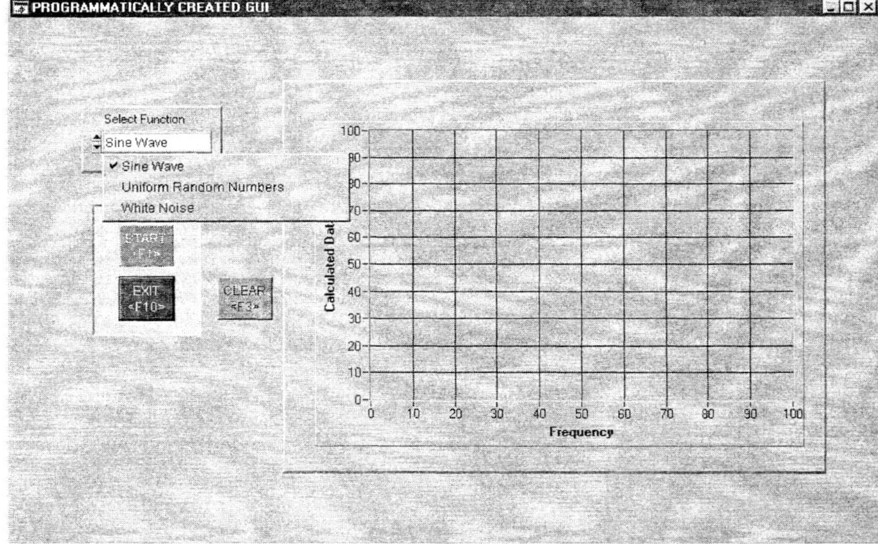

Figure 1–7
Select Function Control Pull-Down List

CLEAR command button on the GUI. When **Sine Wave** and **White Noise** are selected from the **Select Function** ring control, the graph is plotted without displaying the data.

Upon selecting one of these functions, the *RingCB* callback function that was assigned to the ring control is invoked and enables the **START** command button. The listing for the *StartGUICB* function is shown in Figure 1–8.

```
int CVICALLBACK StartGUICB (int panel, int control, int event,
                            void *callbackData, int eventData1, int eventData2)
{
char SelectedLabel[260], buf [500];
int SelectedIndex,i;

        switch (event)
                {
                case EVENT_COMMIT:
                        //Add Label to Graph
                        TEXT_MSG = NewCtrl (GUI_Handle, CTRL_TEXT_MSG, "PLOT FOR: ",
                                                                70, 450);

                        //Obtain Ring item index and label
                        GetCtrlIndex (GUI_Handle, RING_BOX, &SelectedIndex);
                        GetLabelFromIndex(GUI_Handle, RING_BOX, SelectedIndex,
                                                SelectedLabel);
                        //Add selected Ring item to Graph header
                        SELECTED_ITEM = NewCtrl (GUI_Handle,
                                                CTRL_TEXT_MSG,SelectedLabel,70, 520);
                        switch(SelectedIndex)
                        {
                                case 0:  //Sine Wave
                                        CreateDisplayDataPanel();  /* Create panel
                                                        for inputting sine data */
                                        break;

                                case 1: //Uniform Random numbers
                                         //Add Label to Graph
                                        TABLE_MSG = NewCtrl (GUI_Handle,
                                        CTRL_TEXT_MSG, "DATA FOR: ", 70, 450);
                                        //Hide the Graph control
                                        SetCtrlAttribute (GUI_Handle, GRAPH_ID,
                                                        ATTR_VISIBLE, 0);
                                        //Create Text Box control
                                        TEXT_BOX_ID = NewCtrl (GUI_Handle,
                                                CTRL_TEXT_BOX, "", 100, 280);
                                        //Set the text box control mode
                                        SetCtrlAttribute (GUI_Handle, TEXT_BOX_ID,
                                                ATTR_CTRL_MODE,  VAL_INDICATOR);
                                        //Set Text box dimensions
```

Figure 1–8
StartGUICB Function Listing *(continued)*

```
        SetCtrlAttribute (GUI_Handle, TEXT_BOX_ID,
                              ATTR_HEIGHT, 310);
        SetCtrlAttribute (GUI_Handle, TEXT_BOX_ID,
                                 ATTR_WIDTH, 450);
        //Include both scroll bars
        SetCtrlAttribute (GUI_Handle, TEXT_BOX_ID,
                                      ATTR_SCROLL_BARS,
                                 VAL_BOTH_SCROLL_BARS);
        //Set text background color
        SetCtrlAttribute (GUI_Handle, TEXT_BOX_ID,
                         ATTR_TEXT_BGCOLOR,
                                 VAL_BLACK);
        //Set text color to yellow
        SetCtrlAttribute (GUI_Handle, TEXT_BOX_ID,
                                 ATTR_TEXT_COLOR,
                                 VAL_YELLOW);
        //Hide the CLEAR button
        SetCtrlAttribute (GUI_Handle, CLEAR_PLOT_ID,
                                 ATTR_VISIBLE, 0);

        //Create a "PLOT" square command button
        PLOT_UNI_ID = NewCtrl (GUI_Handle,
                        CTRL_SQUARE_COMMAND_BUTTON,
                         "PLOT\n<F4>", 250, 190);
        //Set the width of the command button
        SetCtrlAttribute (GUI_Handle, PLOT_UNI_ID,
                                 ATTR_WIDTH, 40);

        //Create a callback function for the PLOT button
         InstallCtrlCallback (GUI_Handle, PLOT_UNI_ID,
                                 PlotUniformCB, 0);
        //Set the command button color
        SetCtrlAttribute (GUI_Handle, PLOT_UNI_ID,
        ATTR_CMD_BUTTON_COLOR,VAL_OFFWHITE);

        //Set the label color to black
        SetCtrlAttribute (GUI_Handle, PLOT_UNI_ID,
                ATTR_LABEL_COLOR, VAL_BLACK);
        //Set Fonts to "Meta Fonts"
        SetCtrlAttribute (GUI_Handle, PLOT_UNI_ID,
                                 ATTR_LABEL_FONT,
                         VAL_APP_META_FONT);
        //Assign "F4" as shortcut to this button
        SetCtrlAttribute (GUI_Handle, PLOT_UNI_ID,
                                 ATTR_SHORTCUT_KEY,
                                 VAL_F4_VKEY);

        //Create Uniform Random Values
        samples=100;
        seed=20;
        //Display data on Text Box
        Uniform(samples, seed,Y_axis);
```

Figure 1–8
StartGUICB Function Listing *(continued)*

```
94                                  //Display Header
95                                  Fmt(buf,"%s<Samples        Value");
96                                  InsertTextBoxLine(GUI_Handle, TEXT_BOX_ID,-1, buf);
97
98
99                                  for (i=0;i< samples;i++){
100                                   Fmt(buf,"%s<    %d              %f", i,  Y_axis[i]);
101                                   InsertTextBoxLine(GUI_Handle, TEXT_BOX_ID,-1,  buf);
102
103                                 }
104
105                                 break;
106                         case 2:  // White Noise
107                                 amp=5.0;
108                                 samples=50;
109                                 seed=15;
110                                 WhiteNoise(samples, amp, seed, Y_axis);
111                                 PlotHandle = PlotY (GUI_Handle, GRAPH_ID, Y_axis,
112                                                          samples, VAL_DOUBLE,
113                                                          VAL_CONNECTED_POINTS,
114                                                          VAL_ASTERISK, VAL_SOLID, 1,
115                                                                    VAL_WHITE);
116                                 break;
117                         default:
118                                 printf("Error in Ring control selection\n");
119                                 break;
120
121                         } //switch(SelectedIndex)
122                 //Enable the CLEAR command button
123                 SetCtrlAttribute (GUI_Handle, CLEAR_PLOT_ID, ATTR_DIMMED, 0);
124         break;
125                 }
126         return 0;
127 }//StartGUICB
```

Figure 1–8
StartGUICB Function Listing *(continued)*

At line 11 a text message box is created using the *NewCtrl* function. This text message box will display the header "PLOT FOR " on the graph decoration box just above the graph control. The index of the function selected to plot is obtained from ring control at line 15 using the *GetCtrlIndex* function. This index is used as an argument in the *GetLabelFromIndex* function to obtain the label of the item selected.

This label is created and displayed at line 19 as a text message on the graph decoration box next to the "PLOT FOR " text message control. At line 21 the index from the ring control determines the function to plot. When **White Noise** is selected, lines 106-116 are executed, calling the Advanced Analysis Library function *WhiteNoise*.

When **Uniform Random Numbers** is selected from ring control, the program executes at line 28. In this section of code the graph control will be hidden, a new text box control created in the same location as the graph control, and the uniform random numbers displayed on this text box. The **CLEAR** command button is hidden and a **PLOT** command button is created in its place. When the **PLOT** command button is selected, the text box is hidden and the graph control made visible with the plot of the random numbers that were displayed on the text box. Let us look at how this is accomplished by going through the code.

At line 30, a text message label "DATA FOR:" is displayed on the graph decoration box, the graph control is hidden at line 33, and a new text box control is created at line 36 over the graph control. The attributes for this text box are configured at lines 39–55. A square command button with the label "Plot" is created at line 63 over the **CLEAR** command button that is made invisible at line 59. The *PlotUniformCB* callback function is assigned to the **PLOT** command button at line 71. The remaining attributes for the **PLOT** command button are set between lines 74 and 85.

Lines 90–101 are used to set the sample size and the seed, to display the header, and to generate the uniform random numbers using the library function *Uniform* at line 93. The *Uniform* function creates random numbers in the array `Y_axis` that are distributed uniformly between 0 and 1.

At line 123 the **CLEAR** command button is enabled. After the data values are displayed on the text box, you can plot the data on the graph control by selecting the **PLOT** command button.

When the index for the **Sine Wave** function is selected, the user-defined function *CreateDisplayDataPanel* at line 24 is called. This function creates a child panel with the controls where the user can specify the values for the various parameters of the **Sine Wave** plot. The listing for the *CreateDisplayDataPanel* function is shown in Figure 1–9.

The child panel created by the function *CreateDisplayDataPanel* is shown in Figure 1–10 and is titled **SELECT DATA**. Let us look how each of these controls is created and their attributes assigned by examining the *CreateDisplayDataPanel* code listing.

At line 10 the *NewPanel* function creates the child panel with the dimensions and location specified on the parent panel with the panel handle `Child_Handle`. Notice that the first argument of the *NewPanel* function is the name of the parent panel. Thus the new panel will be created on this panel as a child panel.

```
//Create and display the child panel
void CreateDisplayDataPanel(void)
{
#define SAMPLES_VAL 20
#define MAX_SAMPLES 1000

int DONE_ID;

            //Create a child panel
                Child_Handle = NewPanel (GUI_Handle, "SELECT DATA",
                                                        VAL_AUTO_CENTER,
                                                        VAL_AUTO_CENTER,300, 300);

                 InstallPopup(Child_Handle);
                //Create a Numeric control to enter samples
                SAMPLES_ID = NewCtrl (Child_Handle, CTRL_NUMERIC, "Samples", 60, 40);
                //Set data type to integer
                SetCtrlAttribute (Child_Handle, SAMPLES_ID, ATTR_DATA_TYPE,
                                                                VAL_INTEGER);
                //Set the Control Mode
                SetCtrlAttribute (Child_Handle, SAMPLES_ID, ATTR_CTRL_MODE, VAL_HOT);

                //Set number of default samples
                SetCtrlAttribute (Child_Handle, SAMPLES_ID, ATTR_DFLT_VALUE,
                                                        SAMPLES_VAL);
                SetCtrlVal(Child_Handle, SAMPLES_ID, SAMPLES_VAL);
                //Specify minimum value of control
                SetCtrlAttribute (Child_Handle, SAMPLES_ID, ATTR_MIN_VALUE, 1);
                //Specify maximum value of control
                SetCtrlAttribute (Child_Handle, SAMPLES_ID, ATTR_MAX_VALUE,
                                                        MAX_SAMPLES);
                InstallCtrlCallback (Child_Handle, SAMPLES_ID,  SamplesControlCB, 0);

                //Create a Numeric control to enter amplitude
                AMPLITUDE_ID = NewCtrl (Child_Handle, CTRL_NUMERIC, "Amplitude", 60,
                                                                150);
                //Set data type to double
                SetCtrlAttribute (Child_Handle, AMPLITUDE_ID, ATTR_DATA_TYPE,
                                                        VAL_DOUBLE);

                //Set the Control Mode
                SetCtrlAttribute (Child_Handle, AMPLITUDE_ID, ATTR_CTRL_MODE,
                                                                VAL_HOT);
                //Set default amplitude
                SetCtrlAttribute (Child_Handle, AMPLITUDE_ID, ATTR_DFLT_VALUE, 5.0);
                SetCtrlVal(Child_Handle, AMPLITUDE_ID, 5.0);
                //Specify minimum value of control
                SetCtrlAttribute (Child_Handle, AMPLITUDE_ID, ATTR_MIN_VALUE, 0.0);

                //Create a Slider control to enter phase value
                PHASE_ID = NewCtrl (Child_Handle, CTRL_NUMERIC_POINTER_HSLIDE,
```

Figure 1–9
CreateDisplayDataPanel Function Listing *(continued)*

```
                                                        "Phase", 170, 40);
            //Set data type
            SetCtrlAttribute (Child_Handle, PHASE_ID, ATTR_DATA_TYPE,
                                                        VAL_DOUBLE);

            //Set the Control Mode
            SetCtrlAttribute (Child_Handle, PHASE_ID, ATTR_CTRL_MODE, VAL_HOT);
            //Set default
            SetCtrlAttribute (Child_Handle, PHASE_ID, ATTR_DFLT_VALUE, 0.0);
            DefaultCtrl (Child_Handle, PHASE_ID);
            //Specify minimum value of control
            SetCtrlAttribute (Child_Handle, PHASE_ID, ATTR_MIN_VALUE, 0.0);
            //Specify maximum value of control
            SetCtrlAttribute (Child_Handle, PHASE_ID, ATTR_MAX_VALUE, 180.0);

            //Create a Knob control to enter cyles value
            CYCLES_ID = NewCtrl (Child_Handle, CTRL_NUMERIC_KNOB, "Cycles",
                                                                140, 200);
            //Set data type
            SetCtrlAttribute (Child_Handle, CYCLES_ID, ATTR_DATA_TYPE,
                                                        VAL_INTEGER);

            //Set the Control Mode
            SetCtrlAttribute (Child_Handle, CYCLES_ID, ATTR_CTRL_MODE, VAL_HOT);

            //Set default cycles to 2
            SetCtrlAttribute (Child_Handle, CYCLES_ID, ATTR_DFLT_VALUE, 2);

            DefaultCtrl (Child_Handle, CYCLES_ID);
            //Specify minimum value of control
            SetCtrlAttribute (Child_Handle, CYCLES_ID, ATTR_MIN_VALUE, 1);
            //Specify maximum value of control
            SetCtrlAttribute (Child_Handle, CYCLES_ID, ATTR_MAX_VALUE,
                                                        SAMPLES_VAL);
            //Create a "DONE" square command button
            DONE_ID = NewCtrl (Child_Handle, CTRL_SQUARE_COMMAND_BUTTON,
                                                        "DONE", 250, 150);
            //Set the width of the command button
            SetCtrlAttribute (Child_Handle, DONE_ID, ATTR_WIDTH, 30);

            //Create a callback function for the DONE button
            InstallCtrlCallback (Child_Handle, DONE_ID,  DoneGUICB, 0);
            //Set the command button color
            SetCtrlAttribute (Child_Handle, DONE_ID, ATTR_CMD_BUTTON_COLOR,
                                                        VAL_DK_GREEN);
            //Set the label color
            SetCtrlAttribute (Child_Handle, DONE_ID, ATTR_LABEL_COLOR,
                                                        VAL_WHITE);
            //Set Fonts to "Meta Fonts"
            SetCtrlAttribute (Child_Handle, DONE_ID, ATTR_LABEL_FONT,
                                                VAL_APP_META_FONT);
}//CreateDisplayDataPanel
```

Figure 1–9
CreateDisplayDataPanel Function Listing *(continued)*

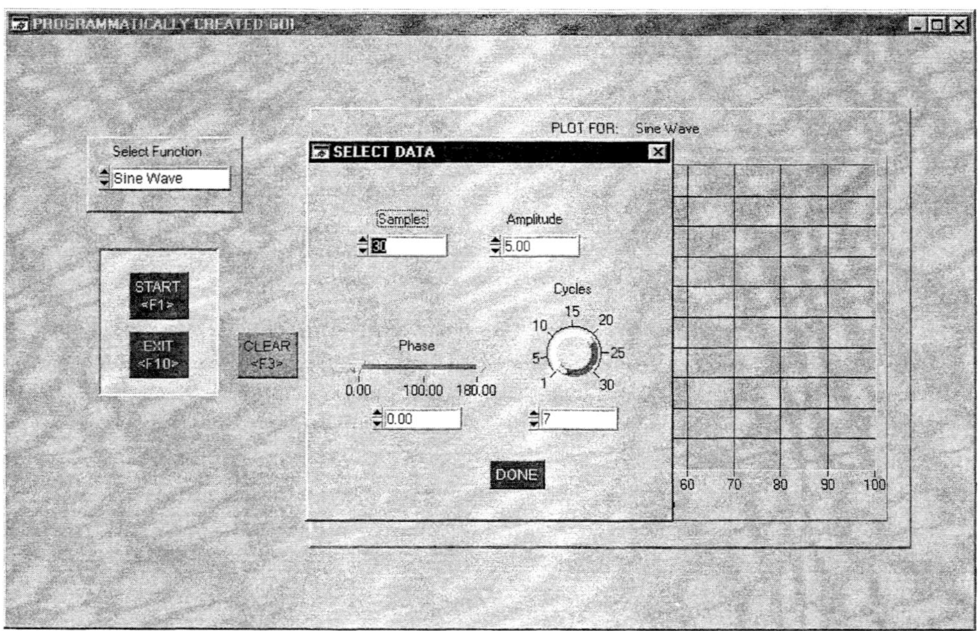

Figure 1–10
SELECT DATA Child Panel

Lines 16–30 create a numeric control with the label `Samples`. This control will receive data of type integer, then set the control mode to **Hot**, the default value to `20`, the minimum value to `1`, and the maximum value to `1000`. At line 32 the callback function *SamplesControlCB* is invoked whenever the value in the **Samples** control is changed. The *SamplesControlCB* function is explained below.

Similarly, at lines 35–48 another numeric control with the attributes specified and the label `Amplitude` is created.

Lines 51–65 create a horizontal slider with the label `Phase`, to enter a phase value between `0.0` and `180.0`.

To enter the number of cycles, a knob control is created at lines 68–84 with the label `Cycles`, the default value is set to 2, and the minimum and maximum values are displayed on **Cycles** control knob.

When all the data has been entered on this panel, you need to click on the **DONE** command button that is created at lines 90–101. This command button is assigned a callback function *DoneGUICB* that accepts the data entered and plots the graph. The listing for *DoneGUICB* function is given in Figure 1–12 and explained in that section.

SamplesControlCB Function

The *SamplesControlCB* function is called when the **Samples** control box on the **SELECT DATA** panel (Figure 1–10) is selected. This function obtains the new value from the **Samples** control and sets this as the new maximum value of the **Cycles** control knob. Setting this value on the **Cycles** knob ensures that the cycles value cannot be greater than the samples value, thus avoiding a possible run-time error. The default value of the **Cycles** control is always initialized to 2. You can change this value from the **Cycles** knob if you desire. This listing for the *SamplesControlCB* function is given in Figure 1–11.

DoneGUICB Function

The *DoneGUICB* function is called when you select the **DONE** command button on the **SELECT DATA** panel (Figure 1–10). The *DoneGUICB* function listing is shown in Figure 1–12.

The values entered in the **SELECT DATA** panel are obtained using the *GetCtrlAttribute* function. Here you could also have used the *GetCtrlVal* library function instead to obtain the same values. Control attributes have

```
//Set the Cycles control based on Samples selected
int CVICALLBACK SamplesControlCB(int panel, int control, int event,
                void *callbackData, int eventData1, int eventData2)

{
        switch (event)
                {
        case EVENT_COMMIT:
                //Get the samples value
                GetCtrlVal(Child_Handle, SAMPLES_ID, &samples);
                //Cycles must be less than samples
                GetCtrlAttribute (Child_Handle, SAMPLES_ID, ATTR_CTRL_VAL, &samples);
                //Set Maximum value of Cycles control to samples value
                SetCtrlAttribute (Child_Handle, CYCLES_ID, ATTR_MAX_VALUE, samples);
                //You can select the cycles by changing the value on the user interface
                SetCtrlVal(Child_Handle, CYCLES_ID,2); //Set Cycles to 2
        break;
    }
        return 0;
}//SamplesControlCB
```

Figure 1–11
SamplesControlCB Function Listing

```
1   //Done entering data
2   int CVICALLBACK DoneGUICB (int panel, int control, int event,
3               void *callbackData, int eventData1, int eventData2)
4
5   {
6           switch (event)
7           {
8           case EVENT_COMMIT:
9           //Obtain the values of the control
10          GetCtrlAttribute (Child_Handle, SAMPLES_ID, ATTR_CTRL_VAL, &samples);
11          GetCtrlAttribute (Child_Handle, AMPLITUDE_ID, ATTR_CTRL_VAL, &amp);
12          GetCtrlAttribute (Child_Handle, PHASE_ID, ATTR_CTRL_VAL, &phase);
13          GetCtrlAttribute (Child_Handle, CYCLES_ID, ATTR_CTRL_VAL, &cycles);
14
15          SinePattern (samples, amp, phase,cycles, Y_axis);//Create sine wave
16          HidePanel(Child_Handle);
17          //Plot Sine Pattern with the selected parameters
18          PlotHandle = PlotY (GUI_Handle, GRAPH_ID, Y_axis, samples, VAL_DOUBLE,
19                      VAL_THIN_LINE, VAL_SIMPLE_DOT, VAL_SOLID, 1, VAL_BLACK);
20          break;
21          }
22      return 0;
23  }// DoneGUICB
```

Figure 1–12
DoneGUICB Function Listing

different data types and different valid ranges, which are given in detail in the *LabWindows/CVI User Interface Reference Manual* and are available through On-line Help.

To retrieve the value of the control, ATTR_CTRL_VAL is used for the **control-Attribute** argument in the *GetCtrlAttribute* function. The values of the controls are obtained using this function at lines 10–13. At line 15 the library function *SinePattern* takes the values entered in the **SELECT DATA** panel as its arguments and creates a data array in the variable Y_axis. The child panel is hidden at line 16 and the data is plotted using line 18 on the graph control on the parent panel using the *PlotY* function. The function *SinePattern* is part of the *CVI* Advanced Analysis Library functions.

The *PlotY* function plots the values of an array on the y-axis against the array indices on the x-axis on the graph control. The **Sine Wave** plot is shown in Figure 1–13 with the following selections from the **SELECT DATA** panel:

```
Samples       25
Amplitude      5.0
Phase          0.0
Cycles         2.0
```

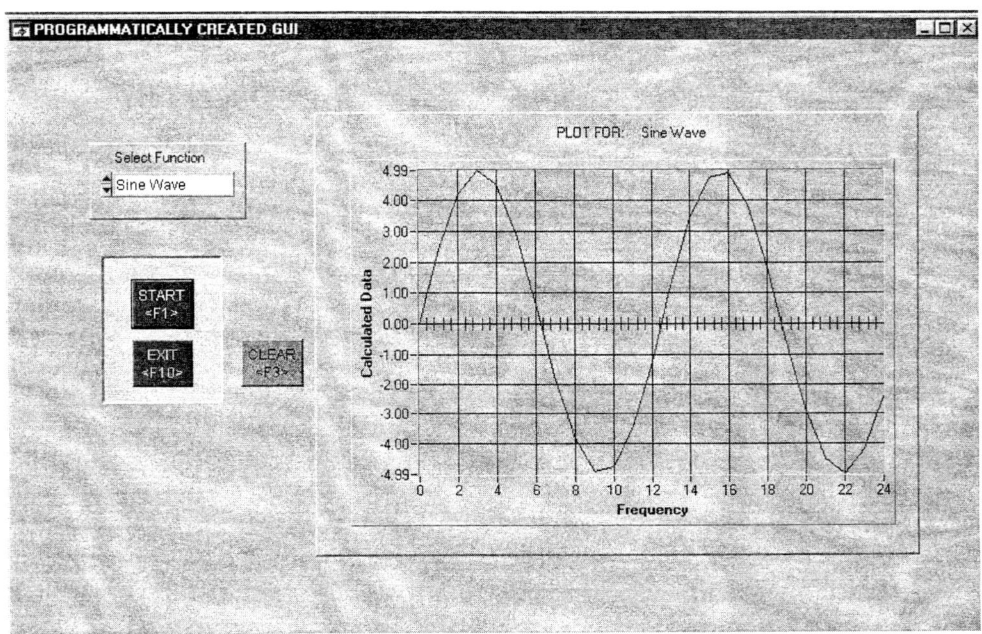

Figure 1–13
Sine Wave Sample Run

ClearGraphCB Function

To delete the plot on the graph control, select the **CLEAR** command button. This will also hide the text message boxes on the graph decoration box and display the text messages when the next selection from the ring control is made. The listing for the callback function *ClearGraphCB* associated with the **CLEAR** command button is shown in Figure 1–14.

At line 8 the *DeleteGraphPlot* library function deletes one or all of the plots from the graph control. You can delete all the plots on the graph control by entering -1 in the third argument of the *DeleteGraphPlot* function. The *RefreshGraph* library function shown on line 10 redraws the plot area immediately. The *SetCtrlAttribute* functions on lines 12 and 13 hide the text message boxes on the graph decoration box in order for the new messages to be displayed.

```
//Delete the last plot
int CVICALLBACK ClearGraphCB (int panel, int control, int event,
                  void *callbackData, int eventData1, int eventData2)
{
        switch (event)
                {
                case EVENT_COMMIT:
                        status = DeleteGraphPlot (GUI_Handle, GRAPH_ID, -1,
                                                                VAL_DELAYED_DRAW);
                        RefreshGraph(GUI_Handle, GRAPH_ID);
                        //Hide the message boxes
                        SetCtrlAttribute (GUI_Handle, TEXT_MSG, ATTR_VISIBLE, 0);
                        SetCtrlAttribute (GUI_Handle, SELECTED_ITEM, ATTR_VISIBLE, 0);
                        break;
                }
        return 0;
}// ClearGraphCB
```

Figure 1–14
ClearGraphCB Function Listing

PlotUniformCB Function

The listing for the callback function *PlotUniformCB* associated with the **PLOT** command button is shown in Figure 1–15.

At line 10 the **CLEAR** command button is made visible and the **PLOT** command and text message for the text box are discarded at lines 12 and 14. The graph control is made visible at line 16 and the text box is hidden at line 18. Line 23 plots the uniform random numbers displayed on the text box using the *PlotY* library function.

ExitGUICB Function

The *ExitGUICB* callback function is selected when the **EXIT** command button is selected. The listing for *ExitGUICB* is shown in Figure 1–16.

Summary

This chapter gave you an overall idea of how to use the *CVI* library functions to create a GUI using the **Source Editor** entirely. In this chapter you learned how to create the GUI programmatically without using the **User Interface Editor**.

```
//Invoked from PLOT command button
int CVICALLBACK PlotUniformCB (int panel, int control, int event,
                void *callbackData, int eventData1, int eventData2)
{
        switch (event)
                {
                case EVENT_COMMIT:
                        //Display the CLEAR button
                        SetCtrlAttribute (GUI_Handle, CLEAR_PLOT_ID, ATTR_VISIBLE, 1);
                        //Delete the PLOT button
                        DiscardCtrl (GUI_Handle, PLOT_UNI_ID);
                        //Delete the Label on Graph
                        DiscardCtrl (GUI_Handle, TABLE_MSG);
                        //Make Graph visible
                        SetCtrlAttribute (GUI_Handle, GRAPH_ID, ATTR_VISIBLE, 1);
                        //Make Text Box invisible
                        SetCtrlAttribute (GUI_Handle, TEXT_BOX_ID, ATTR_VISIBLE, 0);
                        //Plot the Uniform random numbers
                        samples=100;
                        seed=20;
                        Uniform(samples, seed,Y_axis);
                        PlotHandle = PlotY (GUI_Handle, GRAPH_ID, Y_axis, samples,
                                     VAL_DOUBLE, VAL_CONNECTED_POINTS,
                                     VAL_SMALL_X, VAL_SOLID, 1, VAL_DK_RED);
                break;
                }
        return 0;
}//PlotUniformCB
```

Figure 1–15
PlotUniformCB Function Listing

```
//Callback function assigned to "EXIT" command button
int CVICALLBACK ExitGUICB(int panel, int control, int event,
                void *callbackData, int eventData1, int eventData2)
{
        switch (event)
                {
                case EVENT_COMMIT:
                        QuitUserInterface (0);
                        break;
                }
        return 0;
}//ExitGUICB
```

Figure 1–16
ExitGUICB Function Listing

The library functions explaining the various features were shown by means of examples in the code listing. Many aspects of changing control attributes were introduced, opening the way for you to experiment with the various features of the controls and library functions. A blend of creating most of the controls from the **User Interface Editor** and manipulating controls programmatically may seem most desirable, depending on your application.

Library Function Prototypes and Definitions

This section lists alphabetically the *CVI* library functions that were introduced in this chapter.

DeleteGraphPlot Function

The *DeleteGraphPlot* function deletes one or more plots from a graph control. Its prototype is shown below and its arguments described in Table 1–1.

```
int status = DeleteGraphPlot (int panelHandle, int controlID,
                              int plotHandle, int refresh);
```

Table 1–1 DeleteGraphPlot *Function*

Input/ Output	Name	Type	Description
Input	*panelHandle*	integer	panel handle loaded in memory
	controlID	integer	constant name assigned to the control
	plotHandle	integer	handle for the plot to delete; enter –1 to delete all the plots on the graph control
	refresh	integer	selections to decide when to refresh the graph control; see explanation below
Output	*status*	integer	refer to Appendix A in *LabWindows/ CVI User Interface Reference Manual* or see On-line Help for error codes

The *refresh* argument allows you to select when to delete the plot from the graph control. It consists of the following choices:

- VAL_DELAYED_DRAW. If VAL_DELAYED_DRAW is assigned, the plot deleted remains on the graph until one of the following actions takes place:
  - The graph is rescaled.
  - The size of the plot area is changed.
  - The plot area is exposed after hiding or overlapping it.
  - The attribute to plot to screen immediately is set (ATTR_REFRESH_GRAPH =1).
  - The library function *RefreshGraph* is called.
  - Add another plot to the graph control while ATTR_REFRESH_GRAPH =1.
- VAL_IMMEDIATE_DRAW. If VAL_IMMEDIATE_DRAW is assigned to this argument the plot on the graph is redrawn immediately.
- VAL_NO_DRAW. If VAL_NO_DRAW is assigned, the plot deleted remains on the graph until the following action takes place:
  - The graph is rescaled.
  - The size of the plot area is changed.
  - The plot area is exposed after hiding or overlapping it, and ATTR_SMOOTH_UPDATE is 0.

DiscardPanel Function

The *DiscardPanel* function removes from memory the panel displayed and clears it from the screen if visible. Its prototype is shown below and its arguments described in Table 1–2.

```
int status = DiscardPanel (int panelHandle);
```

Table 1–2 DiscardPanel *Function*

Input/ Output	Name	Type	Description
Input	*panelHandle*	integer	panel handle to be removed
Output	*status*	integer	refer to Appendix A in *LabWindows/CVI User Interface Reference Manual* or see On-line Help for error codes

NewCtrl Function

The *NewCtrl* function creates a new control and returns a control identifier that you can use in subsequent function calls. Its prototype is shown below and its arguments described in Table 1–3.

```
int  ControlID= NewCtrl(int panelHandle,
                    int controlStyle, char controlLabel[ ],
                    int controlTop, int controlLeft);
```

controlTop and *controlLeft* positions are referenced from the top left corner of the panel just below the panel's title bar. These coordinates of the control appear before the panel is scrolled.

NewPanel Function

The *NewPanel* function creates a new panel and returns a panel handle that is used in subsequent function calls to refer to this panel. Its prototype is shown below and its arguments described in Table 1–4.

Table 1–3 NewCtrl *Function*

Input/ Output	Name	Type	Description
Input	*panelHandle*	integer	handle for the currently loaded panel on which the control is created
	controlStyle	integer	control style selected
	controlLabel	char[ ]	label for the new control; use " " or 0 for no label
	controlTop	integer	vertical coordinate of the upper left corner of control, not including the label
	controlLeft	integer	horizontal coordinate of the upper left corner of control, not including the label
Output	*ControlID*	integer	returns the ID of the new control; negative values indicate an error; refer to Appendix A in *LabWindows/CVI User Interface Reference Manual* or see On-line Help for error codes

Table 1–4 NewPanel *Function*

Input/ Output	Name	Type	Description
Input	*parentPanelHandle*	integer	handle for the currently loaded panel on which a new child panel is created; zero is used for the top-level panel
	panelTitle	string	title displayed on the panel
	panelVerticalCoord	integer	vertical coordinate where the upper left corner of the panel is placed; see below for further explanation
	panelHorizontalCoord	integer	horizontal coordinate where the upper left corner of the panel is placed; see below for further explanation
	panelHeight	integer	vertical size of the panel without the title bar or panel frame
	panelWidth	integer	horizontal size of the panel without the panel frame
Output	*panelHandle*	integer	value used to refer to this panel; negative values indicate error; refer to Appendix A in *LabWindows/CVI User Interface Reference Manual* or see On-line Help for error codes

```
int  panelHandle= NewPanel(int parentPanelHandle,
            char panelTitle[ ], int panelVerticalCoord,
                     int panelHorizontalCoord, int panelHeight,
                                    int panelWidth);
```

- *panelVerticalCoord*. For the parent panel this is referenced from the top left corner of your screen and is the vertical distance at which to place the upper left corner of the panel.
- *panelHorizontalCoord*. For the parent panel this is referenced from the top left corner of your screen and is the horizontal distance at which to place the upper left corner of the panel.

To center either the parent panel or the child panel, use `VAL_AUTO_CENTER`.

PlotY Function

The *PlotY* function plots an array of y values against its indices along with the x-axis on a graph control. Its prototype is shown below and its arguments described in Table 1–5.

```
int plotHandle = PlotY (int panelHandle, int controlID,
                        void *yArray, int numberOfPoints,
                        int yDataType,int plotStyle,
                        int pointStyle, int lineStyle,
                        int pointFrequency, int color);
```

Table 1–5 PlotY *Function*

Input/ Output	Name	Type	Description
Input	*panelHandle*	integer	panel handle loaded in memory
	controlID	integer	constant name assigned to the graph control
	yArray	void*	array containing the values to plot on the y-axis
	numberOfPoints	integer	number of points to plot; the plot is limited by this value even if there are more elements in the array
	yDataType	integer	select the data type of *yArray* from the following: ■ VAL_CHAR ■ VAL_INTEGER ■ VAL_SHORT_INTEGER ■ VAL_FLOAT ■ VAL_DOUBLE ■ VAL_STRING ■ VAL_UNSIGNED_ ■ SHORT_INTEGER ■ VAL_UNSIGNED_ ■ INTEGER ■ VAL_UNSIGNED_CHAR

(continued)

Table 1–5 PlotY *Function* (continued)

Input/ Output	Name	Type	Description
	plotStyle	integer	curve style for plotting from the following styles: ■ VAL_THIN_LINE ■ VAL_FAT_LINE ■ VAL_CONNECTED_POINTS ■ VAL_SCATTER ■ VAL_THIN_STEPS ■ VAL_FAT_STEPS ■ VAL_VERTICAL_BAR* ■ VAL_HORIZONTAL_BAR* ■ VAL_BASE_ZERO_VERTICAL_BARS* ■ VAL_BASE_ZERO_HORIZONTAL_ BARS* styles marked with * are not valid for strip charts
	pointStyle	integer	point style to use for plot; this determines the type of marker to show on the plot when the plot style is VAL_CONNECTED_ POINTS and VAL_SCATTER; these styles are listed in *LabWindows/CVI User Interface Reference Manual* or see On-line Help for error codes
	lineStyle	integer	line style to use for the plot from the following selection: ■ VAL_SOLID ■ VAL_DASH ■ VAL_DOT ■ VAL-DASH_DOT ■ VAL_DASH_DOT_DOT
	pointFrequency	integer	point interval to draw the marker symbol when the *plotStyle* is VAL_CONNECTED_ POINTS or VAL_SCATTER
	color	integer	color of the curve to plot
Output	*plotHandle*	integer	handle for the plot for possible use in subsequent reference to *DeleteGraphPlot*, *SetPlotAttribute*, and *GetPlotAttribute* functions

RefreshGraph Function

The *RefreshGraph* function redraws the graph. Its prototype is shown below and its arguments described in Table 1–6.

```
int status = RefreshGraph (int panelHandle, int controlID);
```

SetPanelAttribute Function

The *SetPanelAttribute* function sets the value of a particular panel prototype. Its prototype is shown below and its arguments described in Table 1–7.

```
int status= SetPanelAttribute(int panelHandle,
    int panelAttribute, attributeValue[1]);
```

Note that *attributeValue* has different data types that depend on the *panelAttribute* selected and are displayed when you move your cursor to the **Attribute Value** box on the function panel and select <Enter>.

Table 1–6 RefreshGraph *Function*

Input/ Output	Name	Type	Description
Input	*panelHandle*	integer	panel handle loaded in memory
	controlID	integer	constant name assigned to the graph control
Output	*status*	integer	refer to Appendix A in *LabWindows/CVI User Interface Reference Manual* or see On-line Help for error codes

Table 1–7 SetPanelAttribute *Function*

Input/ Output	Name	Type	Description
Input	*panelHandle*	integer	specifier for the panel currently loaded; zero is used for the top-level panel; for the child panel use the child panel handle name
	panelAttribute	integer	panel attribute selected
	attributeValue	[1]depends on attribute selected	value of panel attribute
Output	*status*	integer	refer to Appendix A in *LabWindows/CVI User Interface Reference Manual* or see On-line Help for error codes

2

PLOTTING ON GRAPH CONTROLS

Chapter Highlights

- Graph Attributes and Cursors
- Manual Zooming and Panning
- Creating Graph Legends
- Plotting Geometric Patterns on Graph Control
- Summary
- Library Function Prototypes and Definitions

Graph controls are useful when you want to display your data in a graphical format. This chapter explains the graph control attributes, cursor controls, zooming and panning, and creating graph legends using the toolbox. The ways to draw geometric objects such as lines, arcs, rectangles, and ovals are shown using the library functions. Adding text to the graph control is explained by creating a new metafont based on the existing metafont by changing the point size and other font attributes. You will see how to enter any text string on the graph control to label an item on the graph control or to add the header to the graph control.

Graph Attributes and Cursors

This section discusses how to enhance the features of your graph control using the graph attributes. You will be shown the use of graph cursors to view the values plotted by moving the cursor to different points on your graph. `project2-1.prj` will demonstrate the graph and graph cursor features. Using this project, you will learn to *zoom* the plot area around the point selected. *Zoom* means to change the field of view around a particular point to make the area around the point selected appear bigger or smaller. This feature will be demonstrated by selecting the zoom value from the project GUI control and also by use of the keyboard keys. You will also be introduced on how to *pan* the graph control using the keyboard. *Panning* means to shift the view area around the point selected.

Examine the `project2-1` GUI (Figure 2–1). When you select the **PLOT** command button, the *SinePattern* function is plotted on the graph control using 100 sample points for the X-axis. In the **Zoom Control** box, when the ring control is selected, a pull-down menu gives you the options to select the various zoom ratios to scale the axes of the graph control. This will expand or contract the viewable area selected, depending on your zoom selection.

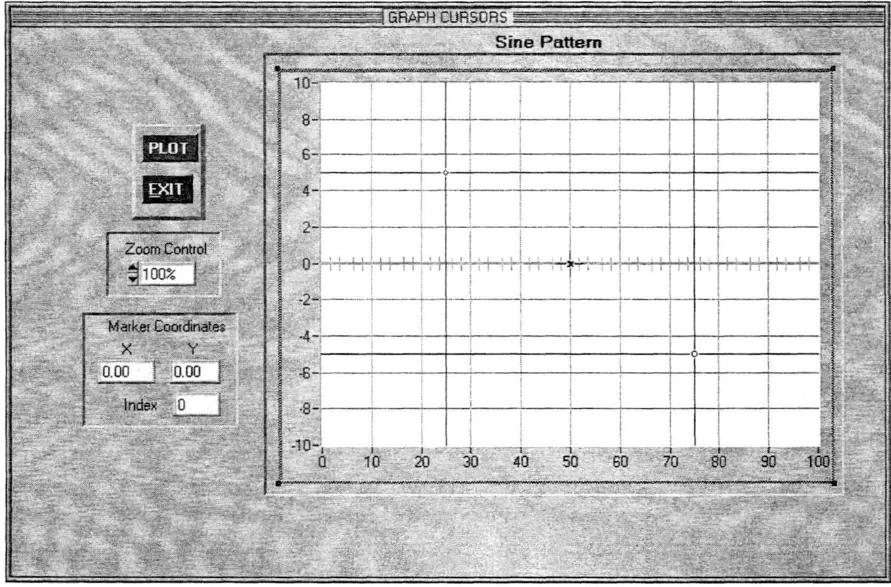

Figure 2–1
`project2-1` GUI

Graph cursor features facilitate better viewing of the data plotted on the graph control. For example, in this graph control you will use three graph cursors. The first cursor is located at the x and y coordinates of 25 and 5.0, respectively, as shown by an open circle on the graph control. An open circle locates the second cursor at the x and y coordinates of 75 and -5.0 respectively. These are the default cursor locations. During execution of the program you will be able to move these cursors using the mouse.

The marker cursor is shown as a cross hair located at the center of the graph control. You can assign a marker cursor to display the current value of the data plotted at a certain point on the graph. The marker cursor can be moved to any specific point on the plot to display the particular value on the **Marker Coordinates** indicator control boxes shown on the GUI.

To understand how to set the graph control cursors, double-click on the graph control and select **Cursors...** from the **Edit Cursors** dialog box shown in Figure 2–2. From this dialog box you can set the number of graph cursors in the **Number of Cursors:** box. The number of cursors is limited only by the memory available. Prior to *CVI* version 5.5 you were limited to using a maximum of 10 cursors. The graph control in project2-1 will use three cursors. By selecting the **Cursor Number:**, you can set the attributes for that particular cursor in the following dialog boxes.

The **Color:** box assigns the color for this graph cursor. The **Mode:** specifies the behavior of the cursor and gives you a choice of setting the cursor to **Free Form** or **Snap to Point**. When using **Free Form** you can move the cursor to any location inside the plot area. **Snap to Point** makes the cursor attach to the nearest data point plotted. **Point Style:** is the cursor shape that is assigned to this cursor on the graph control. You have a choice from among the selections shown in the pull-down menu in Figure 2–2.

The **Cross Hair Style:** pull-down menu shown in Figure 2–3 assigns the available cursor styles. You can experiment with these options and select the cross-hair style you wish to use.

There are two y-axes associated with a graph. By default only the left y-axis is used. You can assign the cursor selected to the right y-axis by checking the **Use Right Y Axis**. To make the cursor operable, check the **Enabled** box.

You can set the attributes for the other graph cursors similarly. Selecting **OK** from the **Edit Cursors** dialog box returns you to the **Edit Graph** dialog box shown in Figure 2–4. **Constant Name:** is the identifier used for the graph control. **Callback Function:** *UpdateMarkerCB* is called whenever an event is generated on the graph control using the mouse.

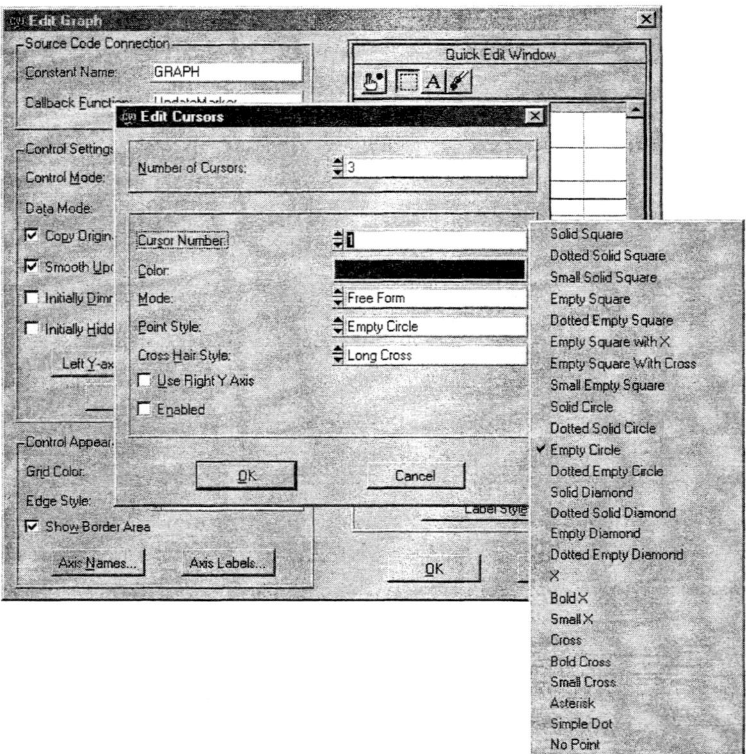

Figure 2–2
Edit Cursors Dialog Box

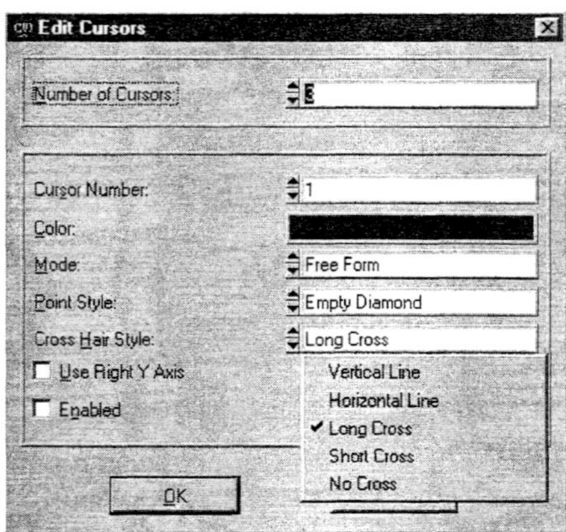

Figure 2–3
Cross Hair Style: Pull-Down Menu

In the **Control Settings** group, **Control Mode:** is set to **Hot** to generate an event on this control. When selected, the **Data Mode:** control brings up a pull-down menu from where you can select **Retain** or **Discard**. This selection allows the data plotted on the graph control to be retained or discarded after drawing new plots.

The graph control can make a copy of the original data plot or keep a pointer to the data. Whenever the graph is scaled, *CVI* uses the original plot data. Erroneous results can occur if a pointer to the data is maintained and the plot is rescaled or the data changed. You will get an error only if you deallocate (or overwrite) the original data array. You should check the **Copy Original Data** box if you will be rescaling the graphs. This option will consume more memory and will also plot the data points more slowly.

To store a copy of the graph in an off-screen bitmap, check the **Smooth Update** box. This will result in fewer flickers when the data is plotted and will allow smoother cursor movements. **Smooth Update** has the disadvantage of using more memory.

Placing a check mark in the **Enable Zooming** box allows you to zoom and pan the data plotted on the graph control using the keyboard keys. You will see this later in the section *Manual Zooming and Panning*.

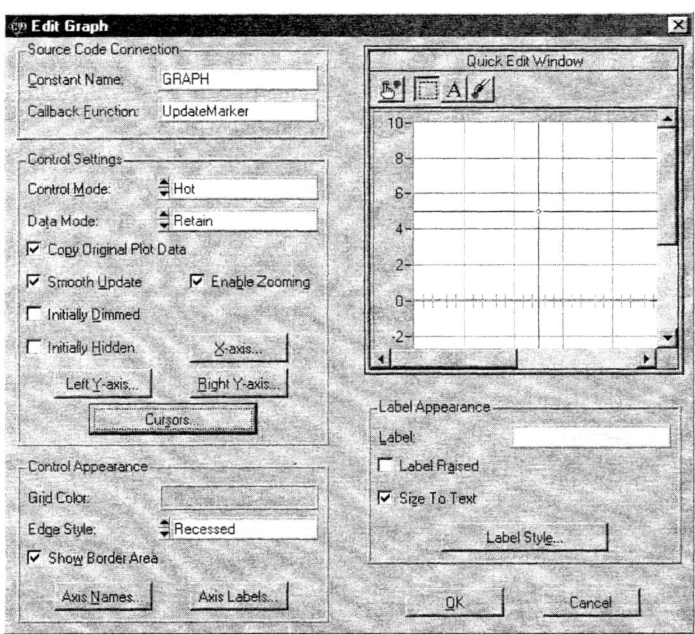

Figure 2–4
Edit Graph Dialog Box

The **X-axis...**, **Left Y-axis...**, and **Right Y-axis...** command buttons shown in Figure 2–4 all have very similar attributes, and when either one of these command buttons is selected it brings up an **Edit Axis Settings** dialog box similar to that shown in Figure 2–5 when the **X-axis...** command button is selected. These axis-setting attributes are described below.

Axis Name: is the name you would like to associate with that axis on the graph control. This label is displayed next to the assigned axis.

Minimum: and **Maximum:** values indicate the range you specify for your value labels on that axis.

Offset: is the number you add to the value label for the axis selected.

Gain: is the factor that you use to multiply the value label. **Gain** is always multiplied with the value label first before **Offset** is added.

By default the **Divisions:** control box is disabled. To enable the **Divisions:** control box, uncheck the **Auto Divisions:** box. In the **Divisions** dialog box you can set the number of tick marks you want displayed on the axis, ranging from 1 to 100. If the **Auto Divisions** box is checked, the tick marks are computed and displayed automatically by *CVI.*

In the **Precision:** box you can enter the significant decimal points you want displayed on the assigned axis or set it to **Auto** in the dialog to be displayed automatically by the program.

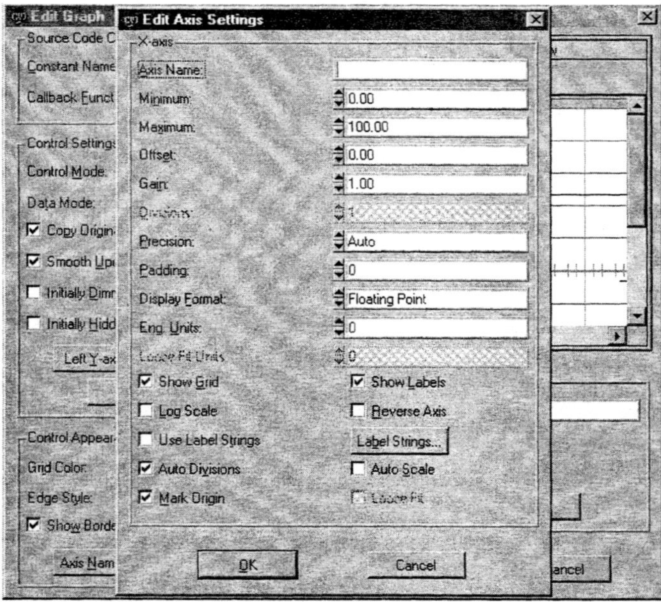

Figure 2–5
Edit Axis Settings Dialog Box

Padding: refers to the number of zeros to be displayed to the left of the value label on the graph axis selected. You may find it of use if your application requires this display.

In the **Display Format:** box you have the option to display the value labels on the axis in **Floating Point**, **Scientific**, or **Engineering** units. **Eng. Units** refers to the significant digits to be displayed on the axis selected and can range from –308 to 308. A positive value shifts the value label to the right of the decimal point. A negative value shifts the value label to the left of the decimal point.

By checking the **Show Grid** box, you can enable the grid markings on the axes selected.

If you want the label values to be displayed along the axis, checkmark the **Show Labels** box.

To display the data on your graph in logarithmic scale, check the **Log Scale** box. This option is acceptable only if the axis selected has a positive range; otherwise, you will be prompted to that effect.

The **Reverse Axis** box, when selected, reverses the orientation of the data plotted. For the x-axis the lowest value appears at the right and the highest value at the left (toward the origin of the x and y axes). For the y-axis the lowest value appears at the top of the graph and the highest value at the bottom (toward the origin of the x and y axes).

If you want to insert the customized labels on the axes instead of the numerical value labels, check the **Use Label Strings** box and enter the label(s) and location of the label(s) using the **Label Strings**... command button.

When you are not sure of the range of the data limits to be plotted on the graph, check the **Auto Scale** box and it will disable the **Minimum:** and **Maximum:** values for the graph control. When the **Auto Scale** box is checked, *CVI* recalculates the limits of the axes, and the range on the axis selected will then be scaled and the graph plotted within the scaled limits. Enabling **Auto Scale** will slow the graph plot if there are many plots on the same graph. This increases the time required to recalculate and remap all the existing plots.

The **Loose Fit** box is enabled when you check the **Auto Scale** box. Selecting this attribute determines how the maximum and minimum values of the axis are calculated when auto scaling is enabled. If the **Loose Fit** box is not checked, the axis is scaled using the minimum and maximum values of the plot. To set the maximum and minimum range of the axis, enable the **Loose Fit** box. The loose fit factor is explained below by means of examples. The **Loose Fit Units:** dialog box is the base 10 logarithm of the loose fit factor. For example, if the **Loose Fit Unit** value is 1, the loose fit factor is 10; if the **Loose Fit Unit** value is 2, the loose fit factor is 100. The smallest multiple of the loose fit factor that is greater than or equal to the largest value of all the plots is used for the maximum value. Similarly, the largest multiple of the loose fit factor that is less than or equal to the smallest value of all plots is used for the minimum value.

Mark Origin, when checked, marks the tick marks of the origin for the axis specified.

Let us look at the features of the *main* function shown in Figure 2–6.

```
/**************************Include files**************************************/
#include <cvirte.h>
#include <userint.h>
#include <ansi_c.h>
#include <analysis.h>
#include "project2-1.h"

/**************** Function prototype ***********************************/

void DisplayCursorPosition(void);

/***********************************************************************/
static int GraphPanelHandle;
double X_StartMin,X_StartMax,Y_StartMin, Y_StartMax;
int ScaleMode;

//Main function
int main (int argc, char *argv[])
{

    if (InitCVIRTE (0, argv, 0) == 0)
        return -1;
    //Load panel and obtain panel handle
    if ((GraphPanelHandle =LoadPanel (0, "project2-1.uir", CURSORS)) < 0)
        return -1;

    //Display the panel
    DisplayPanel (GraphPanelHandle);

    //Store the Starting Axis values for display before any zoom
    GetAxisScalingMode (GraphPanelHandle, CURSORS_GRAPH, VAL_XAXIS, &ScaleMode,
                                                    &X_StartMin, &X_StartMax);

    GetAxisScalingMode (GraphPanelHandle, CURSORS_GRAPH, VAL_LEFT_YAXIS,
                                            &ScaleMode, &Y_StartMin, &Y_StartMax);
    RunUserInterface();

    //Discard panel and release resources
    DiscardPanel (GraphPanelHandle);
    CloseCVIRTE ();

    return 0;
} //main
```

Figure 2–6
project2-1 Header and *main* Function Listing

The *GetAxisScalingMode* functions at lines 31 and 34 obtain the scaling mode and range of the x and y axes when the graph control is loaded initially. You will see its application when it is used in the *SelectZoomCB* function in Figure 2–10. The scaling mode and range are discussed below.

The strip chart features are not discussed here, as they were explained in the book *LabWindows/CVI Programming for Beginners*.

When the **PLOT** command button is selected during execution, the *PlotData* callback function is invoked; its listing is shown in Figure 2–7. The *SinePattern* library function is plotted on the graph control. If you do not have the *CVI* Advanced Analysis Library package that includes the *SinePattern* library function, you can substitute for this function a random number generator or some other function of your choice.

The *PlotData* function calls the *DeleteGraphPlot* library function at line 10 to clear the graph control of any possible previous plots. The *SinePattern* function is plotted using the *PlotY* function at line 15. Notice that library functions requiring the use of a panel handle such as *DeleteGraphPlot* and *PlotY* do not specifically have to use the panel handle (GraphPanelHandle) assigned to the panel when it was loaded at line 28 in the *main* function (Figure 2–6). The variable *panel* is passed by *value* through the callback function and can be used instead. It is advisable to use the panel handle name created in the library function, to avoid possible errors due to ambiguity.

```
1   //Plot SinePattern wave.
2   int CVICALLBACK PlotData (int panel, int control, int event,
3                             void *callbackData, int eventData1, int eventData2)
4   {
5       double Y_axis[100];
6
7       if (event == EVENT_COMMIT)
8       {
9           //Clear the graph
10          DeleteGraphPlot (panel, CURSORS_GRAPH, -1, VAL_DELAYED_DRAW);
11
12          //Create the Sine Pattern array in Y_axis
13          SinePattern (100, 10.0, 0.0, 5, Y_axis);
14          //Plot the graph for 100 points
15          PlotY (panel, CURSORS_GRAPH, Y_axis, 100, VAL_DOUBLE, VAL_THIN_LINE,
16                                     VAL_EMPTY_SQUARE, VAL_SOLID, 1, VAL_RED);
17      }
18
19      return 0;
20  } //PlotData
```

Figure 2–7
PlotData Function Listing

The callback function *UpdateMarkerCB* is invoked when you generate an event on the graph control by moving the marker cursor. This calls the user-defined function *DisplayCursorPosition,* which obtains the marker position and updates the **Marker Coordinates** controls on the GUI (Figure 2–1). The *UpdateMarkerCB* callback function listing is shown in Figure 2–8. The graph control events are processed like any other control events. When you move a cursor on the graph, a commit event is generated.

The listing for the user-defined function *DisplayCursorPosition* is shown in Figure 2–9. As mentioned above, this function obtains the marker cursor's current position and displays it on the **Marker Coordinates** indicator boxes shown in Figure 2–1.

The *GetGraphCursor* library function on line 12 obtains the current position of the graph cursor specified. The `X_Marker_Position` and `Y_Marker_Position` marker positions are displayed on the indicator boxes using the *SetCtrlVal* function at lines 19 and 20. You can obtain the data array index of the current marker cursor location using the library function *GetGraphCursorIndex* at line 16. This array index value is displayed in the indicator box at line 21. *GetGraphCursorIndex* function also returns the plot handle identifier to which the marker is attached.

The **Zoom Control** box on the GUI (Figure 2–1), when selected, displays the pull-down menu with various percentage zooms to change the plotted viewable area on the graph control with respect to the point selected. Selecting a value on this control invokes the *SelectZoomCB* callback function shown in Figure 2–10.

At lines 22–38 the positions of both graph cursors are obtained and their values swapped (if necessary) to obtain positive range values. The *GetCtrlVal* library function at line 43 obtains the value of the ring item selected. **Label/Value Pairs** for the ring control in the **Zoom Control** box is shown in Figure 2–11.

Lines 48–49 obtain the current range values for the X and Y axes. Lines 57–58 ratio the graph control by the user-selected zoom value.

The *SetAxisScalingMode* functions at lines 61 and 63 set the scaling modes and ranges of the X and Y axes using the values calculated at lines 54–58. These values are displayed using the *DisplayCursorPosition* function discussed above.

You can try the various zoom values and see how the graph control reacts. Select `100%` from the **Zoom Control** box to return to the original scaling.

```
//Calls the routine to update graph cursors
int CVICALLBACK UpdateMarkerCB(int panel, int control, int event,
                              void *callbackData, int eventData1,   int eventData2)
{

   if (event == EVENT_COMMIT)
   {
      //Update the marker values on the panel
       DisplayCursorPosition ();
   }
   return 0;
}//UpdateMarkerCB
```

Figure 2–8
UpdateMarkerCB Function Listing

```
//Get cursor position and display on the GUI
void DisplayCursorPosition (void)
{
#define Marker 2

    double X_Marker_Position;
    double Y_Marker_Position;
    int    plothandle;
    int    index;

    // Get Marker's position
    GetGraphCursor (GraphPanelHandle, CURSORS_GRAPH, Marker,
                                        &X_Marker_Position,  &Y_Marker_Position);

    //Obtain plot handle and array index of the point to which the marker is attached
    GetGraphCursorIndex (GraphPanelHandle, CURSORS_GRAPH, Marker, &plothandle,  &index);

    //Update the controls on the panel
    SetCtrlVal (GraphPanelHandle, CURSORS_XREADOUT, X_Marker_Position);
    SetCtrlVal (GraphPanelHandle, CURSORS_YREADOUT, Y_Marker_Position);
    SetCtrlVal (GraphPanelHandle, CURSORS_INDEX, index);
}  //DisplayCursorPosition
```

Figure 2–9
DisplayCursorPosition Function Listing

```
//Invoked when the zoom ring is selected
int CVICALLBACK SelectZoomCB (int panel, int control, int event,
        void *callbackData, int eventData1, int eventData2)
{
#define FirstCursor 1
#define SecondCursor 3

    double X_Pos_First_Cursor;
    double X_Pos_Second_Cursor;
    double Y_Pos_First_Cursor;
    double Y_Pos_Second_Cursor;
    double temp;

    double ZoomValue, ZoomValue_X1, ZoomValue_X2, ZoomValue_Y1,ZoomValue_Y2,
           X_axisMin, X_axisMax, Y_axisMin, Y_axisMax;

    int ZoomIndex, ScalingMode;

    if (event == EVENT_COMMIT)
    {
        //Get the current position of both cursors
        GetGraphCursor (panel, CURSORS_GRAPH, FirstCursor,  &X_Pos_First_Cursor,
                                                            &Y_Pos_First_Cursor);
        GetGraphCursor (panel, CURSORS_GRAPH, SecondCursor, &X_Pos_Second_Cursor,
                                                            &Y_Pos_Second_Cursor);

        // Swap values to obtain positive range values
        if (X_Pos_First_Cursor > X_Pos_Second_Cursor)
        {
            temp = X_Pos_First_Cursor;
            X_Pos_First_Cursor = X_Pos_Second_Cursor;
            X_Pos_Second_Cursor = temp;
        }
        if (Y_Pos_First_Cursor > Y_Pos_Second_Cursor)
        {
            temp = Y_Pos_First_Cursor;
            Y_Pos_First_Cursor = Y_Pos_Second_Cursor;
            Y_Pos_Second_Cursor = temp;
        }

  /**Here the scaling of the graph control is shown using the ring control selection***/
          //Obtain the selected zoom value
          GetCtrlVal(panel, CURSORS_ZOOM_RING, &ZoomValue);
          //Obtain the current X axis minimum and maximum values
          GetAxisScalingMode (panel, CURSORS_GRAPH, VAL_XAXIS, &ScalingMode,
                                                    &X_axisMin, &X_axisMax);
          //Obtain the current Y axis minimum and maximum values
          GetAxisScalingMode (panel, CURSORS_GRAPH, VAL_LEFT_YAXIS,
                                    &ScalingMode, &Y_axisMin, &Y_axisMax);

          //Use the panel min and maximum values for normal mode (100%) scaling
```

Figure 2–10
SelectZoomCB Function Listing *(continued)*

```
            //in main function
            //Ratio the graph control using the zoom factor selected
            ZoomValue_X1= (X_StartMin/ZoomValue);
            ZoomValue_X2= (X_StartMax/ZoomValue);

            ZoomValue_Y1= (Y_StartMin/ZoomValue);
            ZoomValue_Y2= (Y_StartMax/ZoomValue);

            //Set the X and Y axis to the new zoom values
            SetAxisScalingMode (panel, CURSORS_GRAPH, VAL_XAXIS, VAL_MANUAL,
                                                     ZoomValue_X1, ZoomValue_X2);
            SetAxisScalingMode (panel, CURSORS_GRAPH, VAL_LEFT_YAXIS,
                                              VAL_MANUAL,  ZoomValue_Y1, ZoomValue_Y2);
            //Update the marker values on the panel
            DisplayCursorPosition();
        }
    return 0;
} //SelectZoomCB
```

Figure 2–10
SelectZoomCB Function Listing *(continued)*

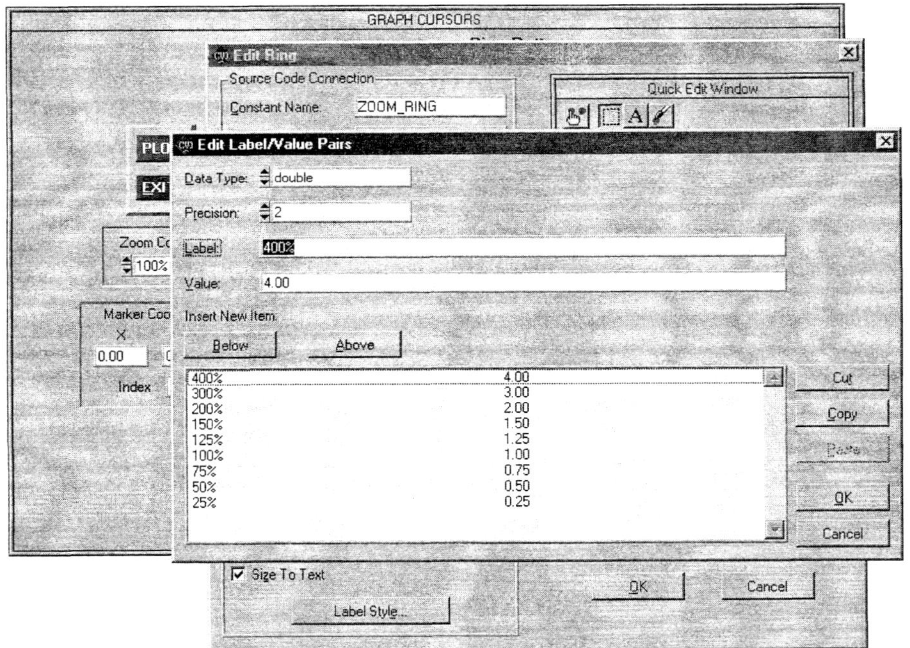

Figure 2–11
Zoom Control Label/Value Pairs

Manual Zooming and Panning

Recall that in Figure 2–4 you had checked the **Enable Zooming** box. This allows you to zoom and pan on the graph using the <Ctrl> and <Shift> keys on the keyboard. Also, for you to use these features the graph control mode must be set to **Normal**, **Hot**, or **Validate**, but not **Indicator**. To zoom on a certain point, move the mouse cursor to that point and hold down the <Ctrl> key and left mouse button. You will continuously zoom in while the mouse button is held down. To stop the zoom, release the mouse button. To zoom on a different area of the graph, drag the mouse and the area around the new mouse location is zoomed when you hold down the <Ctrl> key and the left mouse button.

To zoom out, hold down the <Ctrl> key and the right mouse button. You will zoom out continuously while the mouse button is held down.

To pan, hold down the <Ctrl-Shift> and left mouse keys and move the mouse over a certain point on the graph. Drag the mouse to another point on the graph and the view area starts to scroll to where the mouse is being moved.

After you have played around with the graph, you may wish to restore the graph control to its original state by selecting <Ctrl-spacebar>.

Creating Graph Legends

When you plot multiple graphs on a single graph control you may need to create a graph legend to explain the different plots. To do so, you must first have plots that are distinguishable from other plots on the graph control. You accomplish this by setting the **Plot Style**, **Point Style**, **Line Style**, and **Color** of each plot differently. You specify these attributes when you plot the graph. These features will be explained as we analyze the source code for `project2-2`.

The graph **Legend Control** library functions is included in the instrument driver `legend.fp`. This instrument driver is loaded from the **Instrument** menu in the **Source Editor** window. The graph **Legend Control** instrument driver did not exist for *CVI* versions prior to 5.5. This instrument driver is located in *CVI* folder in the path

```
samples\userint\custrl\legend
```

When creating a project using `legend.fp` it is recommended that the driver be moved to the project directory to avoid giving the full pathname to access it when creating the project. This instrument driver contains functions for opening, configuring, and controlling the legend items in the legend control. The **Legend Control** Function Panel is shown in Figure 2–12.

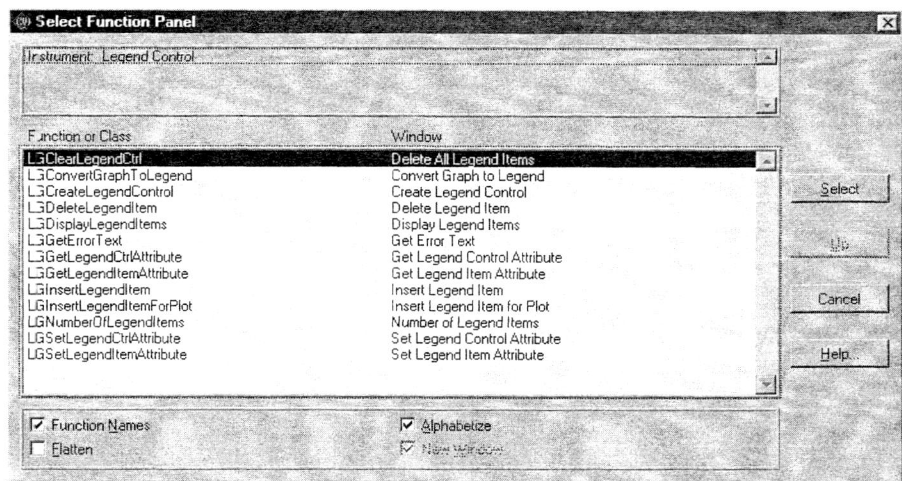

Figure 2–12
Legend Control Function Panel

`project2-2.prj` will build on the application created in `project1-1.prj`. Appropriate command buttons and functions are added to `project1-1.prj` to create the graph legend control and to add/remove item to/from the legend control. The GUI for this project is shown in Figure 2–13. Notice that a **Legend Controls** box has been added with command buttons to demonstrate the various legend functionalities. The command buttons are disabled until the **PLOT** command button is selected and the graphs plotted. In this project you will be creating three *SinePattern* plots offset by phase shifts of 0, 30, and 60 degrees. These plots will have different plot styles, point styles, line styles, and colors to differentiate among the plots. A sample plot will be included in the legend with the appropriate text explaining the plots.

Notice also that in Figure 2–13 a timer labeled **Marker Timer** has been added to the GUI. This timer control has a callback function *ContinuousUpdate* with a time interval of 100 milliseconds. The *ContinuousUpdate* function updates and displays the marker cursor location in the **Marker Coordinates** control boxes as the marker cursor is moved on the graph control.

When the **ADD ALL** command button is selected in the **Legend Controls** group, the *AddLegendCB* callback function is invoked. The listing for the *AddLegendCB* callback function is shown in Figure 2–14. This callback function creates the legend control at the specified position relative to the graph control, sets the attributes of the legend control, and displays the

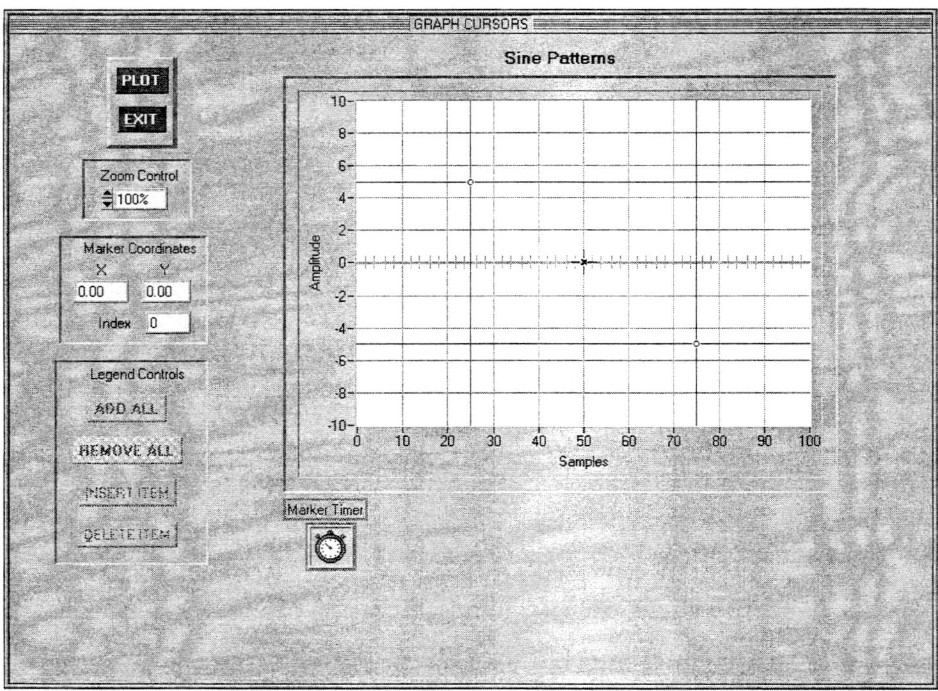

Figure 2–13
`project2-2` GUI

plot samples along with an explanation of the legend items. This function also enables the **REMOVE ALL**, **INSERT ITEM**, and **DELETE ITEM** command buttons. Each of the library functions to create and control the legend is discussed below.

The *LGCreateLegendControl* library function at line 9 creates the legend control. You will see below how to change the legend control attributes using the library function *LGSetLegendCtrlAttribute.*

At line 9 (Figure 2–14) the relative position of the legend control was specified at the center, below the anchor control. Line 12 uses the *LGSetLegendCtrlAttribute* function to move the legend control 25 pixels down from the anchor control.

The *LGSetLegendCtrlAttribute* sets the attributes of the legend control. At line 18 the *LGSetLegendCtlrAttribute* enables autosizing of the legend control. If you prefer to make the legend control border visible, use the *SetCtrlAtrribute* function to enable `ATTR_BORDER_VISIBLE`, as shown on line 15 (Figure 2–14).

```
//Invoked by the ADD ALL button to create the Legends
int CVICALLBACK AddLegendCB (int panel, int control, int event,
          void *callbackData, int eventData1, int eventData2)
{
  switch (event)
          {
          case EVENT_COMMIT:
                  //Create Legend control
                  LGCreateLegendControl (panel, CURSORS_GRAPH,
                                           LG_POS_BELOW_CENTER 1, 1,&PlotLegend);
                  //Move the legend control down 25 pixels from the anchor position
                  LGSetLegendCtrlAttribute (panel, PlotLegend,
                                  LG_ATTR_OFFSET_FROM_ANCHOR, 25);
                  //Make the Legend control border visible
                  SetCtrlAttribute(panel, PlotLegend, ATTR_BORDER_VISIBLE,1);

                  //Automatically re-size the legend box as items are added or deleted
                  LGSetLegendCtrlAttribute (panel, PlotLegend, LG_ATTR_AUTO_SIZE, 1);

                  //Legend item for Sine Plot with no phase offset
                  LGInsertLegendItemForPlot (panel, PlotLegend, 0,
                                  "Phase Offset:  0 degrees",   VAL_BLACK, SinePlot);
                  //Legend item for Sine plot with -30 degrees phase offset
                  LGInsertLegendItemForPlot (panel,PlotLegend , 0,
                              "Phase Offset:-30 degrees", VAL_BLACK, Offset30Plot);

                  //Legend item for Sine plot with -60 degrees phase offset
                  LGInsertLegendItemForPlot (panel, PlotLegend, 0,
                                "Phase Offset:-60 degrees",VAL_BLACK, Offset60Plot);
                  //Set background color the Legend control
                  LGSetLegendCtrlAttribute (panel, PlotLegend,
                                           LG_ATTR_PLOT_BG_COLOR, VAL_CYAN);
                  //Change the Fonts in the Legend box
                  LGSetLegendCtrlAttribute (panel, PlotLegend,
                               LG_ATTR_META_FONT,VAL_SYSTEM_META_FONT);

                  //Enable the INSERT and DELETE buttons
                  SetInputMode (panel, CURSORS_INSERT_ITEM, 1);
                  SetInputMode (panel, CURSORS_DELETE_ITEM, 1);
                  SetInputMode (panel, CURSORS_ADD, 0); /* Disable ADD
                                                               command button */
                  SetInputMode (panel, CURSORS_REMOVE, 1);    /* Enable REMOVE
                                                          command button */
                  break;
          }
  return 0;
} //AddLegendCB
```

Figure 2–14
AddLegendCB Listing

The *LGInsertLegendItemForPlot* library function is used for inserting the legend item in the legend control at the position specified. The legend items for the three *SinePattern* plots are added to the legend control using the

LGInsertLegendItemForPlot function. The appropriate text is used to describe each of the legend items at lines 21–29.

Lines 31–35 set the legend control's background color and the font using the *LGSetLegendCtrlAttribute* functions.

Plotting Geometric Patterns on Graph Control

In this section you will see how to plot certain geometric patterns on the graph control. This will consist of drawing lines between two points, plotting arcs, creating rectangles or ovals, and adding text to these objects. You will be able to modify the *CVI* available metafonts by selecting the point size, and specifying bold, italics, underline, or strike-through attributes to create a new metafont that will be displayed on the graph control. You will be shown how to draw these objects and display the text using a choice of different colors and to fill the object selected with this color or to leave the object unfilled. When used in conjunction with other plots on the graph, these drawing features can enhance your graph considerably, as you will see in `project2-3.prj`. `project2-3` GUI is very similar to the previous two projects explained in this chapter. In this project, using the **Select Plot**, **Select Color**, and **Fill Color** features shown in Figure 2–15 is explained.

When you click on the ring control under **Select Plot**, you have a choice of various plots, as shown in Figure 2–16.

By selecting the appropriate item from the **Select Plot** ring control, you can plot a line, an arc, a rectangle, or an oval between the mouse clicks on the graph control. You will also be able to use the **Plot Text** command on this ring control to write text anywhere on the graph control using the metafont created. **Select Color** under the **Select Plot** ring control box, when selected, pops up with a color palette, as shown in Figure 2–17. Here you can click on any color on the color palette control for use with the object on the graph control.

Load the project in *CVI* to run a demonstration of plotting a line between two points on the graph control. First, click on the **PLOT** command button to plot a sine curve. To plot any object selected on the control, such as a line (in this example), select **Plot Line** from **Select Plot**, and the color of the plotted line from the **Select Color** template.

To draw any object in this project, select the object from **Select Plot**, move the graph marker to the location from where you want to start drawing the object, and right-click the mouse button. To mark the end position of the object, move the graph marker to that point and right-click with the mouse

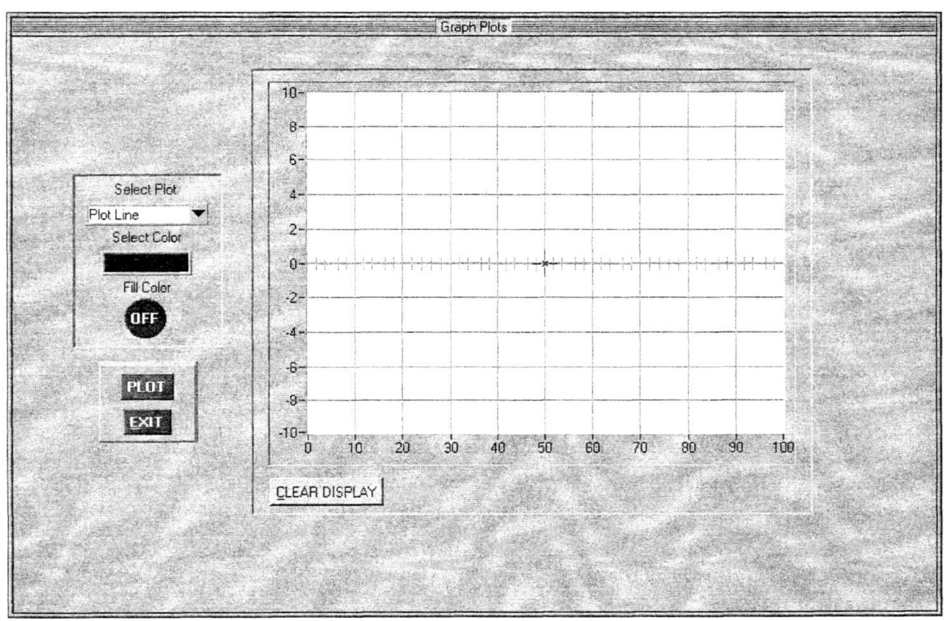

Figure 2–15
`project2-3` **Graph Plots** GUI

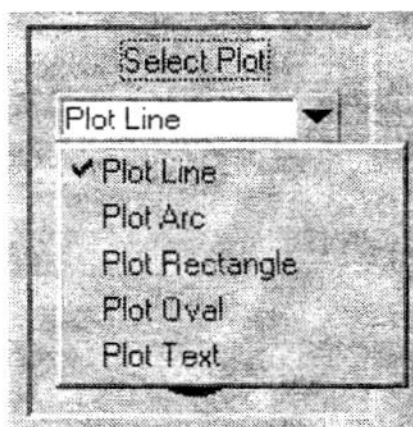

Figure 2–16
Select Plot Ring Control

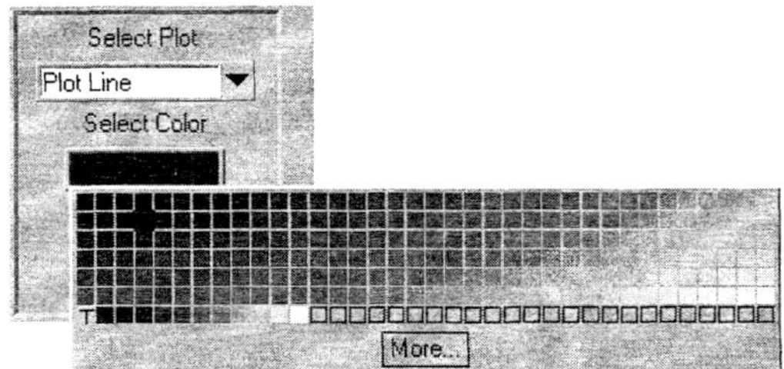

Figure 2–17
Select Color Control

button. Figure 2–18 shows an example of a straight line of the user-defined color drawn between the points selected.

Let us add text to this graph control. From the **Select Plot** ring control, select the **Plot Text** control item. Move the graph cursor to the point where you want the text string displayed on the graph control. The **TEXT ENTRY** GUI (Figure 2–19) is displayed. In this GUI enter the text string in the **Enter Text String:** dialog box that you want displayed at the graph control cursor. Select the preexisting fonts from the **Existing Font:** ring control. This font will be used as the base for creating the new metafont using the font attributes selected from the **Font Attributes** group.

By clicking on the **Existing Font:** ring control you will see a choice of various predefined metafonts, as shown in Figure 2–20. The ring control shows only the metafonts defined in the User Interface Library. You can create your own metafonts by selecting a typeface with a specific choice of point size, normal or bold, italics, underline, strikeout, and so on. Select any metafont, and from the **Font Attributes** group select the other font attributes and color to create the font to display the text message on the graph control.

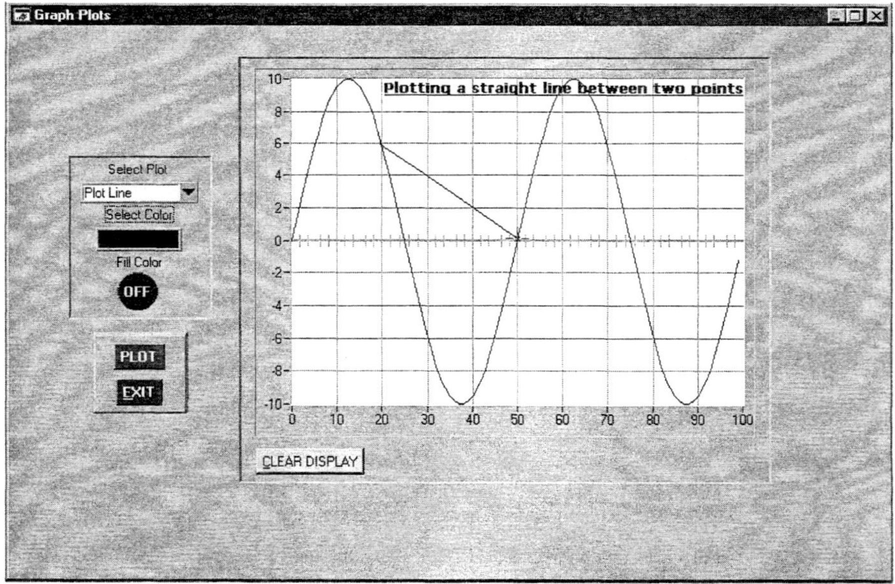

Figure 2–18
Line Plotting and Text Sample

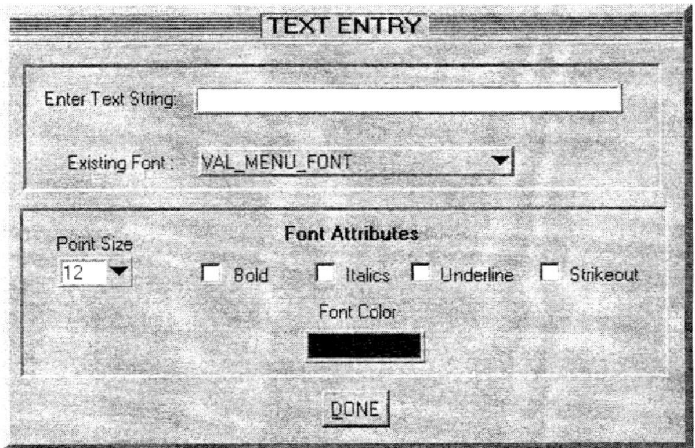

Figure 2–19
TEXT ENTRY and **Font Attributes** GUI

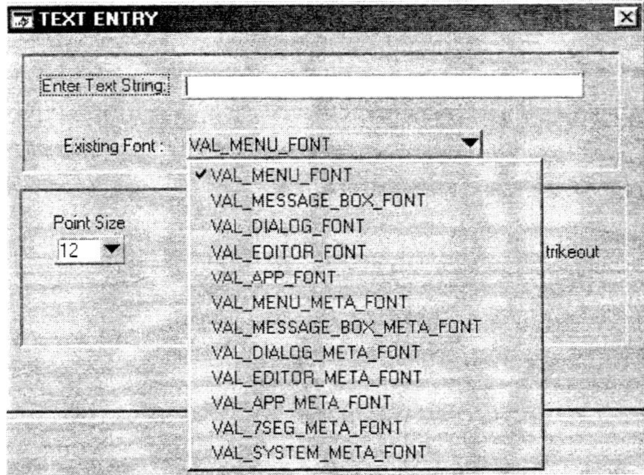

Figure 2–20
Existing Metafont Selections

The text is displayed on the top of the graph control in Figure 2–18 using the metafont created and attributes selected. Examine the source code for the `main` function as shown in Figure 2–21. The `project2-3.uir` GUI is loaded and displayed at lines 30 and 33, respectively. This GUI is shown in Figure 2–15. The panel is discarded at line 36 using the library function *DiscardPanel,* which takes the *panel handle* as its only argument.

```
/******************Include files***************************/
#include <formatio.h>
#include <cvirte.h>
#include <userint.h>
#include <ansi_c.h>
#include <analysis.h>
#include "project2-3.h"
#include "TextEntry.h"
#include "angleselect.h"

/**************** Function prototype ******************/

void DisplayCursorPosition(void);

/*****************************************************/
static int GraphPanelHandle, ArcPanelHandle, TextPanelHandle;
double X_StartMin,X_StartMax,Y_StartMin, Y_StartMax;
static double X1_Position, X2_Position;
static double Y1_Position, Y2_Position;
int SelectedColor, FillColor, StartPlotFlag=0;

#define Marker 1

//Main function
int main (int argc, char *argv[ ])
{
    if (InitCVIRTE (0, argv, 0) == 0)
        return -1;
    //Load panel and obtain panel handle
    if ((GraphPanelHandle =LoadPanel (0, "project2-3.uir", PLOTS)) < 0)
        return -1;
    //Display the panel
    DisplayPanel (GraphPanelHandle);
    RunUserInterface();
    //Discard panel and release resources
    DiscardPanel (GraphPanelHandle);
    CloseCVIRTE ();
    return 0;
} //main
```

Figure 2–21
main Function Source Code Listing

When the graph marker is selected, the *UpdateMarkerCB* callback function is executed, whose listing is shown in Figure 2–22. This function performs the plots for the items selected from the **Select Plot** control. Move the graph marker to the position at which to start drawing the object and right-click the mouse button.

```
//Calls the routine to plot selected shapes on graph control
int CVICALLBACK UpdateMarkerCB (int panel, int control, int event,
                              void *callbackData, int eventData1,
                              int eventData2)
{
    int     Fill;
    int     PlotIndex;

        if (event == EVENT_RIGHT_CLICK)
    {
        GetCtrlIndex (GraphPanelHandle, PLOTS_PLOT_RING, &PlotIndex);
        GetCtrlVal(GraphPanelHandle,PLOTS_COLOR_SELECT,&SelectedColor);
        GetCtrlVal(GraphPanelHandle,PLOTS_FILL_COLOR,&Fill);

        if (!Fill) //Do you want to fill with selected color?
                FillColor =VAL_TRANSPARENT;
        else
                 FillColor= SelectedColor;

        if ( (StartPlotFlag==0) && ( PlotIndex !=4))
        {
        //Get starting Marker's position
        GetGraphCursor (GraphPanelHandle, PLOTS_GRAPH, Marker,
                                                &X1_Position, &Y1_Position);

        StartPlotFlag=1;
    }
    else
    {
        //Get Marker's ending position
        GetGraphCursor (GraphPanelHandle, PLOTS_GRAPH, Marker,
                                              &X2_Position, &Y2_Position);


                switch (PlotIndex)
                {
                 case 0:  //Draw a Line
                        PlotLine (GraphPanelHandle, PLOTS_GRAPH, X1_Position,
                            Y1_Position, X2_Position, Y2_Position, SelectedColor);
                        break;
                 case 1: //Draw an Arc
                        ArcPanelHandle =LoadPanel (GraphPanelHandle,
                                                "AngleSelect.uir", ARC);
                        InstallPopup (ArcPanelHandle);
                break;
                case 2:  //Draw a Rectangle
                        PlotRectangle (GraphPanelHandle, PLOTS_GRAPH, X1_Position,
                                    Y1_Position, X2_Position, Y2_Position,
                                                SelectedColor, FillColor);
                        break;
                case 3:  //Draw an Oval
                        PlotOval(GraphPanelHandle, PLOTS_GRAPH, X1_Position,
```

Figure 2–22
UpdateMarkerCB Source Code Listing *(continued)*

```
53                          Y1_Position, X2_Position, Y2_Position, SelectedColor,
54                                                                  FillColor);
55                      break;
56              case 4:   //Enter Text
57                        TextPanelHandle =LoadPanel (GraphPanelHandle,
58                                                  "TextEntry.uir", TEXT);
59                        // Get Marker's position
60                        GetGraphCursor (GraphPanelHandle, PLOTS_GRAPH, Marker,
61                                                  &X1_Position, &Y1_Position);
62                      InstallPopup (TextPanelHandle);
63                      break;
64              }
65              StartPlotFlag=0;
66          }
67      }
68      return 0;
69  }//UpdateMarkerCB
```

Figure 2–22
UpdateMarkerCB Source Code Listing *(continued)*

Line 11 selects the index of the shape to plot from the **Select Plot** control. The color used for the plot is selected from the color palette in the **Select Color** control at line 12. If you would like to fill the object with the color selected, click on the **Fill Color** toggle button. The value for the **Fill Color** toggle button is obtained at line 15. Based on the **Select Plot** index selected, the appropriate item is plotted using the library functions explained below.

The `StartPlotFlag` is checked to distinguish between the first or second time the mouse is right-clicked. If the mouse is right-clicked the first time and the **Plot Text** is not selected, the library function *GetGraphCursor* is called at line 23 to capture the starting `x` and `y` positions of the graph marker in the variables `X1_Position` and `Y1_Position`. When the graph marker is moved to another position and the right mouse button is clicked again, lines 31–65 are executed and the ending plot position is saved in the variables `X2_Position` and `Y2_Position` using the *GetGraphCursor* at line 31.

To plot a line, the *PlotLine* library function is called at line 38. This function plots the line on the graph control between the starting and ending points using the color selected.

If you want to plot an arc on the graph control, you will need the starting angle and the sweep angle of the arc to define its shape. You will also need to enter the starting and ending coordinates which define the two opposite corners of the rectangle that enclose the arc. When **Arc Plot** is selected from the **Select Plot** control, the **Arc Plot Angles** GUI shown in Figure 2–23 is displayed by means of the *InstallPopup* library function at line 44. You enter the

Begin Angle (starting angle) and the **Arc Angle** (sweep angle) values in this GUI that are used in the *PlotArc* library function. **Begin Angle** and **Arc Angle** are used in the *PlotArc* library function prototype.

When **OK** is selected from **Arc Plot Angles** GUI (Figure 2–23), the *AnglesSelectedCB* callback shown in Figure 2–24 is invoked. The **Begin Angle** and **Arc Angle** values are obtained from the **Arc Plot Angles** GUI and used in the *PlotArc* library function at lines 14–16. The other arguments for this function are what you selected from the GUI in Figure 2–18.

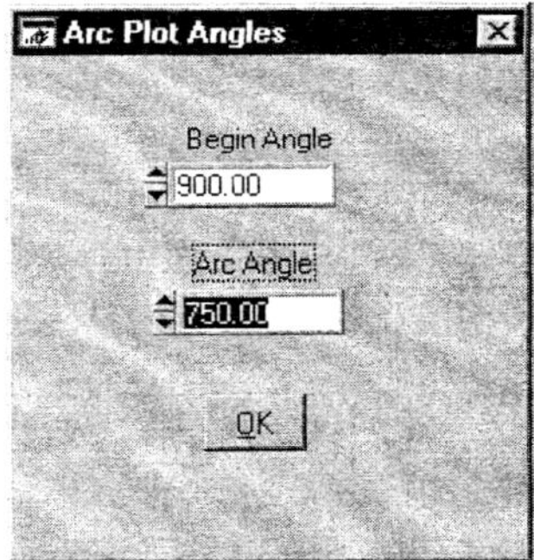

Figure 2–23
Arc Plot Angles GUI

```
1   //Select the angles for the arc
2   int CVICALLBACK AnglesSelectedCB (int panel, int control, int event,
3           void *callbackData, int eventData1, int eventData2)
4   {
5       double BeginAngle, ArcAngle;
6       switch (event)
7     {
8           case EVENT_COMMIT:
9                   //Obtain the Begin and the Arc angles
10                  GetCtrlVal(ArcPanelHandle,ARC_BEGIN_ANGLE,&BeginAngle);
11                  GetCtrlVal(ArcPanelHandle,ARC_BEGIN_ANGLE,&ArcAngle);
12                  RemovePopup(ArcPanelHandle);
13                  //Plot the arc with the selected values
14                  PlotArc (GraphPanelHandle, PLOTS_GRAPH, X1_Position,
15                      Y1_Position, X2_Position, Y2_Position, BeginAngle, ArcAngle,
16                                                  SelectedColor, FillColor);
17                  break;
18      }
19      return 0;
20  }//AnglesSelectedCB
```

Figure 2–24
AnglesSelectedCB Source Code Listing

To plot a rectangle on the graph control, select **Plot Rectangle** from the **Select Plot** control. The *PlotRectangle* library function is called at lines 47–49 in Figure 2–22. You plot the rectangle by selecting its two opposite corners.

Lines 52–54 in Figure 2–22 show the library function *PlotOval* to plot an oval on the graph control. Selecting the two opposite corners of the rectangle will define the bounds to plot the oval.

When you plot the objects on the graph control, you may want to annotate the objects with text. An example of text annotation is shown in Figure 2–25. This figure shows the various plots created using the library functions that we discussed above.

To create text on the graph control, first select **Plot Text** from the **Select Plot** control. Move the graph marker to the place on the graph control where you want to insert the text and right-click using the mouse button. The **TEXT ENTRY** GUI is displayed as shown in Figure 2–26. Enter the text in the **Enter Text String:** box. Select the **Existing Font:** by clicking on this control and choosing from among the available choices shown in Figure 2–20. Select the attributes of the fonts such as **Point Size**, **Bold**, **Italics**, **Underline**, **Strikeout**, and **Color** to create the new font with which to write the text string.

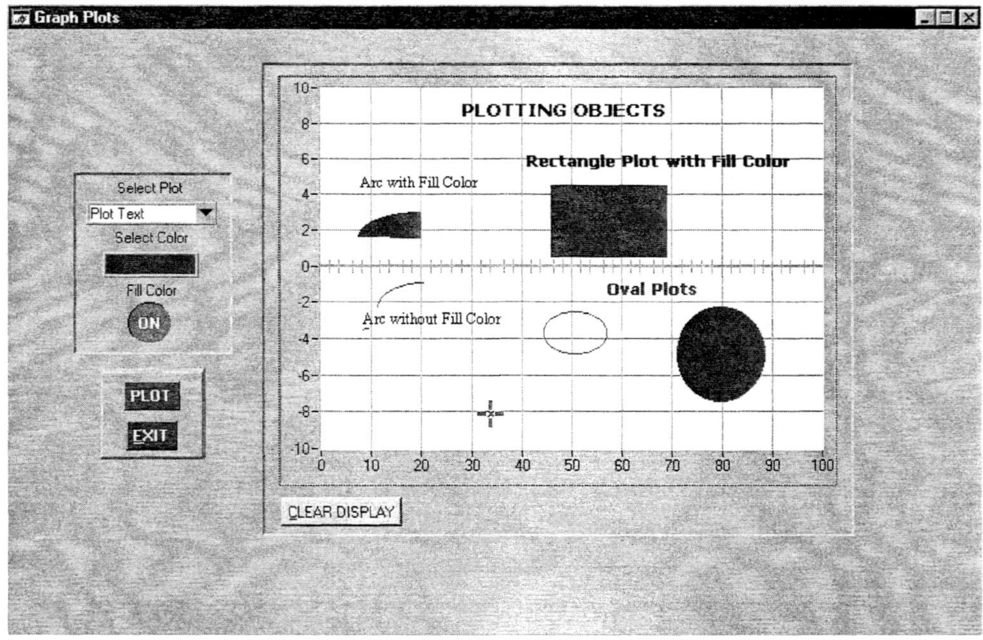

Figure 2–25
Sample of Various Plots

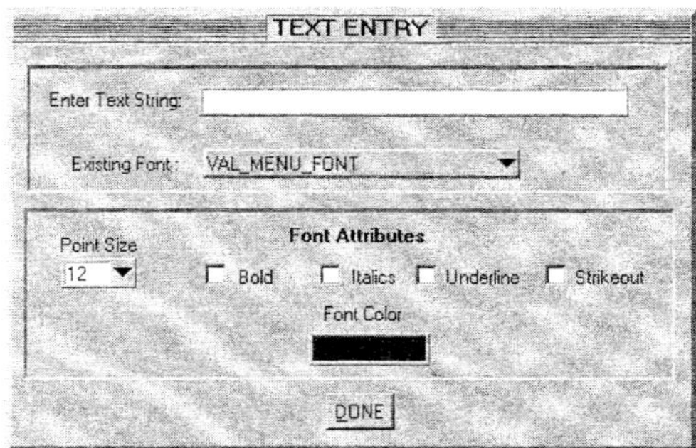

Figure 2–26
TEXT ENTRY GUI

When the **DONE** command button is selected, the *TextOKCB* callback function is called whose listing is shown in Figure 2–27. The library function *CreateMetaFont*, used to create a new metafont based on user selection, is shown on line 35.

At line 43 the *PlotText* library function is used to plot the text on the graph control. This function uses the new metafont created at line 35 to plot the text at the location specified.

In Figure 2–25 various text styles are plotted on the graph, annotating the plotted objects. You can experiment with creating your own metafont, selecting different font attributes from Figure 2–26 and plotting them on the graph control. You will understand the versatility of adding text plot on the graph control by experimenting with the base metafonts and the other attributes of the font. If you have multiple plots on the same graph control, you can use text to highlight the plots if you prefer not to use the legends for the plots.

Summary

This chapter showed you the many features of the graph control by means of projects. You saw how you could use the graph cursors and marker to obtain the value of any point on the plot. You were introduced to the features of zooming and panning both programmatically and manually. Legends are useful when you have multiple plots on the graph control to distinguish among the various plots. You learned how to create legends using the *CVI* toolbox. Plotting objects on the graph control and adding text to them comes in handy in many cases when you do not want to use the legend control but wish to annotate on the graph control.

```
//Selections for Text input
int CVICALLBACK TextOKCB (int panel, int control, int event,
          void *callbackData, int eventData1, int eventData2)
{
char TextBuffer[260], ExistingFontLabel[260], PointSizeLabel[260], NewMetaFont[260]={'\0'};
int FontIndex, FontColor, PointSizeIndex, PointSize, BoldVal, ItalicsVal, UnderlineVal,
      StrikeoutVal;

  switch (event)
  {

          case EVENT_COMMIT:
                  //Get the string entered in the text buffer
                  GetCtrlVal(TextPanelHandle,TEXT_TEXT_STRING,TextBuffer);
                  //Get existing font label
                  GetCtrlIndex(TextPanelHandle,TEXT_RING, &FontIndex);
                  GetLabelFromIndex(TextPanelHandle, TEXT_RING,FontIndex,
                                                          ExistingFontLabel);

                  //Get Point Size index and label.
                  GetCtrlIndex(TextPanelHandle,TEXT_POINT_SIZE, &PointSizeIndex);
                  GetLabelFromIndex(TextPanelHandle, TEXT_POINT_SIZE,
                                                  PointSizeIndex, PointSizeLabel);
                  Scan (PointSizeLabel, "%s>%i", &PointSize); //String to integer
                  //Check if Bold selected
                  GetCtrlVal(TextPanelHandle, TEXT_BOLD, &BoldVal);
                  //Check if Italics selected
                  GetCtrlVal(TextPanelHandle, TEXT_ITALICS, &ItalicsVal);
                  //Check if Underline selected
                  GetCtrlVal(TextPanelHandle, TEXT_UNDERLINE, &UnderlineVal);
                  //Check if Strikeout selected
                  GetCtrlVal(TextPanelHandle, TEXT_STRIKEOUT, &StrikeoutVal);

                  //Create a new meta-font based on the selections
                  CreateMetaFont(NewMetaFont, ExistingFontLabel, PointSize,BoldVal,
                                              ItalicsVal, UnderlineVal,StrikeoutVal);

                  //Obtain Font color
                  GetCtrlVal(TextPanelHandle, TEXT_COLOR_SELECT,&FontColor);

                  HidePanel(TextPanelHandle);
                  //Plot the selected text on the graph control with the created meta-font
                  PlotText(GraphPanelHandle, PLOTS_GRAPH, X1_Position,
                  Y1_Position,TextBuffer,NewMetaFont,FontColor,VAL_TRANSPARENT);
                  break;
  }
  return 0;
}//TextOKCB
```

Figure 2–27
TextOKCB Source Code Listing

Library Function Prototypes and Definitions

This section lists alphabetically the *CVI* library functions that were introduced in this chapter.

CreateMetaFont Function

The *CreateMetaFont* function creates a new metafont based on a predefined font. Its prototype is shown below and its arguments described in Table 2–1.

```
int status = CreateMetaFont (char newMetaFontName[ ],
                char existingFontName[ ], int pointSize,
                  int bold, int italics, int underlined,
                                        int strikeout);
```

Table 2–1 CreateMetaFont *Function*

Input/ Output	Name	Type	Description
Input	*newMetaFontName*	string	new metafont name
	existingFontName	string	base metafont name on which the new metafont is based
	pointSize	integer	point size selected for the new metafont
	bold	integer	if new metafont is bold, select 1, otherwise 0
	italics	integer	if new metafont is italics, select 1, otherwise 0
	underlined	integer	if new metafont is being underlined, select 1, otherwise 0
	strikeout	integer	if new metafont has strikeout text, select 1, otherwise 0
Output	*status*	integer	refer to Appendix A in *LabWindows/CVI User Interface Reference Manual* or see On-line Help for error codes

GetAxisScalingMode Function

The *GetAxisScalingMode* function obtains the scaling mode and the range of one of the graph axes or the y axis of a strip chart. Its prototype is shown below and its arguments described in Table 2–2.

```
int status = GetAxisScalingMode (int panelHandle,
                int controlID, int axis,int *axisScaling,
                                double *min, double *max);
```

The *axis Scaling* argument for the *GetAxisScalingMode* function is explained below. This argument allows you to determine how the axis selected is to be scaled. You have the following choices:

Table 2–2 GetAxisScalingMode *Function*

Input/ Output	Name	Type	Description
Input	*panelHandle*	integer	panel handle loaded in memory
	controlID	integer	constant name assigned to the graph control
	axis	integer	specifies the axis selected to obtain the mode and range from among the following: ■ VAL_XAXIS (graphs only) ■ VAL_LEFT_YAXIS (graphs and strip charts) ■ VAL_RIGHT_YAXIS (graphs only)
Output	*axisScaling*	integer*	scaling mode used for the axis as described below
	min	double*	current minimum value for the axis selected
	max	double*	current maximum value for the axis selected
	status	integer	refer to Appendix A in *LabWindows/CVI User Interface Reference Manual* or see On-line Help for error codes

- VAL_MANUAL. This specifies that the axis is scaled manually, and the *min* and *max* values indicate the low and high ends of the axis range.
- VAL_AUTOSCALE. This specifies that the axis will be scaled automatically by *CVI*, depending on the values plotted. The *min* and *max* values are not used. This option is not available for strip charts.
- VAL_LOCK. This specifies that the axis will be locked and you will not be able to change the scaling. The axis is locked to the current range specified. This option is not available for strip charts.

GetGraphCursor Function

The *GetGraphCursor* function obtains the current position of a specific graph cursor. The position is relative to the current range of the x and y axes. Its prototype is shown below and its arguments described in Table 2–3.

```
int status = GetGraphCursor (int panelHandle,
                 int controlID, int cursorNumber,
                               double *x, double *y);
```

Table 2–3 GetGraphCursor *Function*

Input/ Output	Name	Type	Description
Input	*panelHandle*	integer	panel handle loaded in memory
	controlID	integer	constant name assigned to the graph control
	cursorNumber	integer	cursor number for which the position is required
Output	*x*	double*	current coordinate for the x axis
	y	double*	current coordinate for the y axis
	status	integer	refer to Appendix A in *LabWindows/CVI User Interface Reference Manual* or see On-line Help for error codes

GetGraphCursorIndex Function

The *GetGraphCursorIndex* function obtains the plot handle and array index of the point to which the cursor is currently attached. Its prototype is shown below and its arguments described in Table 2–4.

```
int status = GetGraphCursorIndex (int panelHandle,
                         int controlID, int cursorNumber,
                         int *plotHandle, int  *arrayIndex);
```

LGCreateLegendControl Function

The *LGCreateLegendControl* function creates a legend control. Its prototype is shown below and its arguments described in Table 2–5.

```
int status= LGCreateLegendControl (int panelHandle,
                int anchorControl, int relativePosition,
                              int autoSize, int showSamples,
                                      int *newLegendControl);
```

Table 2–4 GetGraphCursorIndex *Function*

Input/ Output	Name	Type	Description
Input	*panelHandle*	integer	panel handle loaded in memory
	controlID	integer	constant name assigned to the graph control
	cursorNumber	integer	cursor number for which the position is required
Output	*plotHandle*	integer*	plot handle to which the cursor is attached
	arrayIndex	integer*	array index of the data point to which the cursor is attached; –1 is returned if the cursor is attached to the point plot or not attached to the data point
	status	integer	refer to Appendix A in *LabWindows/CVI User Interface Reference Manual* or see On-line Help for error codes

Table 2–5 LGCreateLegendControl *Function*

Input/ Output	**Name**	**Type**	**Description**
Input	*panelHandle*	integer	panel handle ID where the legend control is created
	anchorControl	integer	control ID of the anchor control; the legend control is associated with the anchor control
	relativePosition	integer	position of the legend control relative to the *anchorControl*; valid values are discussed below
	autoSize	integer	selecting "1" resizes the legend control when the legend items are added or removed; "0" disables this feature
	showSamples	integer	selecting "1" enables the sample plot in the legend control; this shows the plot style, point style, line style, and plot color
Output	*newLegendControl*	integer*	identifier of the new legend control created
	status	integer	return value of function; "0" indicates success; negative value indicates error

The *relativePosition* argument indicates where the legend control will be placed in reference to the anchor control. The following are its valid values:

- `LG_POS_DON'T_MOVE` 0
- `LG_POS_ABOVE_LEFT` 1
- `LG_POS_ABOVE_CENTER` 2
- `LG_POS_ABOVE_RIGHT` 3
- `LG_POS_RIGHT_TOP` 4
- `LG_POS_RIGHT_CENTER` 5
- `LG_POS_RIGHT_BOTTOM` 6
- `LG_POS_BELOW_RIGHT` 7
- `LG_POS_BELOW_CENTER` 8
- `LG_POS_RIGHT_LEFT` 9

- LG_POS_LEFT_BOTTOM 10
- LG_POS_LEFT_CENTER 11
- LG_POS_LEFT_TOP 12

Note that if the *anchorControl* is not a valid control, the *relativePosition* parameter is ignored and you must set the position of the legend using the *SetCtrlAttribute* function. If the anchor control is a graph control or strip chart, the legend's background color and border color are taken from the graph or strip chart.

LGInsertLegendItemForPlot Function

The *LGInsertLegendItemForPlot* function inserts a legend in a legend control at the position specified. The plot style, point style, line style, and color of the sample plot are set to the same values as the plot handle specified. Its prototype is shown below and its arguments described in Table 2–6.

```
int status = LGInsertLegendItemForPlot (int panelHandle,
                int legendControl, int insertPosition,
                        char legendText[ ], int textColor,
                                                int plotHandle);
```

Table 2–6 LGInsertLegendItemForPlot *Function*

Input/ Output	Name	Type	Description
Input	*panelHandle*	integer	panel handle ID where the legend control is created
	legendControl	integer	control ID of the legend control
	insertPosition	integer	position at which to insert the legend item
	legendText	char[ ]	text you want displayed with the legend item
	textColor	integer	color of the legend item text
	plotHandle	integer	plot handle with which you will associate this legend item
Output	*status*	integer	return value of function; "0" indicates success; negative value indicates error

LGSetLegendCtrlAttribute Function

The *LGSetLegendCtrlAttribute* function sets an attribute of the legend control. Its prototype is shown below and its arguments described in Table 2–7.

```
int status = LGSetLegendCtrlAttribute (int panelHandle,
                        int legendControl, int attribute,
                                 int attributeValue);
```

The various legend control attributes and their values are listed in Table 2–8.

PlotArc Function

The *PlotArc* function plots an arc on the graph control. Its prototype is shown below and its arguments described in Table 2–9.

```
int plotHandle = PlotArc (int panelHandle,
          int controlID, double starting_X_pos,
          double starting_Y_pos, double ending_X_pos,
                double ending_Y_pos, int beginAngle,
                int arcAngle, int color, int fillColor);
```

Table 2–7 LGSetLegendCtrlAttribute *Function*

Input/ Output	Name	Type	Description
Input	*panelHandle*	integer	panel handle ID where the legend control is created
	legendControl	integer	control ID of the legend control
	attribute	integer	attribute that you want to set; these attributes are discussed in Table 2–8
	attributeValue	depending on attribute	value of the attribute selected; the various attribute values are discussed in Table 2–8
Output	*status*	integer	return value of function; "0" indicates success, negative value indicates error

Table 2–8 Legend Control Attributes

Attribute	Attribute Value Type	Attribute Value
LG_ATTR_META_FONT	string	metafont used when legends are drawn
LG_ATTR_SHOW_SAMPLES	integer	either a "1" or a "0" to specify whether to include a sample plot with the legend
LG_ATTR_REL_POS	integer	position of the legend control relative to the anchor control; same as shown above for *relativePosition* argument in *LGCreateLegendControl*
LG_ATTR_AUTO_SIZE	integer	either a "1" or a "0" to specify whether to resize the legend control automatically when legends are added or deleted
LG_ATTR_OFFSET_FROM_ANCHOR	integer	specifies the number of pixels offset from the anchor control
LG_ATTR_ANCHOR_CTRL	integer	identifier for the anchor control for the legend
LG_ATTR_GRAPH_BG_COLOR	integer	specifies the graph background color; similar to setting it from the *SetCtrlAttribute* function
LG_ATTR_PLOT_BG_COLOR	integer	specifies the plot background color; similar to setting it from the *SetCtrlAttribute* function
LG_ATTR_DELAY_UPDATE	integer	"1" specifies to update the legend control automatically after each attribute change; "0" to not update the change

Table 2–9 PlotArc *Function*

Input/ Output	Name	Type	Description
Input	*panelHandle*	integer	panel handle ID where the legend control is created
	controlID	integer	control ID of the graph control
	starting_X_pos	double	x-coordinate of one corner of the rectangle that encloses the arc
	starting_Y_pos	double	y-coordinate of one corner of the rectangle that encloses the arc
	ending_X_pos	double	x-coordinate of the opposite corner of the rectangle that encloses the arc
	ending_Y_pos	double	y-coordinate of the opposite corner of the rectangle that encloses the arc
	beginAngle	integer	starting angle of the arc in tenths of a degree: range is 0 to 3600; positive value plots the arc counterclockwise and negative values counterclockwise
	arcAngle		sweep angle of the arc in tenths of a degree: range is 0 to 3600; positive value plots the arc counterclockwise and negative values counterclockwise
	color	integer	color selected using the RGB value as a 4-byte integer
	fillColor	integer	color selected to fill the arc
Output	*plotHandle*	integer	handle of the plotted arc

PlotLine Function

The *PlotLine* function plots a line on the graph control. Its prototype is shown below and its arguments described in Table 2–10.

```
int plotHandle = PlotLine (int panelHandle,
           int controlID, double starting_X_pos,
           double starting_Y_pos, double ending_X_pos,
                          double ending_Y_pos, int color);
```

PlotOval Function

The *PlotOval* function plots an oval on the graph control. Its prototype is shown below and its arguments described in Table 2–11.

```
int plotHandle = PlotOval (int panelHandle,
                 int controlID, double starting_X_pos,
                 double starting_Y_pos, double ending_X_pos,
                 double ending_Y_pos, int color, int fillColor);
```

Table 2–10 PlotLine *Function*

Input/ Output	Name	Type	Description
Input	*panelHandle*	integer	panel handle ID where the legend control is created
	controlID	integer	control ID of the graph control
	starting_X_pos	double	starting x-coordinate of the line
	starting_Y_pos	double	starting y-coordinate of the line
	ending_X_pos	double	ending x-coordinate of the line
	ending_Y_pos	double	ending y-coordinate of the line
	color	integer	color selected using the RGB value as a 4-byte integer
Output	*plotHandle*	integer	handle of the plotted line

Table 2–11 PlotOval *Function*

Input/ Output	Name	Type	Description
Input	*panelHandle*	integer	panel handle ID where the legend control is created
	controlID	integer	control ID of the graph control
	starting_X_pos	double	x-coordinate of one corner of the rectangle to define the oval bound
	starting_Y_pos	double	y-coordinate of one corner of the rectangle to define the oval bound
	ending_X_pos	double	x-coordinate of the opposite corner of the rectangle to define the oval bound
	ending_Y_pos	double	y-coordinate of the opposite corner of the rectangle to define the oval bound
	color	integer	color selected using the RGB value as a 4-byte integer
	fillColor	integer	color selected to fill the oval
Output	*plotHandle*	integer	handle of the plotted oval

PlotRectangle Function

The *PlotRectangle* function plots a rectangle on the graph control. Its prototype is shown below and its arguments described in Table 2–12.

```
int plotHandle = PlotRectangle (int panelHandle,
                 int controlID, double starting_X_pos,
                 double starting_Y_pos, double ending_X_pos,
                 double \ending_Y_pos, int color,int fillColor);
```

Table 2–12 PlotRectangle *Function*

Input/ Output	Name	Type	Description
Input	*panelHandle*	integer	panel handle ID where the legend control is created
	controlID	integer	control ID of the graph control
	starting_X_pos	double	x-coordinate of one corner of the rectangle
	starting_Y_pos	double	y-coordinate of one corner of the rectangle
	ending_X_pos	double	x-coordinate of the opposite corner of the rectangle
	ending_Y_pos	double	y-coordinate of the opposite corner of the rectangle
	color	integer	color selected using the RGB value as a 4-byte integer
	fillColor	integer	color selected to fill the rectangle
Output	*plotHandle*	integer	handle of the rectangle plotted

PlotText Function

The *PlotText* function plots a text string on the graph control. Its prototype is shown below and its arguments described in Table 2–13.

```
int plotHandle = PlotText (int panelHandle,
                    int controlID, double X_pos,
                    double Y_pos, char text[ ],
                    int font, int textColor,
                    int backgroundColor);
```

Table 2–13 PlotText *Function*

Input/ Output	Name	Type	Description
Input	*panelHandle*	integer	panel handle ID where the legend control is created
	controlID	integer	control ID of the graph control
	X_pos	double	x-coordinate for text placement
	Y_pos	double	y-coordinate for text placement
	text	string	string to plot
	font	integer	font used to plot the text; could be font created from *CreateMetaFont*
	textColor	integer	color selected using the RGB value as a 4-byte integer
	backgroundColor	integer	background color selected for the text plotted
Output	*plotHandle*	integer	handle of the plot

SetAxisScalingMode Function

The *SetAxisScalingMode* function sets the scaling mode and the range of either one of the axes of a graph, or the Y axis of a strip chart. Its prototype is shown below and its arguments described in Table 2–14.

```
int status = SetAxisScalingMode (int panelHandle,
             int controlID, int axis, int *axisScaling,
                                         double *min, double *max);
```

Table 2–14 SetAxisScalingMode *Function*

Input/ Output	Name	Type	Description
Input	*panelHandle*	integer	panel handle loaded in memory
	controlID	integer	constant name assigned to the graph control
	axis	integer	specifies the axis selected to set the mode and range from among the following: VAL_XAXIS (graphs only) VAL_LEFT_YAXIS (for graphs and strip charts) VAL_RIGHT_YAXIS (graphs only)
Output	*axisScaling*	integer*	same as explained for *GetAxisScaling-Mode* function
	min	double*	minimum value to set for the axis selected
	max	double*	maximum value to set for the axis selected
	status	integer	refer to Appendix A in *LabWindows/CVI User Interface Reference Manual* or see On-line Help for error codes

3

USING DATASOCKET

Chapter Highlights

- Introduction
- Communicating Using DataSocket
- DataSocket Data Files
- Creating a DataSocket Application
- DataSocket Applications
- Accessing the DataSocket Server
- DataSocket Server Manager Configurations
- Summary
- Library Function Prototypes and Definitions

The fundamentals of DataSocket technology to transfer data between applications on a local computer, computers on a network, or to remote computers via the Internet are explained here. To demonstrate the reading and writing of data using DataSocket technology, two projects are created and their functionalities shown. The first project transmits data to the DataSocket Server on a local machine, and another project receives the same transmitted data from the DataSocket Server. The various features of configuring the DataSocket Server Manager are explained. How you can collect data at one location and broadcast to different computers across a local area network (LAN) or publish it on the Web through the Internet is explained.

Introduction

DataSocket is an Internet programming technology that allows you to read, write, and share data between applications on a local computer or a host of computers on a network using a variety of communication technologies. DataSocket technology was developed by National Instruments and included with the *CVI* software starting with version 5.5. You do not have to do anything special to load the DataSocket libraries and the DataSocket Server since they are installed automatically when you install *CVI*.

DataSocket Server is an executable that enables the exchange of data between multiple applications. It receives and stores information from data sources and relays it to other data targets, whether they are on the same computer or on other computers connected through a Transmission Control Protocol (TCP) Ethernet network.

DataSocket provides an easy-to-use high-level interface using the *CVI* DataSocket library functions in which the low-level Transmission Control Protocol/Internet Protocol (TCP/IP) communication protocols are already established for you. The DataSocket technology is built on top of TCP/IP and itself has little overhead. DataSocket manages the TCP that is used to share *live* data between applications on one computer or between one or many computers connected through the network via the Internet and respond to multiple users remotely using the Web browser. Live data (also called *streaming data*) is the data that is not retrieved from a file or any storage media but is transmitted as soon as it is created.

Using the DataSocket technology, all you need is to open a DataSocket connection, write data to that connection, simultaneously read the data on the receiver, and disconnect from the DataSocket when done. There are other technologies, such as TCP/IP and Dynamic Data Exchange (DDE), that accomplish the same objective but require considerable low-level TCP/IP programming and are not suitable for live data exchange. DataSocket technology was designed specifically to transfer the measurement and test data.

Communicating Using DataSocket

To communicate with DataSocket, you specify the data or target source, much like the Web `URL`. The `URL` consists of the data type, the machine name, and the data source to access. Data type is indicated by the prefix in the `URL`, called the URL scheme, access method, or protocol. DataSocket can communicate with several URL schemes, such as http (Hypertext Transfer Protocol), https (encrypted http), ftp (File Transfer Protocol), and Object Linking and Embedding (OLE) for Process Control (OPC).

DataSocket also uses its own protocol, called DataSocket transfer protocol (DSTP), to establish communication between the DataSocket Server, the writer, and the reader. The writer is also referred to as the *publisher* and the reader as the *subscriber*. If you were to specify a URL such as dstp://localhost/test1, it would mean that you want to open a DataSocket transfer protocol connection using the dstp scheme to the machine named localhost and write/receive data to/from the data source test1.

DataSocket consists of a DataSocket application programming interface (API) and the DataSocket Server. You write (publish) the data to the DataSocket Server using the DataSocket API and read (subscribe) the data from the DataSocket Server using the same DataSocket API. The publisher and subscriber are both clients of the DataSocket Server.

To write the code to convert the measurement data to bytes at the sender computer and to reconstruct the string of bytes in a readable format at the receiver computer is taken care of internally by the DataSocket API.

The DataSocket Server, publisher, and subscriber can run on the same machine or on different machines, depending on your application. Running the DataSocket Server on a machine other than the publisher is recommended because you can distribute the various application functions to different machines on the network, thereby improving performance and providing security by isolating network connections from your measurement applications.

DataSocket Data Files

A file scheme is used for reading and writing files on a local machine. The format for this URL scheme is of the form file:../Advprojects/dstest.dsd. This indicates that you would write to or read data from a local file dstest.dsd located in the Advprojects directory on your machine. The data types presently supported by DataSocket are tabbed text files (such as those created in Microsoft Excel), DataSocket Data (DSD) format (*.dsd extension), and sound waves files (*.wav files). The DSD file format is a National Instruments proprietary binary file format used for storing DataSocket waveform data, data values, and data attribute values, including multiple waveforms and arbitrary data attributes, or any data type supported by DataSocket. Presently writing to `.wav` files is not supported, although you can write waveform data in the DSD format. DataSocket can transfer large data files, as there are no limitations on the size of data to transfer, although performance will tend to decrease. The speed of transferring data on the network is limited only by the network bandwidth and the traffic on the network. The speed of transferring data between machines connected by 10BaseT Ethernet can be on the order of 320 kB/s.

Creating a DataSocket Application

To acquaint you with some of the capabilities of DataSocket, let us run the project to transfer live data from a writer (publisher) application to a reader (subscriber) application using the DataSocket Server, all residing on the same computer. This project will create random numbers every second with a time stamp indicating when the random number was generated and count the number of items created. The data generated is displayed on the GUI and written to the data source connected to the DataSocket Server (Figure 3–1). This data is then read from the same data source and connected to the DataSocket Server and displayed on the **READ DATA** GUI (Figure 3–2).

Start *CVI* and load the project `project3Send.prj` from the `AdvProjects` folder. Create and run the executable for this project. The **WRITE DATA** application will start and the **DataSocket Server** will be launched as shown in Figure 3–1. This application will write data to `dstp://localhost/wave`. Here the data is written to `localhost` using the data source `wave`, as shown in the **Target:** control box on the GUI. The `dstp` refers to the data source type being used. This is the default data source when you launch the DataSocket Server.

When you run your DataSocket applications you are using the default DataSocket Server Manager settings. These settings work for most of your applications and may not require any modifications. To change these settings, see the section *DataSocket Server Manager Configurations* later in the chapter.

Before you click on the **CONNECT** command button to transmit data, you need to load `project3Receive.prj` from the same folder, create an executable, and run it. Figure 3–2 will be displayed. This project will read the data sent by `project3Send.prj` to the data source connected to the

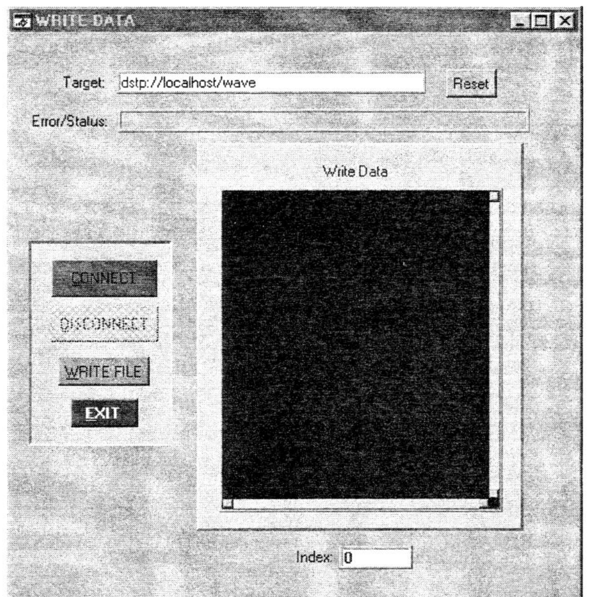

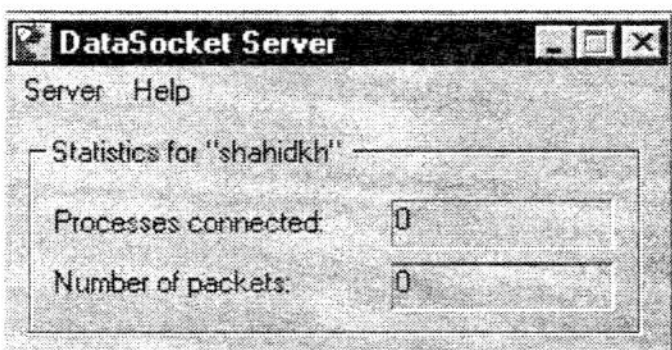

Figure 3–1
WRITE DATA GUI with **DataSocket Server** Launched

DataSocket Server. Notice that the **Target:** control box on **WRITE DATA** GUI and **Source:** control box on **READ DATA** GUI are both pointing to the same data source `dstp://localhost/wave` for both applications to establish communication.

To start communication between the two applications, click first on the **CONNECT** command button on the **WRITE DATA** GUI and then on the **CONNECT** command button on the **READ DATA** GUI. The data written to the DataSocket Server by the **WRITE DATA** project is displayed on the text box with the time stamp on the **WRITE DATA** GUI and transferred simultaneously to the reader application from the DataSocket Server and displayed on the **READ DATA** GUI. A sample run of this is shown in Figure 3–3. The arrow connections that are indicated in this figure represent the direction of data flow. It is important to understand that these two projects communicate through the DataSocket Server only and not directly with each other. `project3Send` writes the data to the DataSocket Server, and `project3Receive` reads the same data from the DataSocket Server.

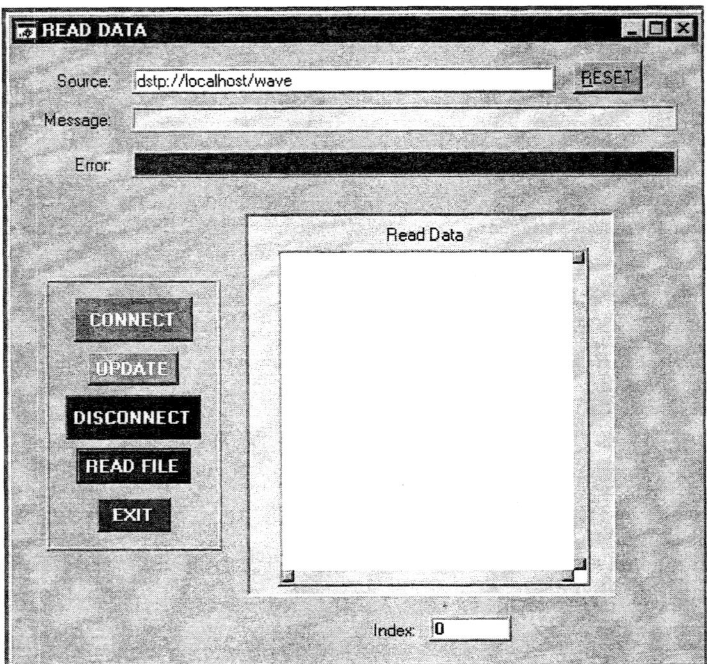

Figure 3–2
READ DATA GUI

To stop the data transfer, first click on the **DISCONNECT** command button on the **READ DATA** GUI and then on the **DISCONNECT** command button on the **WRITE DATA** GUI. You will notice that data is no longer created and updated on both the writer and reader GUIs. To start communication again, click on the **CONNECT** command button on the **WRITE DATA** GUI and after that the **CONNECT** command button on the **READ DATA** GUI. This will retransmit the data as before.

Analyzing the Writer Code

You saw above how data is written to a DataSocket data source in one *CVI* application and read from another *CVI* application connected to the same data source. Now let us analyze the source code used for `project3Send.prj`, the data writer, and explain the DataSocket library functions called in running this application.

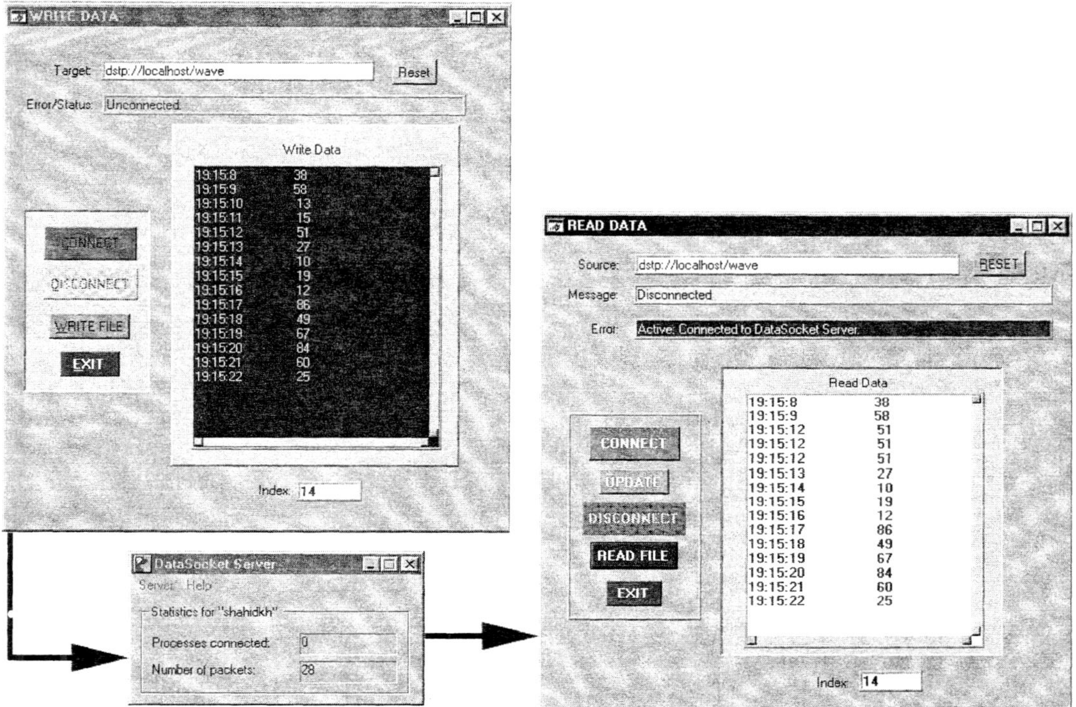

Figure 3–3
`project3` Sample Run Showing Data Path

main Function

Figure 3–4 lists the header and the *main* function for this project. The data is transferred using the *structure* shown on lines 20–24. Two structure members are used in this application. The random number is stored in the structure member `data`. The count of the number of items written and read will be tracked in the structure member `index`, which is a DataSocket attribute of `data`. DataSocket attributes are additional data items that are sent/received in support of the data. These are customized information items that are used to enhance the data sent/received along with the data. Attributes may consist of such information as the timestamp when the data was acquired, test number, operator name, number of data items sent/received, data acquisition rate, and so on. Each attribute has a user-defined name by which you can access it, as you will see later when the *UpdateDSCallback* function (Figure 3–6) is explained.

```
1   #include <cvidef.h>
2   #include <cvirte.h>
3   #include <userint.h>
4   #include <cviauto.h>
5   #include <ansi_c.h>
6   #include <utility.h>
7   #include "dataskt.h"
8   #include <formatio.h>
9   #include "Project3Send.h"
10
11  /* This is the callback function prototype for  DataSocket */
12  void CVICALLBACK UpdateDSCallback (DSHandle DSWriteHandle, int event, void *callbackData);
13  int Checkerror(int ); //Checks error of datasocket functions
14  static int WritepanelHandle;
15  static  DSHandle DSWriteHandle = 0;
16  HRESULT error;
17  char errorBuf[260];
18
19  //Structure to write data
20  typedef struct WriteDataStruct
21  {
22      short data;                 //random number
23      short index;                //counter
24  } WriteDataStruct;
25  WriteDataStruct WriteData;
26
27  //Main function
28  int main (int argc, char *argv[])
29  {
30      if (InitCVIRTE (0, argv, 0) == 0)   /* Needed if linking in external compiler; harmless
31                                                                                   otherwise */
32          return -1;  /* out of memory */
33      if ((WritepanelHandle = LoadPanel (0, "Project3Send.uir", WRT)) < 0)
34          return -1;
35      DisplayPanel (WritepanelHandle);
36     //Start the DataSocket Server
37      error =DS_ControlLocalServer (DSConst_ServerLaunch);
38      Checkerror(error);
39
40      RunUserInterface ();
41      //Close the DataSocket Server
42      error = DS_ControlLocalServer (DSConst_ServerClose);
43      Checkerror(error);
44      return 0;
45  }// main
```

Figure 3–4
`project3Send` Header and *main* Function

DS_ControlLocalServer at lines 37 and 42 is a DataSocket library function that controls the DataSocket Server on the local machine. The function argument indicates the controls you can exercise on this function. Line 37 starts the DataSocket Server using the constant `DSConst_ServerLaunch` as an argument of this function. After the *RunUserInterface* has returned, you close the DataSocket Server at line 42 using the `DSConst_ServerClose` argument in this function. You can hide or display the DataSocket Server using DSConst_ServerHide or DSConst_ServerShow argument in this function, respectively. Checking for errors makes the program more robust and is done throughout this project by the user-defined function *Checkerror(error)*. If there is an error, the library function *DS_GetLibraryErrorString* is called in the *Checkerror(error)* function to convert the integer value passed in error to a meaningful string that is displayed on the GUI.

Connecting Data Source

To start sending data, you need to click on the **CONNECT** command button on the **WRITE DATA** GUI (Figure 3–1). This invokes the *ConnectCB* callback function listed in Figure 3–5. At line 13 *DS_Open* library function creates a DataSocket object and connects it to the data source. The second argument in the *DS_Open* function determines if the DataSocket object is configured for writing, or for reading. When the DataSocket data object is configured for writing, the DataSocket object writes the data to the data source specified in the URL in the function argument. When the DataSocket object is configured for reading, it reads the data from the DataSocket data source specified in the URL argument of the function.

```
1   //Invoked when CONNECT button is selected
2   int CVICALLBACK ConnectCB (int panel, int control, int event,
3           void *callbackData, int eventData1, int eventData2)
4   {
5       char URL[500];
6
7       switch (event) {
8           case EVENT_COMMIT:
9               ResetTextBox(WritepanelHandle,WRT_DSP_DATA, "");
10              //Read the connection from the URL control box
11              GetCtrlVal (WritepanelHandle, WRT_SOURCE, URL);
12              //Open the DataSocket or File connection (depending on the URL)
13              error=DS_Open (URL, DSConst_Write, UpdateDSCallback, NULL, &DSWriteHandle);
14              Checkerror(error);
15              WriteData.index=0; //initialize the index
16
17              //Start the timer to generate  data
18              SetCtrlAttribute(WritepanelHandle, WRT_TIMER ,ATTR_ENABLED,TRUE);
19              //Enable DISCONNECT button
20              SetCtrlAttribute (WritepanelHandle, WRT_DISCONNECT, ATTR_DIMMED, 0);
21              //Disable CONNECT button
22              SetCtrlAttribute (WritepanelHandle, WRT_WRT_CON , ATTR_DIMMED, 1);
23      break;
24      }
25      return 0;
26  }//ConnectCB
```

Figure 3–5
project3Send *ConnectCB* Callback Function

You can update the data written to or read from the data source in either the *AutoUpdate* mode or the *non-AutoUpdate* mode. You can control when you want to send or receive data from the server. When writing data in the AutoUpdate mode the most recent data is sent to the server immediately. When reading data in the AutoUpdate mode, the most recent data is retrieved from the server immediately. Use the non-AutoUpdate mode when you want the data sent or received per your request. Use the non-AutoUpdate mode when you want data synchronized with its attributes. When connecting to other data sources besides DataSocket Servers (`dstp`) or OPC Servers (`opc`) it is advisable to connect in non-AutoUpdate mode. If you are connecting to OPC Servers, non-AutoUpdate mode allows for better error checking and more control over the communication between your application and the OPC Server. However, if you are connecting to multiple items on the OPC Server, your application will run faster in the AutoUpdate mode since the OPC Server sends all the item changes at once.

In the code for the *ConnectCB* function, the second argument, `DSConst_Write`, in the *DS_Open* function configures the data object for updating data manually. Whenever DataSocket data is updated or the status of the connection between a DataSocket object and its DataSocket data source is changed, the DataSocket object invokes and executes the code in the callback function specified in the third argument of the *DS_Open* function (*UpdateDSCallback* in this case). The *UpdateDSCallback* function is discussed in the next section. The status of the DataSocket object can be one of the following conditions:

- The DataSocket object is not connected to any data source or data target.
- The DataSocket object is in the process of transferring data or waiting for an update.
- The DataSocket object has connected to the data source and transferred the data.
- The DataSocket object encountered an error connecting to the data source or target.

The last argument in the *DS_Open* function is the DataSocket handle (`DSWriteHandle`) associated with the DataSocket object. This handle is created by the *DS_Open* function and is used in subsequent function calls when referring to this DataSocket object. Unless there is an error in creating the DataSocket object, this handle must never have a zero value.

Updating the Writer Callback Function

Let us now look at the *UpdateDSCallback* function, which is called from the *DS_Open* library function. The source code for this callback function is listed in Figure 3–6. In this callback function the results of the status update are obtained and displayed on the GUI. Let us first discuss the parameters that are passed to the *UpdateDSCallback* function. `DSWriteHandle` is the DataSocket object handle that generated the event. The `event` parameter specifies the event that invoked this callback function. Recall from above that this function is called when DataSocket object's data value, data attribute, or the status is updated. There are two `event` values that can occur:

- A `DS_EVENT_DATAUPDATED` event occurs when the DataSocket object's data value or any of the data attributes changes.
- A `DS_EVENT_STATUSUPDATED` event occurs when the DataSocket object's status changes.

The `callbackData parameter` contains the value that you pass to this callback function. In the listing shown in Figure 3–6, the `DS_EVENT_STATUS-UPDATED` event is used. The library function *DS_GetLastMessage* returns a string containing a description of either the last communication error, last data update, or last interaction between the DataSocket object and the DataSocket data source. This message is displayed in the **Error/Status:** control box on the GUI (Figure 3–1).

```
//Called from  DS_Open function when data or error is updated
void CVICALLBACK UpdateDSCallback (DSHandle DSWriteHandle, int event, void *callbackData)
{
    char message[1000];

    switch (event) {
        case DS_EVENT_STATUSUPDATED: /* on status updated*/
          //Obtain the last message and update the message box
            error =DS_GetLastMessage (DSWriteHandle, message, 1000);
            Checkerror(error);
            SetCtrlVal (WritepanelHandle, WRT_STATUS, message);
            break;
    }
}//UpdateDSCallback
```

Figure 3–6
UpdateDSCallback Function

Creating and Writing the Data Function

Let us now see how the DataSocket data and its attribute values are being generated and written to the source. Refer to the listing of the *TimerCallback* function in Figure 3–7, which generates a random number every second. The random number ranges in value from 0 to 99 and its value is assigned to the structure member `WriteData.data` at line 12.

```
1  // Create new random data and index
2  int CVICALLBACK TimerCallback (int panel, int control, int event,
3          void *callbackData, int eventData1, int eventData2)
4  {
5
6      int hours, minutes, seconds;
7      char  DataString[50];
8
9      switch (event) {
10         case EVENT_TIMER_TICK:
11           //Generate the random data and write to the structure member
12           WriteData.data = rand() % 100;
13           //Get the system time
14           GetSystemTime(&hours, &minutes, &seconds);
15
16           if (DSWriteHandle)//if handle is valid
17           {
18               //Create a string to write
19               sprintf(DataString, "%-2d:%-2d:%-2d           %-2d", hours, minutes,
20                                                       seconds,WriteData.data);
21               //Display string on the Text Box
22               InsertTextBoxLine(WritepanelHandle,WRT_DSP_DATA,-1, DataString);
23               //Set the data value for the DataSocket object
24               error=DS_SetDataValue (DSWriteHandle, CAVT_CSTRING, &DataString, 50, 0);
25               Checkerror(error);
26               //Set the datasocket attribute value for index
27               error=DS_SetAttrValue (DSWriteHandle, "Index", CAVT_SHORT,
28                                                       &WriteData.index, 0, 0);
29               Checkerror(error);
30               //Synchronize the data value and the attribute value
31               error =DS_Update(DSWriteHandle);
32               Checkerror(error);
33               //Update the Index box
34               SetCtrlVal (WritepanelHandle, WRT_INDX, WriteData.index);
35               WriteData.index++;
36            }
37            else
38            {
39                   MessagePopup ("DATA SOCKET HANDLE ERROR",
40                               "DataSocket Handle is invalid.\nRe-run Project.");
41           exit(-1);  //Terminate project
42     }
43            break;
44      }
45      return 0;
46 } //TimerCallback
```

Figure 3–7
TimerCallback Function

At line 14 a time stamp is obtained using the *GetSystemTime* library function. At line 16 the DataSocket handle is validated. If valid, lines 19–35 are executed; otherwise, line 39 displays a message to indicate that the DataSocket handle is invalid and the program is terminated. The data that is transmitted to the DataSocket server in a string variable `DataString` consisting of the timestamp obtained from the system clock and the random data value assigned to `data`. The `DataString` format is created and displayed on the text box at lines 19–22.

The *DS_SetDataValue* library function sets the data value of the DataSocket object at line 24. The DataSocket object sends the new data value to the data source connected, based on how it is configured in the *DS_Open* function. When configured in `WriteAutoUpdate` mode, the data is sent immediately to the data source. When configured in the `Write` mode, as in this case, the data is sent to the data source only when the *DS_Update* library function is called at line 31. The library function *DS_Update* is used to synchronize the DataSocket object data value with the attribute value when the DataSocket object is configured in `Read` or `Write` modes in the *DS_Open* function. In the `Read` mode the current data value and the attributes are obtained from the data source. In the `Write` mode the current data value and the attributes are sent to the data source. If the DataSocket objects are configured for `ReadAutoUpdate` or `WriteAutoUpdate` mode, use of this function is redundant.

At line 27 the library function *DS_SetAttrValue* sets the value of the data attribute `WriteData.index`. The attribute in this case is the structure member `index` (defined in the program header; see Figure 3–4) that is being written to the data source and displayed on the GUI in the **Index:** control to indicate the number of data values. The `index` is updated every timer tick when the data is generated and written to the data source. Note that the second argument in the *DS_SetAttribute* function contains the name of the data attribute (`Index` in this case) whose value you want to set.

Stop Writing the Data

To stop DataSocket objects from being written to the DataSocket data source, click on the **DISCONNECT** command button from the GUI (Figure 3–1). The *DisconnectCB* function is called and listed in Figure 3–8. The 1-second timer is disabled to stop the sending of data. The library function *DS_DiscardObjHandle* discards the DataSocket object and frees the resources used by it. The *DS_DiscardObjHandle* function contains only one argument, which is the DataSocket handle of the object you want to discard.

```
//Invoked by the DISCONNECT command button to discard the DataSocket object
//and free the resources
int CVICALLBACK DisconnectCB (int panel, int control, int event,
        void *callbackData, int eventData1, int eventData2)
{
    switch (event) {
        case EVENT_COMMIT:
            //Disable timer
            SetCtrlAttribute (WritepanelHandle, WRT_TIMER, ATTR_ENABLED, FALSE);
            if (DSWriteHandle)
            {
                //Discard the DataSocket object and free resources
                error =DS_DiscardObjHandle(DSWriteHandle);
                Checkerror(error);
                DSWriteHandle = 0;
            }

            SetCtrlVal (WritepanelHandle, WRT_STATUS, "Unconnected.");
           //Disable DISCONNECT button
           SetCtrlAttribute (WritepanelHandle, WRT_DISCONNECT, ATTR_DIMMED, 1);
           //Enable CONNECT button
           SetCtrlAttribute (WritepanelHandle, WRT_WRT_CON, ATTR_DIMMED, 0);
      break;
    }
    return 0;
} //DisconnectCB
```

Figure 3–8
DisconnectCB Function

Writing to Data File

Instead of transmitting "live" data, you have the option to record the data to a file and read it later from the DataSocket subscriber. To write data to a data socket file, click on the **WRITE FILE** command button (Figure 3–1). A pop-up panel will appear in which you enter the name of the data file where you want to write the data. Click on the **CONNECT** command button to generate the data and write to the data file.

When you click on the **WRITE FILE** command button on the GUI, the *WriteToFileCB* function listed in Figure 3–9 is called. Between lines 12 and 21 you are asked to enter the file name in the pop-up panel, and the function verifies that the file name entered is not a blank. If a blank entry is made for the file name, the function loops until a file name with one or more characters is entered. The data file will be written to a newly created subdirectory `datafiles` inside the project directory. At line 28 the library function *SetDir* checks if the directory exists with this pathname. If not, it creates the directory inside the project directory using the library function *MakeDir*. To disable run-time error checking, call the function *SetBreakOnLibraryErrors(0)* at line 25 and reset it at line 29. At line 32 the directory path is set to the `datafiles` directory. The format for the URL to write the data to the DataSocket file is of the form

```
file:../directoryName/filename.dsd
```

This string is created at line 32 and is displayed on the **Target:** control box on the **WRITE DATA** GUI. The *DS_Open* function uses this string for the URL argument to inform the DataSocket Server to write the data to this file. The data is written to a data socket file that you entered for the file name in the pop-up panel above. The file name is appended with a `.dsd` extension.

```
1    //Write data to file.  Invoked when the "WRITE FILE" command button is selected
2    int CVICALLBACK WriteToFileCB (int panel, int control, int event,
3              void *callbackData, int eventData1, int eventData2)
4    {
5
6      char  prjDir[261], FileURL[261], dirName[261], DSFileName[261];
7      short len;
8      int nPrevState;
9      switch (event)
10     {
11             case EVENT_COMMIT:
12                     len =0;
13                     do
14                     {
15                         //Enter file name to write data
16                         PromptPopup ("DATASOCKET FILE NAME",
17                   "Enter the file name to write data (no extension)",  DSFileName, 260);
18
19                       len = strlen (DSFileName);
20                       //Check file name entered is at least one character long
21                     } while (len ==0 );
22
23                     SetCtrlVal(WritepanelHandle, WRT_SOURCE, "");
24                     GetProjectDir(prjDir);
25                     nPrevState= SetBreakOnLibraryErrors(0);
26                     sprintf(dirName, "%s\\datafiles",prjDir);
27                     //Make the directory
28                     if ( SetDir(dirName) !=0)  MakeDir (dirName);
29                     SetBreakOnLibraryErrors(nPrevState);
30                     SetDir(dirName);  //Set the path
31                     //Create the path name to write to ".dsd" file
32                     sprintf(FileURL,"file:../datafiles/%s.dsd",DSFileName);
33                     SetCtrlVal(WritepanelHandle, WRT_SOURCE, FileURL);
34     }
35     return 0;
36   } //WriteToFileCB
```

Figure 3–9
WriteToFileCB Function

Analyzing the Reader Code

In the section *Analyzing the Writer Code* you saw how the data and its attribute values are created and written by `project3Send.prj` to the DataSocket data source. The `project3Receive.prj` reads the same DataSocket data and attribute values synchronously from the data source and displays it on the **READ DATA** GUI (Figure 3–2).

main Function

The file header and the *main* function for `project3Receive.c` are shown in Figure 3–10. Notice that in Figure 3–3 the data source name used in the **Source:** control for this GUI is the same as in the **Target**: control box of the **WRITE DATA** GUI, so both DataSocket applications (writer and reader) are pointing to the same data source.

```
#include <cvidef.h>
#include <cviauto.h>
#include <dataskt.h>
#include <formatio.h>
#include <ansi_c.h>
#include <cvirte.h>
#include <userint.h>
#include <utility.h>
#include "project3Receive.h"

// Structure to Read
typedef struct WriteDataStruct
{
    short data;
    short index;
} WriteDataStruct;


int Checkerror(int ); //Checks status of datasocket functions

static WriteDataStruct WriteData;
static int PanelHandle;
static DSHandle DSReadHandle;  //DShandle is the handle used by all of the DataSocket Operations
int hours,minutes, seconds;
HRESULT errorCode;
char Source_String[500], errorBuf[256] , *ReadString;
```

Figure 3–10
`project3Receive` Header and *main* Function *(continued)*

```
//Function prototype
void CVICALLBACK DSCallback(DSHandle DSReadHandle, int event, void *callbackData);

//main function loads the GUI and launches the DataSocket server
int main (int argc, char *argv[])
{
      if (InitCVIRTE (0, argv, 0) == 0)
        return -1;        /* out of memory */
      if ((PanelHandle = LoadPanel (0, "project3Receive.uir", DS_READ)) < 0)
        return -1;
      DisplayPanel (PanelHandle);

      RunUserInterface ();
      return 0;
} //main
```

Figure 3–10
`project3Receive` Header and *main* Function *(continued)*

Connecting the Reader

To connect the reader to the same data source as the writer, click on the **CONNECT** command button on the **READ DATA** GUI (Figure 3–2) and the *ConnectReadCB* callback function is invoked. The listing for this function is shown in Figure 3–11. This function obtains the string in the **Source:** control box from the GUI and uses it as an argument in the *DS_Open* library function at line 23. The DataSocket object is configured to read in AutoUpdate mode to receive the data as soon as the data value or the attribute value changes. The DataSocket handle `DSReadHandle` is created by the *DS_Open* library function, and the *DSCallback* function is called when the data value or data attributes are updated or when there is a DataSocket status change. The connection status between a DataSocket object and its DataSocket data source is obtained from the *DS_GetStatus* library function at line 27 and displayed on the GUI using the statement at line 45.

Updating the Reader

Figure 3–12 lists the code for the *DSCallback* function. This function receives the data and the data attribute value from the data source as soon as the data is updated. The data string sent by the writer is received and displayed on the text box and the data attribute displayed in the **Index:** control if there are no errors in receiving the data.

```
//Invoked when the CONNECT button is selected.
int CVICALLBACK ConnectReadCB (int panel, int control, int event,
          void *callbackData, int eventData1, int eventData2)
{
  char buf [260]= {"\0"};
  DSEnum_Status Status;

  switch (event)
  {
          case EVENT_COMMIT:
                  ResetTextBox (PanelHandle,DS_READ_RD_DSP, "");
                  if (DSReadHandle)
                  {
                          errorCode =DS_DiscardObjHandle(DSReadHandle);
                          Checkerror(errorCode);
                          DSReadHandle=0;
                  }

                  //Read the Source
                  GetCtrlVal(PanelHandle, DS_READ_SOURCE, Source_String);

                  //Creates a DataSocket object and connects it to the data source
                  errorCode=DS_Open (Source_String, DSConst_ReadAutoUpdate,
                                                   DSCallback, NULL, &DSReadHandle);

                  Checkerror(errorCode);
                  errorCode=DS_GetStatus (DSReadHandle,&Status);
                  Checkerror(errorCode);
                  switch (Status)
                  {
                          case 1:
                                   sprintf(buf, "Unconnected");
                                   break;
                          case 2:
                                   sprintf(buf, "Connection Active");
                                   break;
                          case 3:
                                   sprintf(buf, "Connection Idle");
                                   break;
                          case 4:
                                   sprintf(buf, "Connection Error");
                                   break;
                  }
                  //Display DS Status
                  SetCtrlVal(PanelHandle, DS_READ_MSG ,buf);

  }
  return 0;
} //ConnectReadCB
```

Figure 3–11
ConnectReadCB Function

```
//This callback function is executed whenever the data or the status changes
void CVICALLBACK DSCallback(DSHandle DSReadHandle, int event, void *callbackData)
{
  char msg[1000];
  unsigned  int StringLen;


  switch(event)
  {
        case DS_EVENT_DATAUPDATED:
                errorCode = DS_GetDataType (DSReadHandle, NULL, &StringLen, NULL);

                if (StringLen ==0)  //No data received
                {
                 MessagePopup ("DATASOCKET ERROR",
                      "No data received. Connect Write DataSocket first");
                 //Wait a while for the operator to enable the DataSocket Writer
                               Delay (3.0);
                 //Disable DISCONNECT button
                 SetCtrlAttribute (PanelHandle, DS_READ_DISCONNECT,
                                                          ATTR_DIMMED, 1);
                 //Enable CONNECT button
                 SetCtrlAttribute (PanelHandle, DS_READ_CONNECT ,
                                                          ATTR_DIMMED, 0);
                    return;
                }
                else
                    ReadString= malloc(StringLen +1);


                //Get the string from DataSocket Writer
                errorCode = DS_GetDataValue (DSReadHandle, CAVT_CSTRING,
                                          ReadString, StringLen+1, 0, 0);
                Checkerror(errorCode);

                //Get the index value from DataSocket Writer
                errorCode = DS_GetAttrValue (DSReadHandle, "Index", CAVT_SHORT,
                                    &WriteData.index, sizeof(short), 0, 0);
                Checkerror(errorCode);

                //No errors, write data
                if (errorCode>=0)
                {
                   InsertTextBoxLine(PanelHandle,DS_READ_RD_DSP, -1,  ReadString);

                   SetCtrlVal(PanelHandle, DS_READ_INDEX,WriteData.index);
                }

                //Enable DISCONNECT button
                SetCtrlAttribute (PanelHandle, DS_READ_DISCONNECT,
                                                          ATTR_DIMMED, 0);
                //Disable CONNECT button
```

Figure 3–12
DSCallback Function *(continued)*

```
53                  SetCtrlAttribute (PanelHandle, DS_READ_CONNECT ,
54                                                                  ATTR_DIMMED, 1);
55
56            break;
57
58            case DS_EVENT_STATUSUPDATED:
59                  //DataSocket message
60                  errorCode=DS_GetLastMessage(DSReadHandle, msg,1000);
61                  Checkerror(errorCode);
62                  SetCtrlVal(PanelHandle, DS_READ_ERROR_MSG ,msg);
63                  break;
64            }
65
66      return;
67   } //DSCallback
```

Figure 3–12
DSCallback Function *(continued)*

The *DSCallback* function executes the code depending on the `event` passed in the *DSCallback* function argument. When the `DS_EVENT_DATAUPDATED` event is passed in this function, the code between lines 13 and 54 is executed. When the `DS_EVENT_STATUSUPDATED` event is passed, the code between lines 58 and 63 is executed.

When the `DS_EVENT_STATUSUPDATED` event is passed to the *DSCallback* function, at line 60 the last status message for the DataSocket object is retrieved using the library function *DS_GetLastMessage*. This message is then displayed on the **Message:** control box on the GUI at line 62.

Now let us look at the code when the data or the data attribute values are updated and the event `DS_EVENT_DATAUPDATED` is passed to this function. At line 11 the *DS_GetDataType* library function is used to obtain the type of the DataSocket object's data. This function is used here to obtain the length of the string, and memory is allocated for `ReadString` using the ANSI C library function *malloc()* at line 28. Lines 13–25 test the data received by checking the size of the string. If the string is of length zero, a pop-up panel informs the user to click on the **CONNECT** command button on the **WRITE DATA** GUI since the reader is not receiving data. If the data is received from the DataSocket server, it is obtained in the `ReadString` buffer using the library function *DS_GetDataValue* at line 32. The DataSocket object's attribute value is obtained in the variable `WriteData.index` using the library function *DS_GetAttrValue* function at line 37. Note that the name of the data attribute, `Index` in this case, is passed in this function. This tells the function the name of the data attribute whose value is to be retrieved. This data attribute value is dependent on when the connection to the data source is made and how the DataSocket is configured for reading the data. The *DS_GetAttrValue* function performs the tasks of calling the DataSocket library function *DS_GetAttrHandle*, which obtains the handle of the attribute, and using this handle in the *DS_GetDataValue* function to obtain the data value of the attribute.

Reading the Data File

As you saw above in the section *Writing to Data File*, you can write data to a DataSocket local file using the **WRITE FILE** command button in the **WRITE DATA** GUI (Figure 3–1). To read the data from the data socket file, click on the **READ FILE** command button on the **READ DATA** GUI (Figure 3–2). The file select pop-up panel will display all the file names with the `.dsd` extension. Click on the same file name you entered in the **WRITE DATA** GUI to read back the data from this file. Click on the **CONNECT** button on the **READ DATA** GUI. The data will be read from the file and displayed in the text box.

Let us look at the code for the *ReadFileCB* function that is called when the **READ FILE** command button on the **READ DATA** GUI is selected. The listing for the *ReadFileCB* function is shown in Figure 3–13. This function creates the same path with a `file:..` prefix to where the data socket file was written in the **WRITE DATA** GUI in the format

```
file:..//directoryName/filename.dsd
```

This string is written to the **Source:** control box on the **READ DATA** GUI. When the **CONNECT** command button on this GUI is selected, this string from the **Source:** control box is retrieved and entered in the *DS_Open* library function and the connection established with this data source. While the writer is writing the data to the file constantly, you can click on **UPDATE** command button on the **READ DATA** GUI to obtain the latest value written to the data socket file.

```
/Read DataSocket file data
int CVICALLBACK ReadFileCB (int panel, int control, int event,
        void *callbackData, int eventData1, int eventData2)
{
       char ProjectDir[261], SelectedDataSocketPath[261],FileSource[261], dirName[261],
                      Directory[261], Path[261], SelectedFile[261];
       short FileSelected;

       switch (event)
       {
         case EVENT_COMMIT:
                  GetProjectDir(ProjectDir);
                  sprintf(Directory, "%s\\datafiles", ProjectDir);
                  FileSelected = FileSelectPopup (Directory, "*.dsd", "","SELECT
                                              DATASOCKET FILE", VAL_LOAD_BUTTON, 0, 1, 1, 0,
                                                                SelectedDataSocketPath);

                  if (FileSelected !=0)// file is selected
                  {
                          //Get the file name
                          SplitPath (SelectedDataSocketPath, NULL, dirName,
                                                                           SelectedFile);
                          //Create the pathname from where to read the file
                          sprintf(Path, "%s\\datafiles",ProjectDir);
                          SetDir(Path);  //Set the path

                          sprintf(FileSource,"file:../datafiles/%s",SelectedFile);
                          SetCtrlVal(PanelHandle, DS_READ_SOURCE, FileSource);
                  }
                  else
                          SetCtrlVal(PanelHandle, DS_READ_SOURCE, "");

         break;
       }
       return 0;
} //ReadFileCB
```

Figure 3–13
ReadFileCB Function

Manual Update Function

Selecting the **UPDATE** command button calls the *UpdateCB* callback function shown in Figure 3–14. The *UpdateCB* callback function calls the *DS_Update* library function, and the last data value and attribute written to the data socket file are updated. These two projects introduced you to the basics of DataSocket communications. You can build on these basic concepts and library functions to create more elaborate applications to transfer data to other computers on the network.

```
//Update Data
int CVICALLBACK UpdateCB (int panel, int control, int event,
        void *callbackData, int eventData1, int eventData2)
{
      switch (event)
      {
        case EVENT_COMMIT:
                errorCode=DS_Update(DSReadHandle);
                Checkerror(errorCode);
                break;
      }
      return 0;
} //UpdateCB
```

Figure 3–14
UpdateCB Function

DataSocket Applications

You saw above how the DataSocket could be used for transferring data between two applications on the same computer. Ideally, you would like to publish data from one computer to multiple computers on the network. If you prefer, your data can be published within your company's domain using an intranet or to any place in the world using the Internet. An example of such an application would be to run long tests in a laboratory for which you would like to view the data periodically at a remote location. You can set up the computer conducting the test to stream the test data to your computer at a remote location on the network. This same application can be expanded to have multiple clients view the same live test data at different locations connected across the network.

You can send your live test data to another computer for analysis, prepare the test results report, and publish it to yet another computer. In this way you can distribute various tasks across the network. You do not have to overload one computer with all the tasks and assign only process-intensive tasks to the computers with faster processors. You can keep the test and measurement computer inside the Internet firewall by assigning the publishing task to a different computer, preventing clients from gaining access to the test and measurement computer and keeping your test data secure. A *firewall* is a security measure to prevent unauthorized Internet users from accessing private networks connected to the Internet, such as intranets. Firewalls can be implemented in hardware, software, or both.

DataSocket technology allows you to restrict access to data by administrating security and permissions. It allows you to share confidential measurement data over the Internet while preventing access from unauthorized viewers.

DataSocket is an ActiveX control, allowing you to create data-sharing applications in ActiveX containers such as LabWindows/CVI 6.0 and higher versions, Visual Basic, Visual C++, and Borland Delphi. Suppose that you want to create a DataSocket client Web page with which users can connect to a DataSocket Server, retrieve data, and display the data on the Web page. You can do the following:

- Create a reader component using ActiveX controls in Visual Basic.
- Create the HTML file to build a reader Web page.
- View the reader Web page with an Internet browser.
- Connect the reader Web page to the DataSocket Server.
- Read and display the data on the Web page.

DataSocket can also be used to post data and reports to clients across the Internet using a Web server once created in HyperText Markup Language (HTML) format. This can be advantageous even if your clients are using different operating systems on their machines since Web servers publish data in a format that is available on most of the common computing platforms. For example, you can publish your reports on a Web server that is running on Windows 2000, and post it to the client machine, which may be using a Linux or Macintosh operating system. Using the Web-based tools, you can embed animation, graphics, and even sound into your Web page to add enhancement to your report.

DataSocket finds its use in monitoring the process variables on the production floor and sending the live data to various computers located on the network. This facilitates in keeping the production manager, who may be physically located in a different part of the plant, apprised of the current productivity data. If the floor productivity is not meeting the factory goals, the manager can take immediate action to solve the problem. This data can be used to create a periodic report and to send it to upper-level management through the network.

DataSocket technology has made it possible to take remote measurements used for monitoring the environment in locations that may be inaccessible or hazardous to human beings. For example, you can set up a measurement device in such a location and link it to the client computers located in a safe area, using an Ethernet connection. The client computer can be used to publish the acquired data from the remote measurement equipment to various computers on the network and to flag the receiver of any environment and test parameters that are out of range.

The sample programs given in the *CVI* `samples\datasocket` folder contain many examples of DataSocket applications. You can run the `weather.prj` to understand how to receive live data using the Internet. Other examples of interactive data transfer using the Internet are in the `datasocket\armadillo` folder. You are encouraged to run these sample programs to gain insight into what can be accomplished using the DataSocket technology.

Accessing the DataSocket Server

As mentioned above, the purpose of the DataSocket Server is to facilitate information exchange between the data source and data target(s). In most cases, when you run a DataSocket application your software launches the DataSocket Server. To launch the DataSocket Server from outside your application, select the path **Start>>Programs>>National Instruments>>DataSocket>>DataSocket Server** (see Figure 3–15). The DataSocket Server is launched in the default configuration, or if you had modified it, in the settings previously selected. The default configurations of the DataSocket Server are explained in the section *DataSocket Server Manager Configurations*.

Figure 3–16 shows an instance of the DataSocket Server transferring data between connected processes. This figure shows the number of clients (processes) connected and the number of data packets transmitted/received so far.

DataSocket Server Manager Configurations

In some cases you may want to add or delete data items, change the host name, or set up the permissions group. To configure the **DataSocket Server Manager**, select the **DataSocket Server Manager** in Figure 3–15. The **DataSocket Server Manager** window will be displayed (Figure 3–17). The **DataSocket Server Manager** is an executable that is used for changing the settings of the DataSocket Server. An explanation of each of these configuration items is given below.

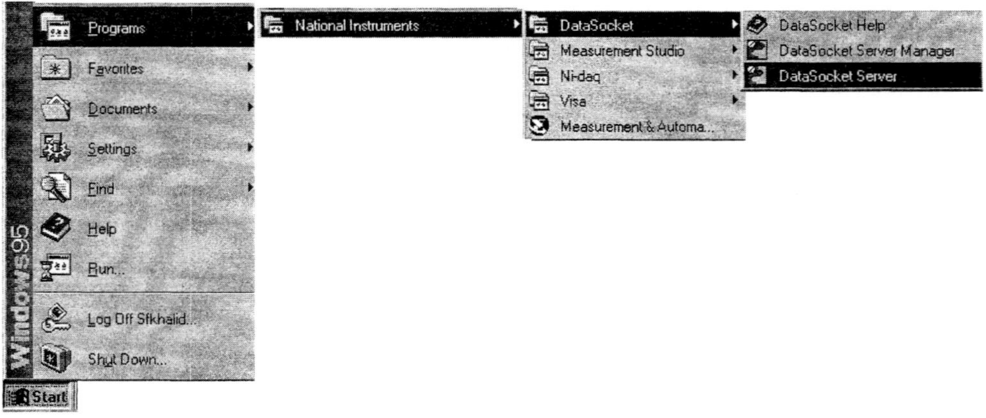

Figure 3–15
DataSocket Server Path

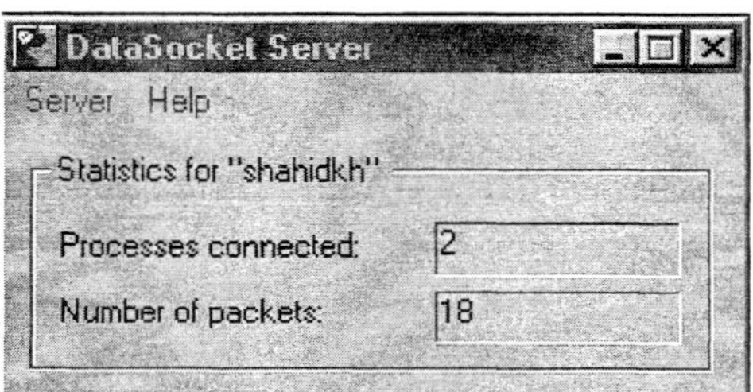

Figure 3–16
DataSocket Server

Server Settings

The **Server Settings** group shown in Figure 3–17 consists of **MaxConnections** and **MaxItems**.

- **MaxConnections**. **MaxConnections** is the maximum number of clients that can be connected simultaneously. You can choose a value from 1 to 1000. The default value is 50. Allowing permission to connect too many clients can slow down the server.
- **MaxItems**. **MaxItems** is the maximum number of items that the DataSocket Server can create dynamically. You are allowed a value between 1 and 1000. The default value is 200. This number does not include the items defined in the section *Predefined Data Items* below.

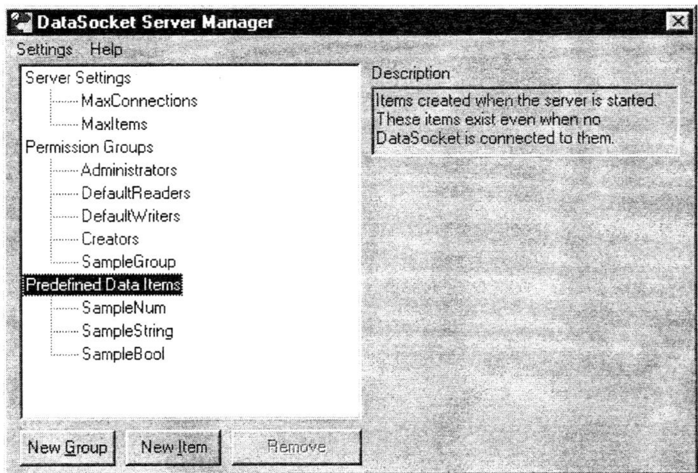

Figure 3–17
DataSocket Server Manager Configuration

Permissions Groups

The **Permissions Groups** shown in Figure 3–17 consists of the following five items: **Administrators**, **DefaultReaders**, **DefaultWriters**, **Creators**, and **Sample Group**. These items are defined below.

- **Administrators**. The **Administrators** group is reserved for future use and is intended to retrieve management information from the DataSocket Server.
- **DefaultReaders**. **DefaultReaders** consists of default host(s) that can read all dynamically created data items from the DataSocket Server. The default host is **everyhost**, allowing any computer connected to read the data item. To add a user to this group, select the **Add Host** button from Figure 3–18. Enter the name of the host in the dialog window at the bottom in this figure. In this case, `newhost` has been created in Figure 3–18.
- **DefaultWriters**. **DefaultWriters** consists of default host(s) that can write all dynamically created data items to the DataSocket Server. By default the host is only **localhost**. For the remote computer(s) to write data to the DataSocket Server, you must change this host. To add a user to this group, select the **Add Host** button as shown for **DefaultReaders** above.

- **Creators**. The **Creators** can create new data items on the DataSocket Server. The default host is **localhost**. Select **Add Host** to add another host, as shown above.
- **Sample Group**. This is a sample user-defined group. You can create your own custom group by clicking on the **NewGroup** button and entering the **Description**, **Name** of the group, and **Hosts** in the dialog boxes. Figure 3–19 shows the new group `MyGroup` that we have created with hosts `localhost, labwin.com`, and `mycomputer.sk.com`.

You use **localhost** and **everyhost** to refer to special types of hosts. The **DataSocket Server Manager** uses an IP address of **127.0.0.1** for **localhost** to refer to your local machine, and **everyhost** is used for communication with any machine on the Internet. After the **Permissions Groups** have defined the hosts that belong to a certain group, you can assign permissions for the group through the **Predefined Data Items** discussed next.

Predefined Data Items

Predefined Data Items contain the default data items created when the DataSocket Server is started. Provided that they have been assigned the appropriate permissions in the **Creators** group, DataSocket clients create the data items dynamically. These data items exist only while the DataSocket is connected. The **Predefined Data Items** lists four sample predefined data items: **SampleNum**, **SampleString**, **SampleBool**, and **SampleShort**. You can

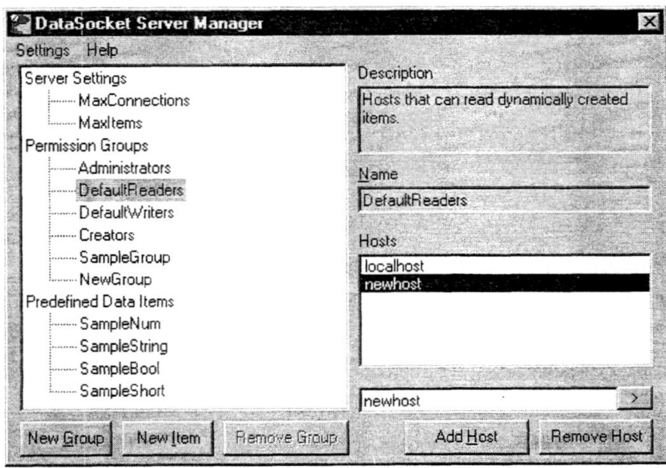

Figure 3–18
Setting Up Permission Groups for **DefaultReaders**

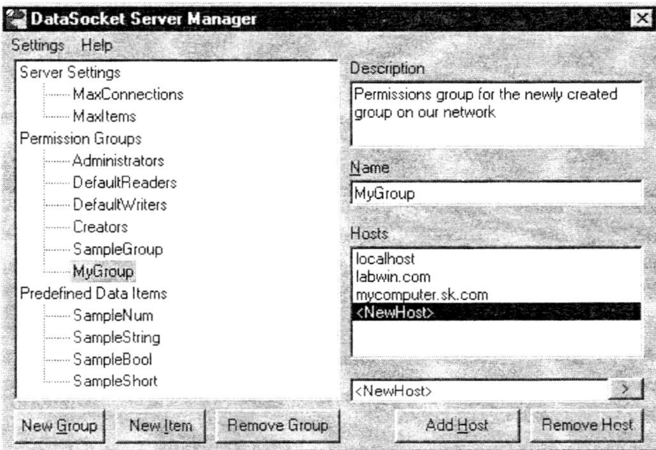

Figure 3–19
`MyGroup` Permission Group

create your own data items and define which groups have access and the kind of access they have to that data item(s). To add a new data item, click on the **NewItem** button. A dialog window appears, as shown in Figure 3–20.

Enter a short description in the **Description** box to describe the data item you are creating. In the **Name** box enter a descriptive name for the data item. Click on the **Read Access:** box to assign the appropriate read permissions from

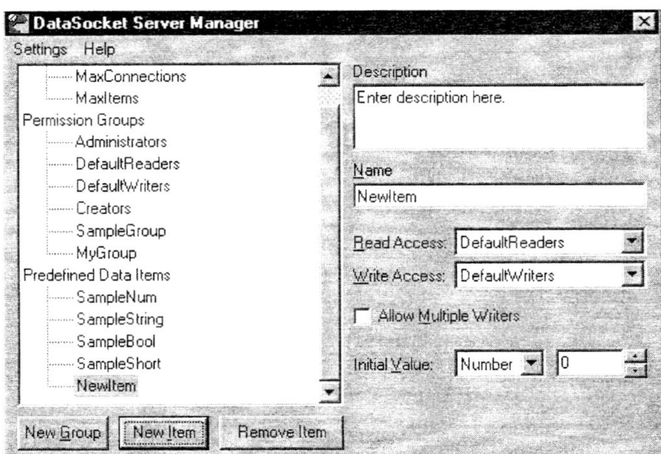

Figure 3–20
NewItem Creation Dialog Window

the pull-down list. Similarly, click on the **Write Access:** box to assign the write permissions for this data item. If you want multiple writers to be able to connect to the DataSocket Server, check the box next to **Allow Multiple Writers**. The box next to the **Initial Value:** gives you a choice from among **Boolean**, **Number**, and **String**. It is useful to assign an initial value to the data item, as it aids in determining if your program is communicating with the DataSocket Server by checking the value of the data item returned. For example, if a DataSocket Reader is connected and the DataSocket Writer is not connected, DataSocket returns a zero or an empty string to the Reader. If the default value that you specified in the **Initial Value:** dialog box is returned, you know that the DataSocket connection is good.

Caution: Changes made to the DataSocket Server using the DataSocket Server Manager are implemented only when you close the DataSocket Server Manager and restart the DataSocket Server.

There are two caveats regarding the DataSocket Server Manager.

1. You cannot configure the DataSocket Server Manager from a remote computer.
2. You cannot set DataSocket Server Manager configurations programmatically.

Summary

In this chapter you learned how to set up a simple application to transmit and receive data using the DataSocket technology on a local computer. You were shown the functions required to open a DataSocket connection, to transmit data, and to receive data through another application. All the DataSocket library functions used in the examples were explained. You were introduced to configuring the DataSocket object in various write and read modes and the advantages of using each. You were also given a taste of writing data to and reading data from a DataSocket file. You learned how to configure the items in the DataSocket Server Manager. Many areas where DataSocket technology is being used and how it can be used to your advantage were mentioned. DataSocket technology is in its inception stage and is expanding as the growth of data dissemination across the network and Internet continues to increase rapidly.

Library Function Prototypes and Definitions

This section lists alphabetically the *CVI* library functions that were introduced in this chapter.

The *HRESULT* data type is defined in the SDK header file `winnt.h`. HRESULT is the ActiveX data type used to report errors detected by the DataSocket object or any operating system network. *DSHandle* is of integer type defined in the data socket header file `dataskt.h`.

DS_ControlLocalServer Function

This function is used to start, close, display, or hide the DataSocket Server application on a local machine. The *DS_ControlLocalServer* function prototype is shown here and its arguments explained in Table 3–1.

```
HRESULT status = DS_ControlLocalServer (DSEnum_ServerOps
                                                  DSControl);
```

DSEnum_ServerOps is an enumerated type defined in the data socket header file `dataskt.h`.

DS_GetAttrValue Function

The data value of a DataSocket objects's attribute is obtained with this function. The *DS_GetAttrValue* function prototype is shown here and its arguments explained in Table 3–2.

```
HRESULT status = DS_GetDataValue (DSHandle DSHandle,
                        char *attributeName, unsigned int type,
                        void *value, unsigned int size,
                        unsigned int *dimension1,
                                       unsigned int *dimension2);
```

Table 3–1 DS_ControlLocalServer *Function*

Input/ Output	Name	Type	Description
Input	*DSControl*	DSEnum_ ServerOps	DataSocket Server application on local machine; the following are valid constants: ■ `DSConst_ServerLaunch`. Start the local DataSocket Server. ■ `DSConst_ServerHide`. Hide the local DataSocket Server window ■ `DSConst_ServerShow`. Display the local DataSocket Server ket window. ■ `DSConst_ServerClose`. Close the local DataSocket Server.
Output	*status*	HRESULT	refer to DataSocket library error codes; negative value represents error

Table 3–2 DS_GetAttrValue *Function*

Input/ Output	Name	Type	Description
Input	*DSHandle*	DSHandle	DataSocket object's handle
	attribute Name	char*	string to specify the attribute name
	type	unsigned integer	constant that indicates the type of DataSocket object's data value; the data value returned could be one of the following types: ■ `CAVT_DOUBLE` ■ `CAVT_FLOAT` ■ `CAVT_LONG` ■ `CAVT_SHORT` ■ `CAVT_UCHAR` ■ `CAVT_CSTRING` ■ `CAVT_BOOL` ■ `CAVT_VARIANT`

(continued)

Table 3–2 DS_GetAttrValue *Function (continued)*

Input/ Output	Name	Type	Description
			to pass an array type, you can combine CAVT_ARRAY with one of the data types above (except CAVT_VARIANT) using the OR operator (|); for example, for an array of float values, pass CAVT_FLOAT | CAVT_ARRAY
	size	unsigned integer	if *type* is an array or string, pass the size in bytes of the *value* buffer; for scaler types this argument is meaningless
Output	*value*	void*	for *type* array or string, pass the variable of large enough size to hold the data value; for scalar types, pass the address of the variable to receive the data value
	dimension1	unsigned integer*	type of DataSocket object's data value returned as follows, if ■ String—length of string ■ One-dimensional array—size of first dimension of array ■ Two-dimensional array—size of first dimension of array ■ Others—not defined pass NULL if you do not want this value
	dimension2	unsigned integer*	returns the size of the second dimension of the array if the DataSocket object's data value is a two-dimensional array, otherwise it is undefined; pass NULL if you do not want this value
	status	HRESULT	refer to DataSocket library error codes; negative value represents error

DS_GetDataType Function

The type of DataSocket object data value is obtained with this function. The *DS_GetDataType* function prototype is shown here and its arguments explained in Table 3–3.

```
HRESULT status = DS_GetDataType (DSHandle DSHandle,
                    unsigned int *type, unsigned int dimension1,
                                        unsigned int dimension2);
```

Table 3–3 DS_GetDataType *Function*

Input/ Output	Name	Type	Description
Input	*DSHandle*	DSHandle	DataSocket object's handle
Output	*type*	unsigned integer*	constant indicating the type of DataSocket object's data value; the data value returned could be of one of the following types: ■ CAVT_DOUBLE ■ CAVT_FLOAT ■ CAVT_LONG ■ CAVT_SHORT ■ CAVT_UCHAR ■ CAVT_CSTRING ■ CAVT_BOOL ■ CAVT_VARIANT array can also be of one of the types above (except VARIANT); pass NULL if you do not want this value
	dimension1	unsigned integer	type of DataSocket object's data value returned as follows, if ■ String—length of string ■ One-dimensional array—size of first dimension of array ■ Two-dimensional array—size of first dimension of array ■ Others—not defined pass NULL if you do not want this value

(continued)

Table 3–3 DS_GetDataType *Function (continued)*

Input/ Output	Name	Type	Description
	dimension2	unsigned integer	returns the size of the second dimension of the array if the DataSocket object's data value is a two-dimensional array, otherwise it is undefined; pass NULL if you do not want this value
	status	HRESULT	refer to DataSocket library error codes; negative value represents error

DS_GetDataValue Function

The value of the DataSocket object data is obtained with this function. The *DS_GetDataValue* function prototype is shown here and its arguments explained in Table 3–4.

```
HRESULT status = DS_GetDataValue (DSHandle DSHandle,
                        unsigned int type, void *value,
                              unsigned int size,
                                    unsigned int *dimension1,
                                         unsigned int *dimension2);
```

DS_GetLastMessage Function

The status message of the DataSocket object is obtained with this function. The status message consists of one of the following:

- A description of the last error in communication between the DataSocket object and the DataSocket data source
- A description of the last step taken in connecting the DataSocket object to the DataSocket data source
- A description of the last data update between the DataSocket object and the DataSocket data source

Table 3–4 DS_GetDataValue *Function*

Input/ Output	Name	Type	Description
Input	*DSHandle*	DSHandle	DataSocket object's handle
	type	unsigned integer	constant that indicates the type of DataSocket object's data value; the data value returned could be of one of the following types: ■ CAVT_DOUBLE ■ CAVT_FLOAT ■ CAVT_LONG ■ CAVT_SHORT ■ CAVT_UCHAR ■ CAVT_CSTRING ■ CAVT_BOOL ■ CAVT_VARIANT array can also be of one of the types above (except VARIANT); see description in *DS_SetAttrValue* function to specify an array type
	size	unsigned integer	if *type* is an array or string, pass the size in bytes of the *value* buffer; for scaler types this argument is meaningless
Output	*value*	void*	for *type* array or string, pass the variable of large enough size to hold the data value; for scalar types, pass the address of the variable to receive the data value
	dimension1	unsigned integer*	type of DataSocket object's data value returned as follows, if ■ String—length of string ■ One-dimensional array—size of first dimension of array ■ Two-dimensional array—size of first dimension of array ■ Others—not defined pass NULL if you do not want this value

(continued)

Table 3–4 DS_GetDataValue *Function (continued)*

Input/ Output	Name	Type	Description
	dimension2	unsigned integer*	returns the size of the second dimension of the array if the DataSocket object's data value is a two-dimensional array, otherwise it is undefined; pass `NULL` if you do not want this value
	status	HRESULT	refer to DataSocket library error codes; negative value represents error

The *DS_GetLastMessage* function prototype is shown here and its arguments explained in Table 3–5.

```
HRESULT status = DS_GetLastMessage (DSHandle DataSocketHandle,
                 char buffer [ ], unsigned int bufferSize);
```

DS_GetLibraryErrorString Function

The *DS_GetLibraryErrorString* function converts the error code returned by one of the DataSocket Library functions and converts it to a meaningful string. The *DS_GetLibraryErrorString* function prototype is shown here and its arguments explained in Table 3–6.

```
void DS_GetLibraryErrorString (HRESULT errorCode,
              char errBuffer[],unsigned int errorBufferSize);
```

Table 3–5 DS_GetLastMessage *Function*

Input/ Output	Name	Type	Description
Input	*DataSocket Handle*	DSHandle	handle for the DataSocket object
	bufferSize	unsigned integer	size in bytes of buffer returning the message
Output	*buffer*	char[ ]	buffer containing the message
	status	HRESULT	refer to DataSocket library error codes; negative value represents error

Table 3–6 DS_GetLibraryErrorString *Function*

Input/ Output	Name	Type	Description
Input	*errorCode*	HRESULT	error code returned from one of the DataSocket library functions
	errorBufferSize	unsigned integer	size in bytes of *errBuffer*
Output	*errBuffer*	char[ ]	contains the error message

DS_GetStatus Function

The *DS_GetStatus* function returns the status of the connection between a DataSocket object and its DataSocket data source. When the DataSocket object's status changes, it calls the callback you specified in the call to *DS_Open* to create the DataSocket object. The DataSocket object passes the value DS_EVENT_ONSTATUSUPDATED in the event parameter of your callback. The *DS_GetStatus* function prototype is shown here and its arguments explained in Table 3–7.

```
HRESULT status = DS_GetStatus (DSHandle DS_Handle,
                               DSEnum_Status *DS_Status);
```

DSEnum_Status is an enumerated type defined in the DataSocket header file `dataskt.h`.

Table 3–7 DS_GetStatus *Function*

Input/ Output	Name	Type	Description
Input	*DataSocket Handle*	DSHandle	handle for the DataSocket object
	bufferSize	unsigned integer	size in bytes of buffer returning the message
Output	*DS_Status*	DSEnum_ Status*	status between the Datasocket object and the DataSocket source; for the status values returned, see On-line Help
	status	HRESULT	refer to DataSocket library error codes; negative value represents error

DS_Open Function

The *DS_Open* function creates a DataSocket object and connects it to a data source. Each DataSocket object is connected to a single DataSocket data source. The *DS_Open* function prototype is shown here and its arguments explained in Table 3–8.

```
HRESULT status = DS_Open (char *SOURCE,
    DSEnum_AccessMode accessMode, DSCallbackPtr eventFunction,
                    void *callbackData, DSHandle *DSHandle);
```

Table 3–8 DS_Open *Function*

Input/ Output	Name	Type	Description
Input	*SOURCE*	char*	data source for the DataSocket object connection; this argument contains the protocol and the data source path; the following protocols can be used: `http`, `ftp`, `dstp`, `file`, and `opc`
	access Mode	DSEnum_ Access-Mode	used to configure the DataSocket object's connection in the write or read mode; this specifies whether the data is being updated automatically or manually; you can specify the following constants: ■ `DSConst_Read`. The DataSocket object is configured as a reader and data is updated manually when the *DS_Update* function is called. ■ `DSConst_ReadAutoUpdate`. The DataSocket object is configured as a reader and data is updated automatically when the data or any of the data attributes are changed. ■ `DSConst_Write`. The DataSocket object is configured as a writer and data is updated manually when the *DS_Update* function is called.

(continued)

Table 3–8 DS_Open *Function (continued)*

Input/ Output	Name	Type	Description
			■ DSConst_WriteAutoUpdate. The DataSocket object is configured as a writer and data is updated automatically when the data or any of the data attributes are changed
	event Function	DS Callback-Ptr	callback function that is executed if either the DataSocket object's data value, the data attributes, or the status is changed; If NULL is passed, then no events are received and you do not have to create this callback function
	callbackData	void*	specify the value that you want DataSocket object to pass to your callback function specified in the *EventFunction* argument
Output	*DSHandle*	DSHandle*	handle used for identifying the DataSocket object
	status	HRESULT	refer to DataSocket library error codes; negative value represents error

DSEnum_AccessMode is an enumerated type defined in the DataSocket header file `dataskt.h`.

DS_SetAttrValue Function

The *DS_SetAttrValue* function sets the data value of the attribute you specify when the DataSocket object is configured for writing. The *DS_SetAttrValue* function prototype is shown here and its arguments explained in Table 3–9.

```
HRESULT status = DS_SetAttrValue (DSHandle DataSocketHandle,
                     char *attributeName, unsigned int type,
                       void *value, unsigned int dimension1,
                                unsigned int dimension2);
```

Table 3–9 DS_SetAttrValue *Function*

Input/ Output	Name	Type	Description
Input	*DataSocket Handle*	DSHandle	handle for the DataSocket object
	attribute Name	char*	string specifying the attribute name; this function creates the attribute name if the name does not already exist
	type	unsigned integer	type of data passed for the *value* argument; the following are valid data types: ■ CAVT_DOUBLE ■ CAVT_FLOAT ■ CAVT_LONG ■ CAVT_SHORT ■ CAVT_UCHAR ■ CAVT_CSTRING ■ CAVT_BOOL ■ CAVT_VARIANT to pass an array type, see the *DS_GetAttrValue* function
	value	void*	for scalar values, pass the address of the variable containing the data value; for string or an array, pass the buffer containing the data; if *type* argument is CAVT_CSTRING | CAVT_ARRAY, pass an array of elements of type char*
	dimension1	unsigned integer	if the array is not passed in the *value* argument, enter 0; if a one-dimensional array is passed in the *value* argument, enter the number of elements in the first dimension of the array; if a two-dimensional array is passed in the *value* argument, enter the number of elements in the first dimension of the array

(continued)

Table 3–9 DS_SetAttrValue *Function (continued)*

Input/ Output	Name	Type	Description
	dimension2	unsigned integer	if the array is not passed in the *value* argument or is a one-dimensional array, enter 0; if a two-dimensional array is passed in the *value* argument, enter the number of elements in the second dimension of the array
Output	*status*	HRESULT	refer to DataSocket library error codes; negative value represents error

DS_SetDataValue Function

The *DS_SetDataValue* function sets the data value of the DataSocket object you specify when the DataSocket object is configured for writing. The *DS_SetDataValue* function prototype is shown here and its arguments explained in Table 3–10.

```
HRESULT status = DS_SetDataValue (DSHandle DataSocketHandle,
                                  unsigned int type, void *value,
                                        unsigned int dimension1,
                                       unsigned int dimension2);
```

DS_Update Function

The *DS_Update* function causes the DataSocket object to synchronize its data value and attributes with the DataSocket data source to which the DataSocket object is connected. If the DataSocket object is configured for the read mode, it obtains the current data value and attributes from the DataSocket data source. If the DataSocket object is configured for the write mode, it sends the current data value and attributes to the DataSocket data source. This function has no effect on DataSocket objects that are configured for ReadAutoUpdate mode or WriteAutoUpdate mode. The *DS_Update* function prototype is shown here and its arguments explained in Table 3–11.

```
HRESULT status = DS_Update (DSHandle DataSocketHandle);
```

Table 3–10 DS_SetDataValue *Function*

Input/ Output	Name	Type	Description
Input	*DataSocket Handle*	DSHandle	handle for the DataSocket object
	type	unsigned integer	type of data passed for the *value* argument; the following are valid data types: ■ CAVT_DOUBLE ■ CAVT_FLOAT ■ CAVT_LONG ■ CAVT_SHORT ■ CAVT_UCHAR ■ CAVT_CSTRING ■ CAVT_BOOL ■ CAVT_VARIANT to pass an array type, see the *DS_GetAttrValue* function
	value	void*	for scalar values, pass the address of the variable containing the data value; for a string or an array, pass the buffer containing the data if the *type* argument is CAVT_ CSTRING | CAVT_ARRAY, pass an array of elements of type char*
	dimension1	unsigned integer	if the array is not passed in the *value* argument, enter 0; if a one-dimensional array is passed in the *value* argument, enter the number of elements in the first dimension of the array; if a two-dimensional array is passed in the *value* argument, enter the number of elements in the first dimension of the array

(continued)

Table 3–10 DS_SetDataValue *Function (continued)*

Input/ Output	Name	Type	Description
	dimension2	unsigned integer	if the array is not passed in the *value* argument or is a one-dimensional array, enter 0; if a two-dimensional array is passed in the *value* argument, enter the number of elements in the second dimension of the array
Output	*status*	HRESULT	refer to DataSocket library error codes; negative value represents error

MakeDir Function

The *MakeDir* function creates a new directory with the name you specify. This function is part of the *CVI* Utility Library and is listed under the Directory Utilities. The *MakeDir* function prototype is shown here and its arguments explained in Table 3–12.

```
result = MakeDir (char DirectoryName[]);
```

SetBreakOnLibraryErrors Function

The *SetBreakOnLibraryErrors* function displays the run-time error if debugging is enabled and suspends execution when a *CVI* library function reports an error. This function is part of the *CVI* Utilities Library. The *SetBreakOn-LibraryErrors* function prototype is shown here and its arguments explained in Table 3–13.

```
result = SetBreakOnLibraryErrors (int New_State);
```

Table 3–11 DS_Update *Function*

Input/ Output	Name	Type	Description
Input	*DataSocket Handle*	DSHandle	handle for the DataSocket object
Output	*status*	HRESULT	refer to DataSocket library error codes; negative value represents error

Table 3–12 MakeDir *Function*

Input/ Output	Name	Type	Description
Input	*DirectoryName*	string	name of directory you want to create
Output	*result*	integer	results of creating the directory could be from among the following: `0` success `-1` one of the path components not found `-3` general I/O error occurred `-4` insufficient memory to complete the operation `-5` invalid path `-6` access denied `-8` disk is full `-9` directory or file already exists with same pathname

Table 3–13 SetBreakOnLibraryErrors *Function*

Input/ Output	Name	Type	Description
Input	*New_State*	integer	if debugging is enabled, display or inhibit the run-time error dialog box `0` inhibit run-time dialog box `1` display run-time dialog box if debugging is disabled, this parameter has no effect
Output	*result*	integer	previous state of the state of the "break on library errors" option; if debugging is disabled, 0 is always returned `1` enabled `0` disabled

SetDir Function

The *SetDir* function sets the directory to the specified working directory that you specify. This function is part of the *CVI* Utility Library and is listed under the Directory Utilities. The *SetDir* function prototype is shown here and its arguments explained in Table 3–14.

```
result = SetDir (char DirectoryName[]);
```

Table 3–14 SetDir *Function*

Input/ Output	Name	Type	Description
Input	*DirectoryName*	string	directory name you want to set
Output	*result*	integer	results of creating the directory could be from among the following: 0 success -1 specified directory not found or is out of memory

4

TABLE CONTROL

Chapter Highlights

- Introduction
- Table Control Basics
- Browsing the Table Control Dialog Windows
- Table Control Project
- Examining the Project Code
- Summary
- Library Function Prototypes and Definitions

This chapter will introduce you to the features of table control. Table control did not exist in *CVI* versions prior to 5.5. You will be shown the capabilities of table control to display data in the cells that comprise the rows and columns. You will learn the features of the various buttons on table control's property windows and you will be walked through the various attribute dialog windows. You are shown how to sort the columns in ascending and descending order, search for a cell value, and paste it to the system clipboard. At your request the system clipboard data is then displayed on table control. You are shown how to set the cell, row, and column attributes and to enable/disable the grid lines on table control.

Introduction

Table control lets you display and control data in rows and columns much like a spreadsheet. Using table control you have the ability to control and manipulate an individual cell, a range of cells, or rows and columns. You have the ability to change the value and attributes of the individual cells. You can sort and display the data on table control, search the data value on a particular cell, and perform many database operations just as you would on a spreadsheet application. Right-click on table control and you have the option through a built-in pop-up menu to go to a target cell, find a cell in the range selected, or sort the cell for the range selected. You can also customize this table control pop-up menu to create your own menu items. These same features can be accomplished programmatically through the use of a variety of table control library functions, as you will see later when we discuss the project. Using ActiveX you are able to interface with spreadsheet programs, but the advantage of using table control is that you can embed it in your GUI and manipulate data from your customized command buttons and software. The interface to table control is faster than using the ActiveX interface since it uses fewer Windows resources.

A sample table control is shown in Figure 4–1. The table control shown here consists of numeric cells, string cells, and picture cells, all of which are explained in the next section.

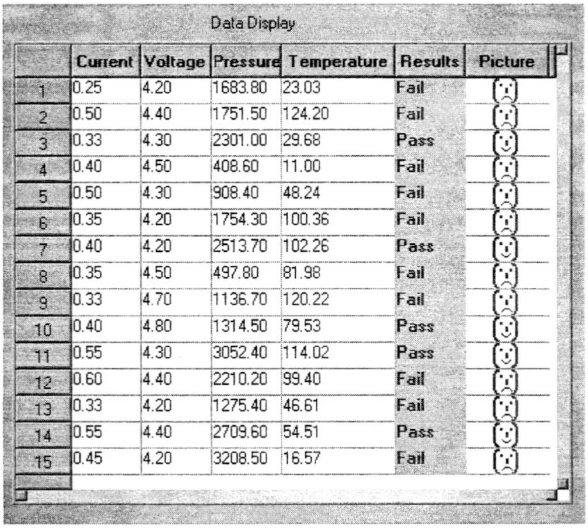

Data Display

	Current	Voltage	Pressure	Temperature	Results	Picture
1	0.25	4.20	1683.80	23.03	Fail	
2	0.50	4.40	1751.50	124.20	Fail	
3	0.33	4.30	2301.00	29.68	Pass	
4	0.40	4.50	408.60	11.00	Fail	
5	0.50	4.30	908.40	48.24	Fail	
6	0.35	4.20	1754.30	100.36	Fail	
7	0.40	4.20	2513.70	102.26	Pass	
8	0.35	4.50	497.80	81.98	Fail	
9	0.33	4.70	1136.70	120.22	Fail	
10	0.40	4.80	1314.50	79.53	Pass	
11	0.55	4.30	3052.40	114.02	Pass	
12	0.60	4.40	2210.20	99.40	Fail	
13	0.33	4.20	1275.40	46.61	Fail	
14	0.55	4.40	2709.60	54.51	Pass	
15	0.45	4.20	3208.50	16.57	Fail	

Figure 4–1
Sample Table Control

Table Control Basics

The rows and columns of a table control are made up of a collection of individual cells. The table control cells can be of the following data types: *numeric*, *string*, and *picture*. The *numeric cell* can contain data values of integer, short integer, float, double, unsigned short integer, and unsigned integer numeric data types. The *string cell* consists of `char` and `unsigned char` types that hold the text data. The *picture cell* can contain images in the following image file formats: BMP, DIB, RLE, ICO, PCX, WMF, and EMF.

Table Control States

You can operate table control in either of two states: *Selection State* or *Edit State*. The Selection State is the default state, in which by using the keyboard or the mouse you can move around the table control and make different cells active but cannot change the data in any of the cells. The Edit State allows you to change the data of the individual cell unless the cell is dimmed, configured as an indicator control, or is a picture cell. To go to the Edit State from the Selection State, double-click on the cell you want to make active or press <F2> on the cell. To return to the Selection State, press <Esc> or click on another cell on the table control.

Moving Around in the Table Control

To move between cells on a table control is similar to navigating in any of your favorite spreadsheet programs. For instance, if you want to move one cell to the left, press the left arrow key once. To move to the right, press the right arrow key. To move up, press the up arrow key, and to move down, press the down arrow key. Similarly, you can jump to the first cell in the current row by selecting the <Home> key, and to the last cell in the current row by selecting the <End> key. The <Page Up> and <Page Down> keys scroll the table up or down one page. The first cell of the first row of the table control is made active when you select <Ctrl-Home>, and the last cell of the table is activated when you select <Ctrl-End>.

In the Edit State, you can move to the cell immediately below the current cell by selecting the <Enter> key. If you are in the last cell of the column, pressing the <Enter> key moves you to the first cell in the next column. If you are in the last cell of the table, pressing <Enter> will take you to the Selection

State. Also note that in the Edit State, using the <Ctrl> key in conjunction with the arrow key will take you to the adjoining cell in the direction of the arrow key selected.

To select a range of cells using the keyboard, hold down the <Shift> key in conjunction with the other keys mentioned above. Using the mouse, you can select the entire row or column by clicking on the label for that row or column. To select the entire table, click on the upper left corner of the table. You can select a range of cells by holding down the left mouse button and moving the mouse over the cells to select.

Resizing Rows and Columns

You can resize the row height and column width of table control if the `ATTR_ENABLE_ROW_SIZING` and/or `ATTR_ENABLE_COLUMN_SIZING` attributes are enabled from the library function *SetCtrlAttribute*. Each row or column is sizable if the `ATTR_SIZE_MODE` attribute is set to `VAL_USE_EXPLICIT_SIZE` in the library functions *SetTableRowAttribute* and/or *SetTableColumnAttribute* for that particular row or column. You can resize by dragging the line separating the adjoining rows or columns (or adjoining cells) by using your mouse cursor. When you move the mouse cursor to the separator line, a double-sided arrow is formed which you can move to resize the row or column. If you double-click on the separator line while resizing the width of the row or the column, it is adjusted automatically to accommodate for the biggest cell size (numeric, string, or picture) in that row or column.

Using the System Clipboard

In the Selection State you can copy an individual cell or a range of cells from the table to the system clipboard by highlighting the cell and selecting <Ctrl-C>. You can paste the selection to the table by selecting <Ctrl-V>. If multiple cells are selected, the <Tab> ASCII character separates data between each of the cells. The linefeed ASCII character separates the data between each of the rows. When the data on the clipboard includes empty cells or invalid data, the paste operation makes no change to the table control cells receiving the data. The cells cannot be copied on the system clipboard if the range selected includes picture cells. However, if only a single picture cell is selected, it is copied to the clipboard in a bitmap format.

Table Control Events

Like most other controls, the table control callback function can process the following *User Interface Library* events:

```
EVENT_LEFT_CLICK                  EVENT_GOT_FOCUS
EVENT_LEFT_DOUBLE_CLICK           EVENT_LOST_FOCUS
EVENT_RIGHT_CLICK                 EVENT_DISCARD
EVENT_RIGHT_DOUBLE_CLICK          EVENT_COMMIT
EVENT_KEYPRESS                    EVENT_VAL_CHANGED
```

Browsing the Table Control Dialog Windows

To create the table control, select **Create>>Table** from the **User Interface Editor**. Double-click on the table control and the **Edit Table** dialog window in Figure 4–2 will be displayed. To understand some of the features of table control, let us look at the features of this window.

Some of the control items in the **Source Code Connection** group and **Label Appearance** group are the same, as you have seen on other controls. In the **Control Settings** group the **Table Mode:** dialog box defines how the new cells receive their initial, default attribute values when new row(s) or new column(s) is/are inserted in the table control.

When you click on the **Table Mode:** dialog box, you have a choice of setting the table control to one of the three table attributes shown in the pull-down menu in Figure 4–3.

Table Mode is meant to govern attribute inheritance of new cells from their parent row, their parent column, or the table itself. Each column, each row, and the table contain a "master" cell configuration. When you set the cell attributes on a row, a column, or the table using *SetTableRowAttribute*, *SetTableColumnAttribute*, or *SetCtrlAttribute*, respectively, you are setting properties on the corresponding "master" cell. When you create a new cell on the table, regardless of which function you use, the initial state of the new state is copied from the appropriate master cell, depending on which mode the table is in: column, row, or grid (table). **Column** mode is the default setting for **Table Mode.**

The same **Table Modes** can be configured programmatically. To inherit the attributes from the column cells, set the table attribute ATTR_TABLE_MODE to VAL_COLUMN in the library function *SetCtrlAttribute*. To use the attributes

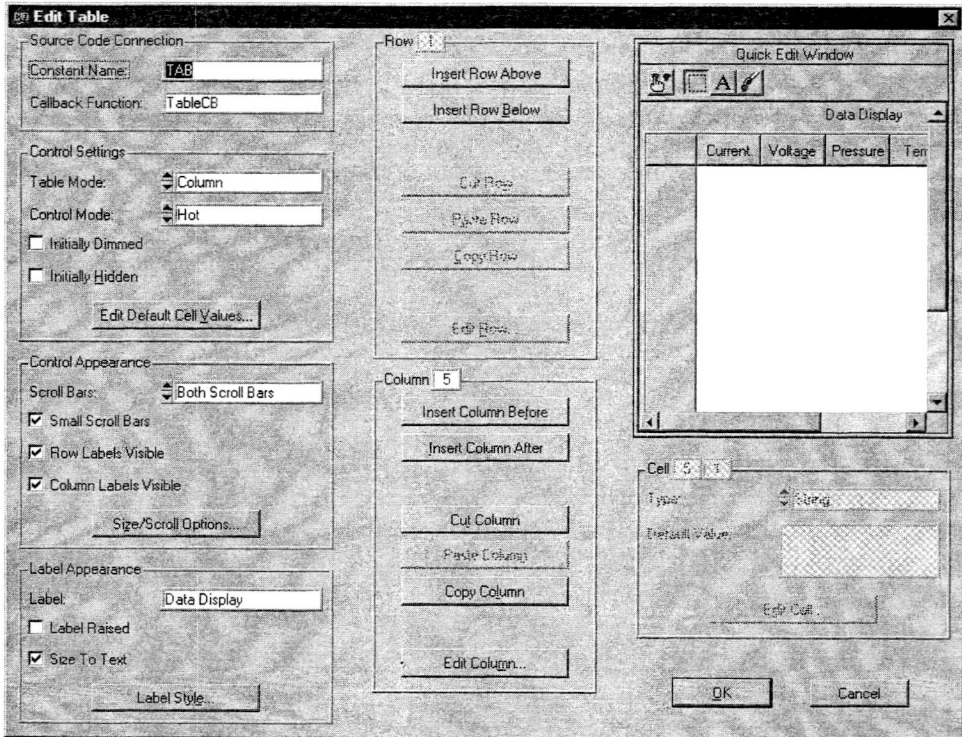

Figure 4–2
Edit Table Control

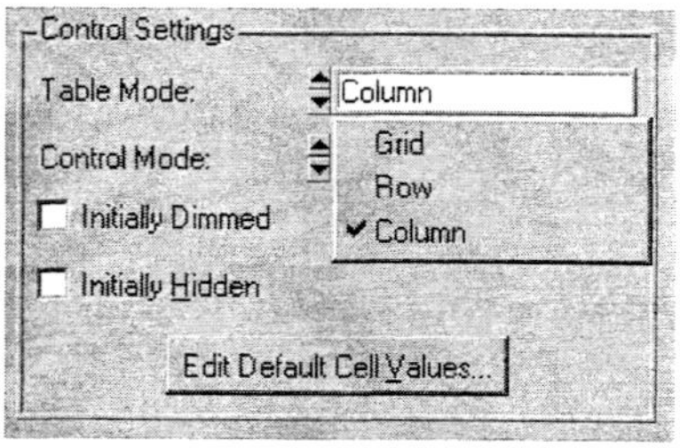

Control Settings
Table Mode:
Column
Control Mode:
Grid
Row
Column
Initially Dimmed
Initially Hidden
Edit Default Cell Values...

Figure 4–3
Table Mode: Dialog Box

from the row cells, set the table attribute ATTR_TABLE_MODE to VAL_ROW in the library function *SetCtrlAttribute,* and to use the cell attributes from the table, set the table attribute ATTR_TABLE_MODE to VAL_GRID in the library function *SetCtrlAttribute.*

The **Row** group shown in the top middle section of Figure 4–2 is shown separately in Figure 4–4. This group contains the controls to create the rows on the table, set the default cell attributes, control the appearance of the data inside the individual cell, and set the row label appearance for the selected row. To set any of these row attributes, you have to insert a row on the table control by selecting either the **Insert Row Above** or the **Insert Row Below** command button from this **Row** group. When you insert a row, the box next to the **Row** label is enabled with the number of the row for which you can set the attributes. To set the attributes of a different row, just enter a different number in the **Row** box (top of Figure 4–4), provided that a row number exists.

When you select the **Edit Row...** command button shown at the bottom of Figure 4–4, the **Edit Row** dialog window is displayed as shown in Figure 4–5. From this window you can set the **Size Mode:** to determine the method to set the height of the row. This is equivalent to setting the `ATTR_SIZE_MODE` attribute from the library function *SetTableRowAttribute*, as explained in the section *Resizing Rows and Columns*.

The **Size Mode** pull-down menu allows you the following choices:

- `Use Explicit Size`
- `Size To Cell Text`
- `Size To Cell Images`
- `Size To Cell Images And Text`

When you select the **Size Mode** as `Use Explicit Size`, the actual row height is the value entered in the **Height:** dialog box below the **Size Mode:**

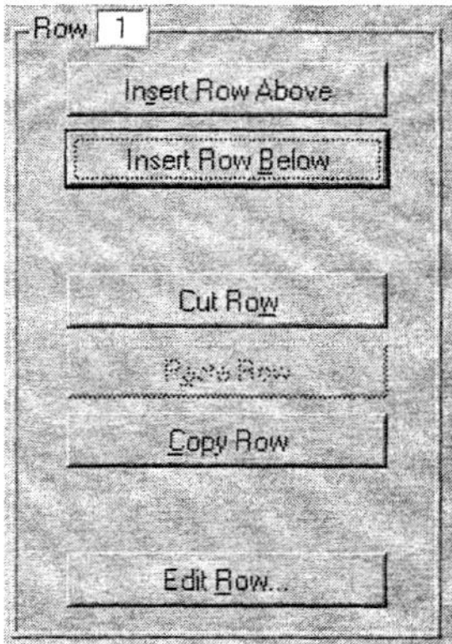

Figure 4–4
Row Controls

dialog box. In this case the ATTR_SIZE_MODE constant is set to VAL_USE_EXPLICIT_SIZE and the **Height** attribute, ATTR_ROW_HEIGHT, contains the actual value of the row height. Similarly, when the **Size Mode** is selected for Size To Cell Text, the largest height of text in all numeric or string cells in the row is used. In this case the ATTR_SIZE_MODE is set to VAL_SIZE_TO_CELL_TEXT. When the **Size Mode** is chosen as Size To Cell Images, the largest height of the images in all picture cells in the row is used, and the ATTR_SIZE_MODE constant is set to VAL_SIZE_TO_CELL_IMAGE. When Size To Cell Images And Text is selected from the **Size Mode** dialog box, the height of the row is set to the larger value of either Size To Cell Text or the Size To Cell Images. In this case, the ATTR_SIZE_MODE is set to VAL_SIZE_TO_CELL_IMAGE_AND_TEXT.

Selecting the **Edit Default Cell Values...** command button (Figure 4–5) will bring up the **Edit Default Cell Values** dialog window (Figure 4–6) to set the row cell default values. If you have set **Table Mode** to something other than **Row**, *CVI* will display a warning message to change **Table Mode** to **Row** since you will be setting the attributes of the row "master" cells here. This will be of use only when **Table Mode** is set to **Row**.

You can select the **Type:** of data for the cell from among numeric, picture, or string. You can also set attributes for each cell, using the library function *SetTableCellAttribute* or a range of cells using the library function *SetTable-*

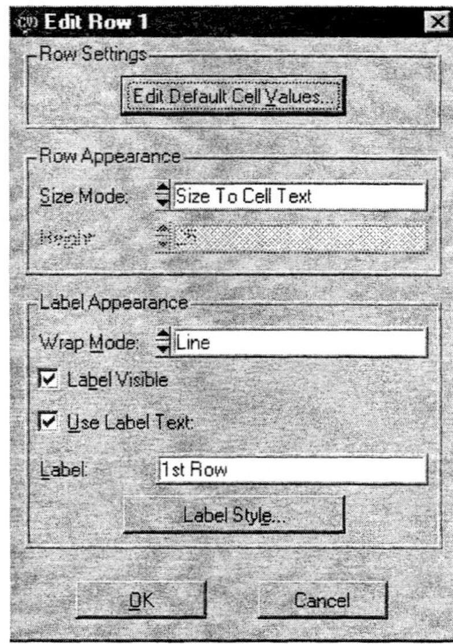

Figure 4–5
Edit Row Window

CellRangeAttribute. You specify the ATTR_CELL_TYPE attribute to be VAL_CELL_NUMERIC, VAL_CELL_STRING, or VAL_CELL_PICTURE for the cell or range of cells to contain numeric, text, or image data, respectively.

The **Mode:** of the cell is the same as you have seen for any other controls. You can set the mode of the cell or range of cells by selecting **Normal**, **Hot**, **Validate**, or **Indicator** mode.

From the **Cell Appearance** group you can set the horizontal and vertical grid colors of the cells, the background color, enable the horizontal and vertical grid lines, and set the text style of the cells in the row.

If the cell is of numeric type, using the **Numeric Attributes** group you can set the attributes shown in this group (Figure 4–6, top right). When you place a check mark in the box next to the **Show Inc/Dec Arrows** check box, the increment/decrement arrows are visible only for the cell that is edited and when the table is in the Edit State.

Similarly, if the cell type is of string type, attributes in the **String Attributes** group can be selected (Figure 4–6, bottom right). The **Wrap Mode:** in the **String Attributes** group gives you a choice of how you want to wrap the text on the row. Clicking on the box next to **Wrap Mode:** gives you the following choices:

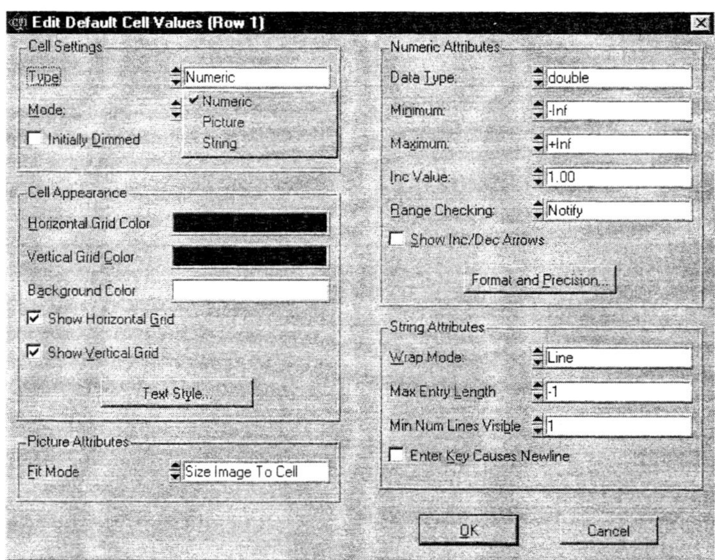

Figure 4–6
Edit Default Cell Values (Row) Window

- **Character**
- **Word**
- **Line**

If you select **Character**, the text is wrapped to a new line after the last character that fits the line. For **Word**, the text is wrapped to a new line after the end of the last word that fits on the line. **Line** wrap mode is the default and puts the text on a new line only after a new line character (\n).

In the **Min Num Lines Visible** control box, specify the number of lines of text to take into account when performing the calculations associated with the VAL_SIZE_TO_CELL_TEXT value of ATTR_SIZE_MODE. When a row is autosized to the text (VAL_SIZE_TO_CELL_TEXT or VAL_SIZE_TO_CELL_IMAGE_AND_TEXT), the table adjusts automatically to the maximum "text height" of all cells in the row. Only the font used in each cell in the row determines the "height" of each cell. However, you might decide that for a given cell you want to display more than one line of text on a given cell, which would cause the text to be chopped off. In order to take this factor into account when calculating the height of the row, you decide how many lines of text you wish to display at a given cell. Note that this attribute is used only for the purpose of calculating the row height. It doesn't prevent users from displaying however many lines they want at any given cell.

The **Column** group is shown in the lower-middle section of Figure 4–2, below the **Row** group. This is shown in Figure 4–7 for clarity. This grouping contains the controls to create the columns on the table and to set its attributes. To set any of these column attributes, you have to insert a column on the table control by selecting either the **Insert Column Above** or the **Insert Column Below** command button from this **Column** group. The column number for which you are setting the attributes is displayed next to the **Column** label (top of Figure 4–7). To set the attributes of another column, enter the column number in the box next to **Column**, if that column exists.

When you select the **Edit Column...** command button (from Figure 4–7), the **Edit Column** dialog window is displayed (see Figure 4–8).

Selecting the **Edit Default Cell Values...** command button will bring up the **Edit Default Cell Values** dialog window (Figure 4–9) to set the column cell default values for the column selected. Again, if you have set **Table Mode** to something other than **Column**, *CVI* will display a warning message to change **Table Mode** to **Column** since you will be setting the attributes of the column "master" cells here. This will be of use only when **Table Mode** is set to **Column**. This is equivalent to setting the `ATTR_SIZE_MODE` attribute

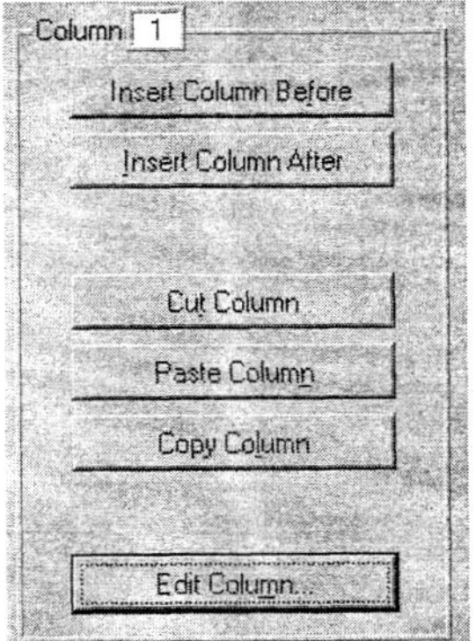

Figure 4–7
Column Controls

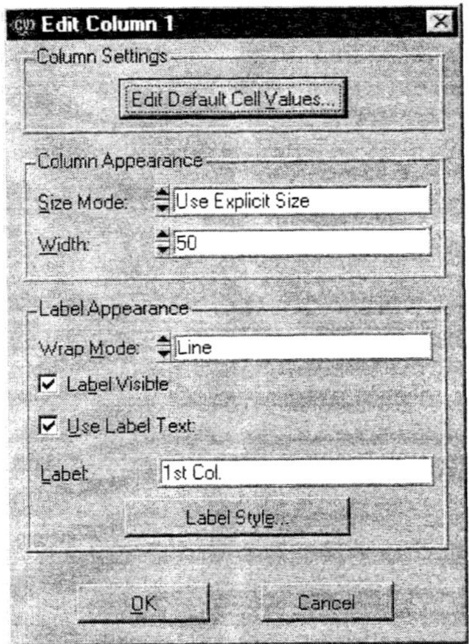

Figure 4–8
Edit Column Window

in the library function *SetTableColumnAttribute* to one of the following **Size Mode** choices:

- `Use Explicit Size`
- `Size To Cell Images`

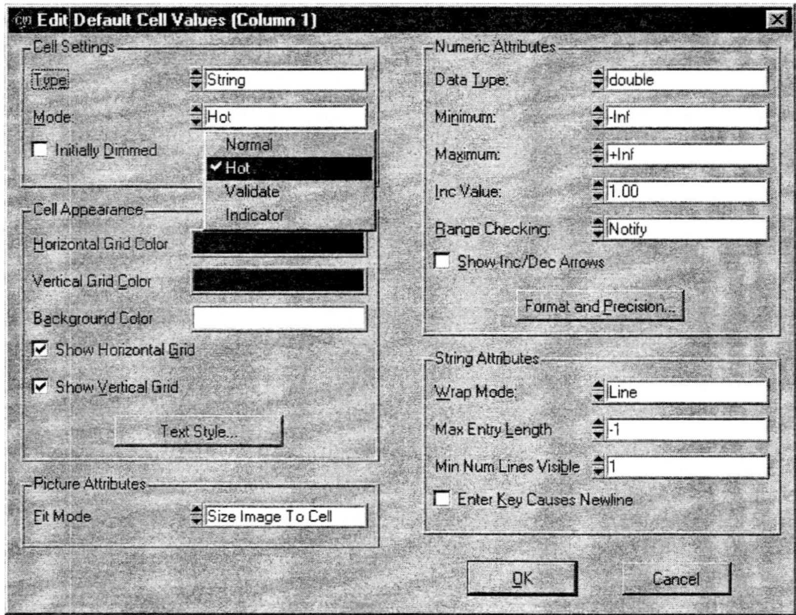

Figure 4–9
Edit Default Cell Values (Column) Window

When you select **Size Mode** as `Use Explicit Size`, the actual column width is the value entered in the **Width:** dialog box below the **Size Mode:** dialog box. In this case the `ATTR_SIZE_MODE` constant is set to `VAL_USE_EXPLICIT_SIZE` and the **Width** attribute, `ATTR_COLUMN_WIDTH`, contains the actual value of the column width. Similarly, when **Size Mode** is selected as `Size To Cell Images`, the largest width of the images in all picture cells in the column is used. The rest of the explanation for editing the default column cell values is the same as explained in **Edit Default Cell Values...** for the rows above.

You can configure the table's "master" cell using the **Edit Default Cell Values...** command button (Figure 4–2). Enter the values in the dialog window that is displayed (Figure 4–10). The dialog boxes are similar to the **Edit Default Cell Values** explained for Rows and Columns above.

The **Size/Scroll Options...** command button (Figure 4–2) affects the appearance of the table at the time it is loaded. When you select this command button the **Size/Scroll Options** dialog window appears (Figure 4–11). Here you can set the number of rows and columns that you want visible on the table control, starting with the row and column numbers selected. The **Auto**

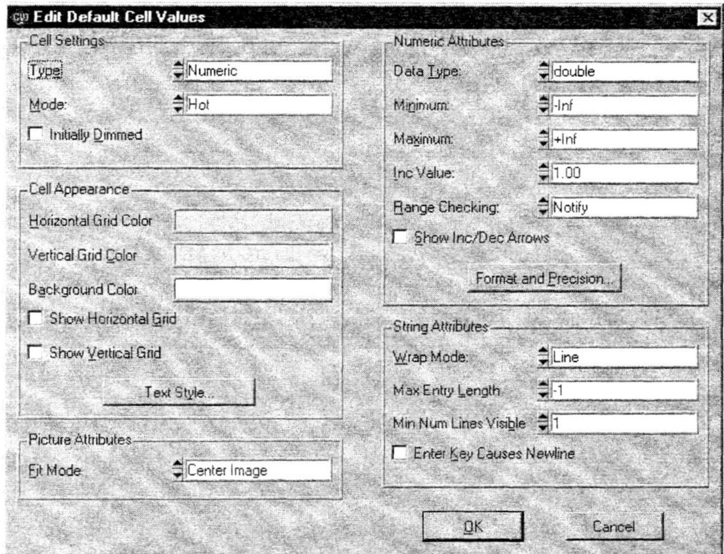

Figure 4–10
Edit Default Cell Values Window

Size (when loaded) check box assures that the same number of rows and columns are displayed when table control is created on the GUI, as when displayed when the program is executed. The row height of each row may depend on the font you are using in each cell. It is possible that the font used may be different on the development machine than on the execution machine because the system fonts can vary in typeface and size. This could result in rows "growing" or "shrinking" from the time they are created on the user interface to when loaded at execution. To avoid this situation so that the table control displays the same number of settings as when created, you should check mark the **Auto Size (when loaded)** check box.

There is a convenient feature in table control where you can change the attributes of the individual cells or a group of cells using the **Quick Edit Window** shown on the top right side of the **Edit Table** dialog window (Figure 4–2). To do this you must first set **Control Mode:** in the **Edit Table** dialog window to a mode other than **Indicator**. If there are no rows displayed in the **Quick Edit Window**, add the rows by clicking on the **Insert Row Above** or **Insert Row Below** command buttons. In the **Quick Edit Window** click on the **Operate Mode** icon on the toolbar (hand icon) to select the individual cell to change the

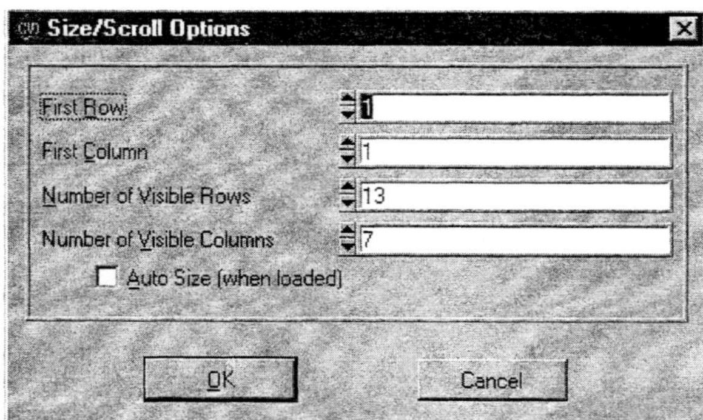

Figure 4–11
Size/Scroll Options Window

cell's type, default value, or other attributes by selecting the **Edit Cell...** command button. When you select multiple cells, the **Edit Cell...** command button changes to **Edit Cell Group...**, where you can edit the attributes of a range of cells instead of just a single cell.

Table Control Project

You have seen above how to create the table control and how to set its attributes. Many other table control items will be explained in more detail by looking at the project source code and understanding the associated library functions.

Load `project4.prj`. This project has been created to show you a variety of table control features using many of the table control library functions. A sample run for this project was shown in Figure 4–1. The GUI for this project (`project4.uir`) is shown in Figure 4–12. To start the application, click on the **LOAD DATA** command button. In this project data is loaded from two ASCII data files. One file loads the data in the **Current** column and the other in the **Voltage** column. The data for the remaining columns is generated using random numbers. Pass and fail values are randomly assigned to the data. The picture cell selects a "happy" face if the data passes or a "sad" face if the data fails. Here we are not interested in the actual data values or the criteria for pass/fail but in showing you how to manipulate the data on the table control.

Once the data is displayed on the table control, you can sort on any of the columns by selecting any column label from the ring control under **Select Sort Column** in the **SORT** group. Any selection in this ring control undims the ring control **Sort Results as:** to enable you to select either **Ascending** or **Descending** from the ring control.

The search feature is demonstrated by clicking on any cell whose value you want to search. The cell value selected is shown bold and the rows containing the cell value(s) searched are highlighted using a certain color when you select the **SEARCH** command button. When you search on a different cell value, that cell is shown bold and the rows containing the cell value searched are again highlighted using a different color. The search results are written to the system clipboard and pasted to a separate table when requested using the **DISPLAY** command button.

For this project **Table Mode** is set to **Column**. Six columns are created for this project, and the default attributes values of the cell are created using the **Edit Table** properties window(s).

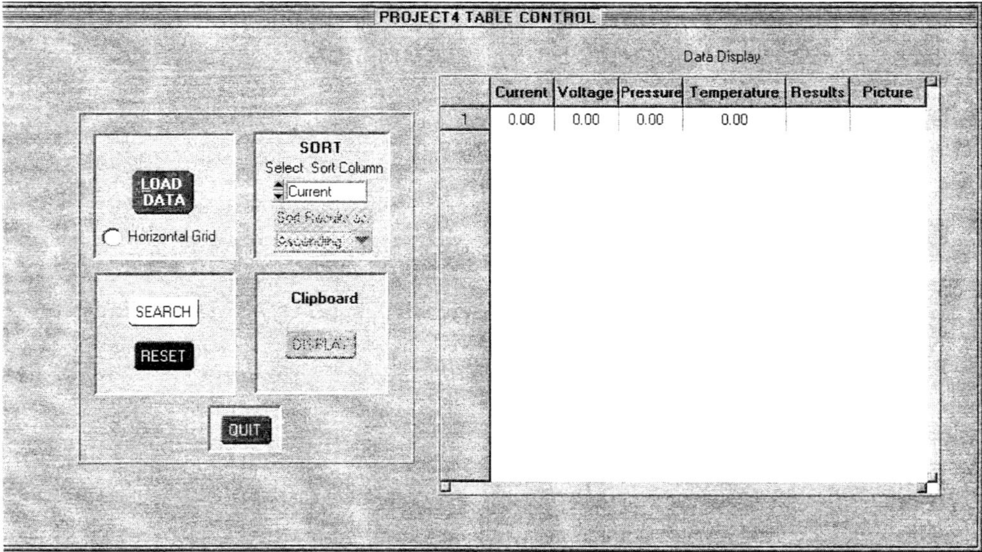

Figure 4–12
`project4` Table Control GUI

Examining the Project Code

Header and *main* Function

The *main* function loads and displays the `project4.uir` GUI. The `project4SearchTable.uir` file will be loaded, which will contain the results of the data searched, as you will see later, and the bitmap images are read from files that are displayed in the picture cells of the table control. The header information and *main* function for the `project4` source code file are listed in Figure 4–13.

A table control contains a pop-up menu consisting of menu items that are displayed on the table control when you right-click on it. This facilitates the user to navigate through the table using the default menu items: **Goto**, **Find**, and **Sort**. You can also add your own customized menu items. You have the option to hide all or some of the menu items or to create new customized menu items as needed. To create a new menu item, use the library function *NewCtrlMenuItem*. Whenever the new menu item is selected from this pop-up menu, *CVI* calls the corresponding callback function. To remove the built-in menu items included as defaults, use the *HideBuiltInCtrlMenuItem* library function shown at line 35 to hide the **Find** built-in menu item on the table control since we will be demonstrating the search feature from the **SEARCH** command button. If you want to display any of the previously hidden menu items, you can use *ShowBuillnCtrlMenuItem*. Figure 4–14 shows the default pop-up menu items when you right-click on the table control. As expected, the **Find** menu item is hidden. Note that the **Sort** menu item is disabled, as neither row or column is selected on the table control.

The GUI to display the results searched (`project4SearchTable.uir`) is created at line 39. The *GetBitmapFromFile* library function at lines 43 and 44 reads the bitmap images from the files `smiley.ico` and `frown.ico` and creates a bitmap object. This image is displayed in picture cells on the table control under the **Picture** column. The bitmap files `smiley.ico` and `frown.ico` were created using the *CVI* **Icon Editor** utility.

Load Data Function

To start the project, select **LOAD DATA** from the GUI to load the data on table control. This will activate the function *LoadDataCB* function and will load the data in a table control from the two ASCII files in the **Current** and **Voltage** columns, and random data in the other columns. The code in the *LoadDataCB* callback function is listed in Figure 4–15.

```
#include <formatio.h>
#include <ansi_c.h>
#include <cvirte.h>
#include <userint.h>
#include "project4.h"
#include "Project4SearchTable.h"
#define NMBR_OF_ROWS 15   //Total number of rows in the table

//Define the columns of the table control
#define    CURRENT_COL      1
#define    VOLTAGE_COL      2
#define    PRESSURE_COL     3
#define    TEMP_COL         4
#define    RESULTS_COL      5
#define    PICTURE_COL      6

static int TablePanel,SearchedPanel, SearchedRows=0;
int status, DataLoaded=1, visible, smilePicture,
      frownPicture,  SelectedColumnIndex;


//Function prototype
int HighlightRow(Point NewCell, int SearchNumber);

//Main
int main (int argc, char *argv[])
{
  if (InitCVIRTE (0, argv, 0) == 0)
          return -1;       /* out of memory */
  if ((TablePanel = LoadPanel (0, "project4.uir", PNL)) < 0)
          return -1;

  //Hide the right click option to Find item.
  //We will do the search programmatically
  HideBuiltInCtrlMenuItem (TablePanel, PNL_TAB, VAL_SEARCH);

  DisplayPanel (TablePanel);
  //Panel to display searched data
  if ((SearchedPanel = LoadPanel (PNL, "project4SearchTable.uir", SRCH_PNL )) < 0)
          return -1;

  //Load the bitmap files for happy and sad faces
  GetBitmapFromFile ("smiley.ico", &smilePicture);
  GetBitmapFromFile ("frown.ico", &frownPicture);

  RunUserInterface ();
  DiscardPanel (TablePanel);
  return 0;
}//main
```

Figure 4–13
project4 Header and *main* Function

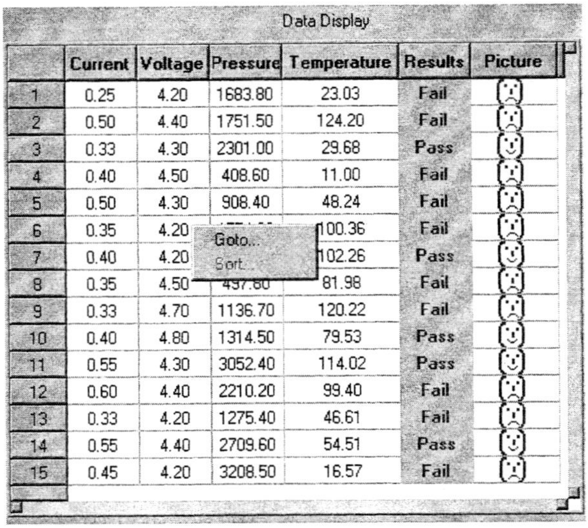

Data Display

	Current	Voltage	Pressure	Temperature	Results	Picture
1	0.25	4.20	1683.80	23.03	Fail	
2	0.50	4.40	1751.50	124.20	Fail	
3	0.33	4.30	2301.00	29.68	Pass	
4	0.40	4.50	408.60	11.00	Fail	
5	0.50	4.30	908.40	48.24	Fail	
6	0.35	4.20	[illegible]	100.36	Fail	
7	0.40	4.20	[illegible]	102.26	Pass	
8	0.35	4.50	[illegible]	81.98	Fail	
9	0.33	4.70	1136.70	120.22	Fail	
10	0.40	4.80	1314.50	79.53	Pass	
11	0.55	4.30	3052.40	114.02	Pass	
12	0.60	4.40	2210.20	99.40	Fail	
13	0.33	4.20	1275.40	46.61	Fail	
14	0.55	4.40	2709.60	54.51	Pass	
15	0.45	4.20	3208.50	16.57	Fail	

Goto...
Sort...

Figure 4–14
Pop-up Menu for Table Control

Before displaying the data on the table control, the library function *DeleteTableRows* at line 12 deletes all the rows on the table if any rows are created from a previous data load. You can specify the *DeleteTableRows* function to delete the specified number of rows or all the rows in the table control. Here all the rows are deleted.

The library function *SetTableColumnAttribute* at line 14 sets the column attribute in a table control. Here the background color attribute of the master cell of the **Results** column is set. All cells that are created under this column will later inherit this background color attribute. The *InsertTableRows* library function at line 21 inserts new rows in the table control at the one-based index specified. In this code, the *InsertTableRows* function is inside a *for* loop that is called for the number of rows (`NMBR_OF_ROWS`) in the table, creating a new row at the end of the table for each iteration of the *for* loop. The indices of the existing rows are incremented automatically for the rows beyond where the row(s) are inserted. This function also creates a new cell for each column in the table. The same could have been achieved if the *InsertTableRows* function was placed before the start of the *for* loop and `NMBR_OF_ROWS` entered in the number of rows argument for this function. This will create a table with `NMBR_OF_ROWS` rows and insert NULL data in the cells. Either approach would be acceptable for inserting rows in a table control. Note that it is more efficient to insert as many rows or columns that need to be created inside a single *InsertTableRows* function.

```
//Create and display data on Table. Invoked from "LOAD DATA" command button
int CVICALLBACK LoadDataCB (int panel, int control, int event,
          void *callbackData, int eventData1, int eventData2)
{
  short int index, result;
  double volts_array[ NMBR_OF_ROWS+1], pressure, temperature,
              current_array[ NMBR_OF_ROWS+1];
  switch (event)
  {
          case EVENT_COMMIT:
                  //If rows exit then delete them before loading the data again
                  if (DataLoaded==0) DeleteTableRows (TablePanel, PNL_TAB, 1, -1);
                  //Set background color of "Results" column
                  SetTableColumnAttribute (TablePanel, PNL_TAB, RESULTS_COL,
                                           ATTR_TEXT_BGCOLOR, VAL_LT_GRAY);

                  //Create and load data
                  for (index=1; index <= NMBR_OF_ROWS; index++)
                  {

                          InsertTableRows (TablePanel, PNL_TAB, -1, 1,
                                           VAL_USE_MASTER_CELL_TYPE);

                          //Read the simulated data from the ASCII files
                          //The data from the file is input to current_array
                          FileToArray ("Current.txt", current_array, VAL_DOUBLE,
                                           NMBR_OF_ROWS, 1,
                                           VAL_GROUPS_TOGETHER,
                                           VAL_GROUPS_AS_COLUMNS, VAL_ASCII);
                          //The data from the file is input to volts_array
                          FileToArray ("Volts.txt", volts_array, VAL_DOUBLE,
                                                   NMBR_OF_ROWS, 1,
                                                   VAL_GROUPS_TOGETHER,
                                                   VAL_GROUPS_AS_COLUMNS, VAL_ASCII);

                          //Create random data for pressure and temperature
                           pressure = rand ()/10.0;
                           temperature = rand ()/250.0;

                           //Display results on the table
                           SetTableCellVal(TablePanel,
                                           PNL_TAB,MakePoint(CURRENT_COL,index),
                                                             current_array[index-1]);
                           SetTableCellVal(TablePanel,
                                           PNL_TAB,MakePoint(VOLTAGE_COL,index),
                                                            volts_array[index-1]);
                           SetTableCellVal(TablePanel,
                                    PNL_TAB,MakePoint(PRESSURE_COL,index), pressure);

                           SetTableCellVal(TablePanel,
                                                   PNL_TAB,MakePoint(TEMP_COL,index),
                                                                     temperature);
```

Figure 4–15
LoadDataCB Function Listing *(continued)*

```
                        result=rand() % 2;  //generate odd and even number
                        if (result == 0)//pass
                        {
                                //Display the "Pass" string in green and bold attributes
                                SetTableCellVal(TablePanel,
                                  PNL_TAB,MakePoint(RESULTS_COL,index),
                                                                "Pass");
                                SetTableCellAttribute (TablePanel, PNL_TAB,
                                     MakePoint(RESULTS_COL,index),
                                        ATTR_TEXT_COLOR, VAL_DK_GREEN);
                                SetTableCellAttribute (TablePanel, PNL_TAB,
                                     MakePoint(RESULTS_COL,index),
                                        ATTR_TEXT_BOLD, 1);
                                SetTableCellVal(TablePanel,
                                        PNL_TAB,MakePoint(PICTURE_COL,index),
                                                        smilePicture);
                        }
                        else  //fail
                        {
                                //Display the "Fail" string in red and bold attributes
                                SetTableCellVal(TablePanel,
                                  PNL_TAB,MakePoint(RESULTS_COL,index),"Fail");

                                SetTableCellAttribute (TablePanel, PNL_TAB,
                                                MakePoint(RESULTS_COL,index),
                                                        ATTR_TEXT_COLOR, VAL_RED);
                                SetTableCellAttribute (TablePanel, PNL_TAB,
                                                MakePoint(RESULTS_COL,index),
                                                       ATTR_TEXT_BOLD, 1);
                                SetTableCellVal(TablePanel,
                                        PNL_TAB,MakePoint(PICTURE_COL,index),
                                                               frownPicture);


                        }

                } //for

                //Make horizontal grid control visible
                SetCtrlAttribute (TablePanel, PNL_H_GRID, ATTR_VISIBLE, 1);
                SetCtrlVal(TablePanel, PNL_H_GRID,0);

                //Dim the "LOAD" command button
                SetCtrlAttribute (TablePanel, PNL_LOAD, ATTR_DIMMED, 1);
                DataLoaded =0;  //Rows are created on the table
                break;

    }

    return 0;
} //LoadDataCB
```

Figure 4–15
LoadDataCB Function Listing *(continued)*

The *FileToArray* library function at lines 26 and 31 writes the data from the files `Current.txt` and `Volts.txt` to the arrays `current_array` and `volts_array`, respectively. The purpose of the *FileToArray* function is to read the data from a file into an array. The data read into an array must be of the same data type. This function opens the file, reads it, and then closes it after the data has been read. At lines 41 and 44 the *SetTableCellVal* function loads the data from these arrays into the appropriate column cells for each row. In some applications it will be more efficient to use the library function *SetTableCellRangeVals* to replace the values in a cell range by values from an array. You can also use *FillTableCellRange* if you want to replace the same value in all the cells in the cell range. Either of these functions can be used in place of the library function *SetTableCellVal*.

Notice that the third argument in *SetTableCellVal* function is a library function *MakePoint*. The *MakePoint* function will be used frequently when addressing the cells of the table control and needs to be explained. A cell on a table control is always referenced by a *Point* structure. The *Point* structure is defined as follows:

```
typedef struct
{
    int x;
    int y;
} Point;
```

The one-based column index of the cell is passed in the `x` field of the structure and the one-based row index to the `y` field of the structure. For example, if you are referencing a cell at row 4 and column 6, you will use

```
Point Cell;  //Declare variable of type Point
Cell.x =6;   //Column index
Cell.y =4;   //Row index
```

An easier way to reference the cell without having to declare a variable is to use the *MakePoint* function. This function returns a *Point* structure using the `x` and `y` coordinate values you supply in the function argument, such as *MakePoint(x,y)* . In the example above you would use `MakePoint(6,4)` to refer to the cell located at column 6 and row 4.

Notice that at lines 41, 44, 47, and 50 the data values are placed in cells located in the **Current**, **Voltage**, **Pressure**, and **Temperature** columns, respectively, for all the rows of the table. The rows of the table control are indexed by the *for* loop between lines 18 and 88.

Before we go any further you should be aware that similar to a *MakePoint* function there is a *MakeRect* library function. The *MakeRect* function is used to address the cell range of the table control and is explained in the section *Library Function Prototypes and Definitions* at the end of the chapter.

The random number `result` at line 53 determines the string "Pass" or "Fail" in the **Results** column cell for each row. Based on the **Results**, either a "happy" or a "sad" face image is entered in the cell in the **Picture** column for each row.

The *SetTableCellAttribute* library function sets the attribute for the cell selected. In the code, the *SetTableCellAttribute* function at lines 60, 63, 66, and 76 sets the text color and the fonts for the cells in the **Results** column.

Line 91 makes the **Horizontal Grid** radio button visible. The grid lines are drawn only when you click on the **Horizontal Grid** radio button (see Figure 4–12).

Selecting Columns to Sort

Once the data is displayed on the table control you can select the column label for which you want to sort the data by selecting from the **Select Sort Column** control (see Figure 4–16).

Selecting any column label invokes the *SortColumnSelectCB* function, whose listing is shown in Figure 4–17. Lines 10–15 check if the data is loaded before sorting the data. If data has not been loaded on the table, a message via a pop-up panel is displayed, reminding you to load the data before performing the sort. If the data is loaded, the **Sort Results as:** ring control is enabled at line 19 and the index of the item to sort is obtained from the **Select Sort Column** ring control at line 20.

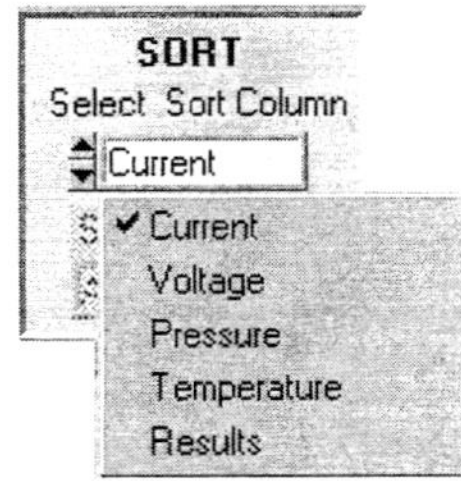

Figure 4–16
Select Sort Column Ring Control

```
//Select the column to sort
int CVICALLBACK SortColumnSelectCB (int panel, int control, int event,
          void *callbackData, int eventData1, int eventData2)
{

  switch (event)
  {
    case EVENT_COMMIT:
    //Check if data loaded
     if (DataLoaded==1)
     {
          //Inform the user
          MessagePopup ("LOAD DATA FIRST",
                           "No data loaded.  Load data before sorting");
          return 0;
     }

     //Un-dim Ascending/Descending control
     SetCtrlAttribute (TablePanel, PNL_RNG, ATTR_DIMMED, 0);
     GetCtrlIndex (TablePanel, PNL_SEL_COL, &SelectedColumnIndex);

  break;
  }
  return 0;
} //SortColumnSelectCB
```

Figure 4–17
SortColumnSelectCB Function Listing

Sort Ascending/Descending

You can sort in either **Ascending** or **Descending** order by choosing from the **Sort Results as:** ring control (Figure 4–18), which calls the *SortResultsCB* function, whose listing is shown in Figure 4–19.

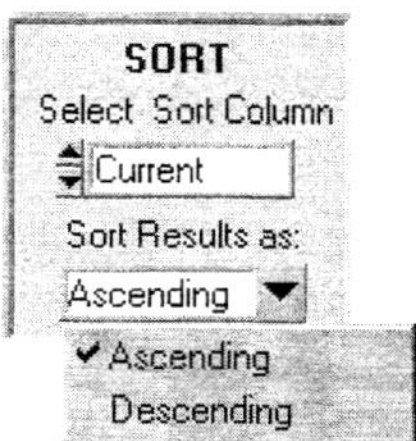

Figure 4–18
Sort Results as: Ring Control

```
//Sort using the Ring selection
int CVICALLBACK SortResultsCB (int panel, int control, int event,
        void *callbackData, int eventData1, int eventData2)
{
    int direction[5];
    static int SortFlag[5]={2,2,2,2,2};
    switch (event)
    {
     case EVENT_COMMIT:
     //Get the ring index for ascending or descending sort
     GetCtrlIndex (TablePanel, PNL_RNG, &direction[SelectedColumnIndex]);
     //Check if already sorted in the direction selected
     if( (direction[SelectedColumnIndex] ==0 &&
                                 SortFlag[SelectedColumnIndex]==0)  )  //Ascending
     {
        //Inform the user
        MessagePopup ("ALREADY SORTED IN THIS DIRECTION",
                     "Sorted in this direction. Try a different direction");
        SortFlag[SelectedColumnIndex]=0;
        return 0;
     }
    else if ( direction[SelectedColumnIndex] ==1 &&
                                 SortFlag[SelectedColumnIndex]==1 )  //Descending
     {
       //Inform the user
       MessagePopup ("ALREADY SORTED THIS DIRECTION",
                     "Sorted in this direction. Try a different direction");
       SortFlag[SelectedColumnIndex]=1;
       return 0;
     }

     //Sort in the direction specified in the ring control
     SortTableCells (TablePanel, PNL_TAB, VAL_TABLE_ENTIRE_RANGE,
                                  VAL_COLUMN_MAJOR, SelectedColumnIndex + 1,
                                      direction[SelectedColumnIndex], 0, 0);

   //Set flag to determine the direction presently sorted
   if (direction[SelectedColumnIndex] ==0)
       SortFlag[SelectedColumnIndex]=0;  //Ascend sort Flag
   else
       SortFlag[SelectedColumnIndex]=1;  //Descend sort flag

   break;
  }
 return 0;
} //SortResultsCB
```

Figure 4–19
SortResultsCB Function Listing

The direction of sort (ascending or descending) is obtained at line 11 from the ring control in the **Sort Results as:** box. Between lines 13 and 30, if the column selected is already sorted in the direction selected, the program apprises you to this fact via a pop-up panel, otherwise conducts the sort in the selected direction. The column is sorted using the library function *SortTableCells* at line 33. This function sorts a cell range using the row or column as the sort key in the sorting order selected.

Instead of writing your own software to sort the data table control, table control has a built-in pop-up menu that can sort on either rows or columns. Using the left mouse button, click on the column label of the table control, the column selected is highlighted. Right-click on the column selected and a pop-up menu appears. Selecting the **Sort...** menu item from this pop-up menu brings up the dialog window, from where you can make your sort selections. Figure 4–20 is displayed when the label in the **Temperature** column in the table control is highlighted and the **Sort...** menu item is selected from the pop-up menu. In the **Sort Table Cells** dialog window that is displayed, you have a choice of sorting the rows or columns in ascending or descending order.

The purpose of writing the customized sort callback functions was to show you the various table control library functions that are available for your use and to give you an understanding of the working of these functions. The major difference between the programmatic sort and the built-in sort is that the programmatic sort allows you to truly customize the sort by

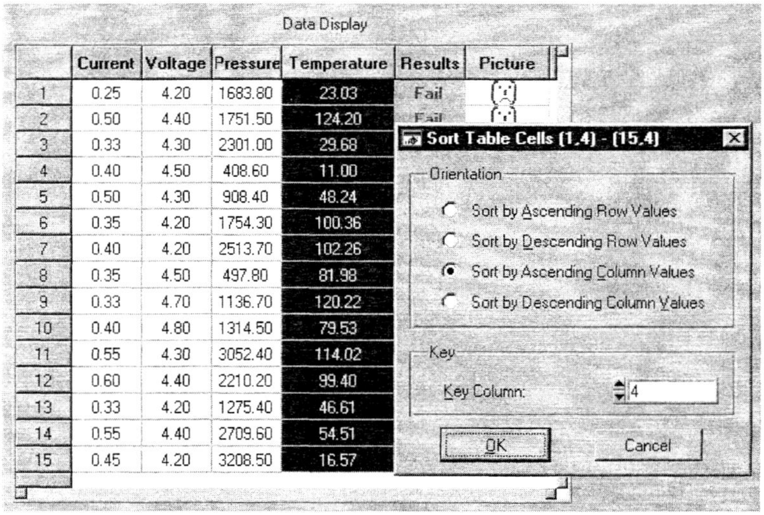
Data Display
Current
Voltage
Pressure
Temperature
Results
Picture
1 0.25 4.20 1683.80 23.03 Fail
2 0.50 4.40 1751.50 124.20 Fail
3 0.33 4.30 2301.00 29.68
4 0.40 4.50 408.60 11.00
5 0.50 4.30 908.40 48.24
6 0.35 4.20 1754.30 100.36
7 0.40 4.20 2513.70 102.26
8 0.35 4.50 497.80 81.98
9 0.33 4.70 1136.70 120.22
10 0.40 4.80 1314.50 79.53
11 0.55 4.30 3052.40 114.02
12 0.60 4.40 2210.20 99.40
13 0.33 4.20 1275.40 46.61
14 0.55 4.40 2709.60 54.51
15 0.45 4.20 3208.50 16.57
Sort Table Cells (1,4) - (15,4)
Orientation
Sort by Ascending Row Values
Sort by Descending Row Values
Sort by Ascending Column Values
Sort by Descending Column Values
Key
Key Column:
4
OK
Cancel

Figure 4–20
Sorting Using the Built-in Menu Table Control

overriding the comparison function in *SortTableCells* by passing your customized function in the `sortingFunction` argument of *SortTableCells* (Table 4–25). Using your own sorting function gives you the flexibility to compare complex data instead of just scalar data.

Search Function

The search function searches a cell value in the column specified, and if the cell value is found, shows the cells selected in bold and highlights all the rows containing that cell value. The row(s) containing the cell(s) searched is/are placed on a system clipboard to be displayed later when the **DISPLAY** command button is selected from the GUI. When you make another selection on the table, the rows containing that cell value are highlighted using a different color and the rows are pasted on the system clipboard again.

To search for a cell value, click on the cell with the left mouse button whose value you want to search. Click the **SEARCH** command button on the GUI in Figure 4–12 to start the search. The *SearchCB* function is called and its listing is shown in Figure 4–21. The first item this routine checks for is to determine if you have selected a cell on the table control. The library function *GetActiveTableCell* at line 15 returns the *Point* structure of the cell, indicating the one-based row and one-based column of the cell selected. The *GetTableCellVal* at line 18 obtains the value of the cell selected. If no cell is selected on the table control (indicated by the cell value being `0`) and you clicked on the **SEARCH** command button, a pop-up message is displayed at line 23 asking you first to select a cell. Otherwise, the cell selected is obtained at line 29 and is shown in bold using the library function *SetTableCellAttribute* at line 30. The cell is shown in bold to indicate the cell selected for searching. Lines 34–40 determine if the cell searched is a picture cell, and if so, displays a pop-up panel indicating that it cannot search on this cell type. In this function the code is set to search on either a numeric or a string cell. Line 41 determines if the cell selected is a string by checking the column number of the cell selected. If a string is to be searched, the library function *GetTableCellVal* obtains the search value in a string variable `SearchString` at line 44; otherwise, the value is contained in a numeric variable `SearchValue` at lines 48 and 49.

The library function *GetNumTableRows* at line 52 determines the number of rows in the table to search. The *SetActiveTableCell* library function sets the cell selected as active and obtains the cell structure at line 57. The *GetActiveTableCell* library function at line 59 assigns a variable `NewCell` to the cell selected. This cell structure is used in the *GetTableCellVal* library function at line 62 or

```
//Search on selected cell value
int CVICALLBACK SearchCB (int panel, int control, int event,
          void *callbackData, int eventData1, int eventData2)
{
  Point NewCell, SearchCell;
  double SearchValue, NewValue, value;
  int RowsInTable,row, SearchFlag=0;
  static int SearchNumber=0;
  char SearchString[10], NewStringValue[10];

  switch (event)
  {
          case EVENT_COMMIT:
                  //Get active cell
                  GetActiveTableCell (TablePanel, PNL_TAB, &SearchCell);

                  //Get Cell value
                  GetTableCellVal (TablePanel, PNL_TAB, SearchCell, &value);

                  if (value ==0.0) //no active cell
                  //Popup message if no cell selected
                  {
                          MessagePopup ("NO SEARCH CELL SELECTED",
                                           "Select cell on table before searching");
                          return 0;
                  }

                  //Set the selected cell and highlight it
                  GetActiveTableCell (TablePanel, PNL_TAB, &SearchCell);
                  SetTableCellAttribute (TablePanel, PNL_TAB, SearchCell,
                                                         ATTR_TEXT_BOLD, 1);


                  if (SearchCell.x == PICTURE_COL)  //Searching on picture value
                  {
                       MessagePopup ("PICTURE SEARCH DISABLED",
                     "Code not set up to search on picture cell.\nSelect another
                                                                       column.");
                          return 0;
                  }
                  if (SearchCell.x == RESULTS_COL)  //Searching on string value

                          //Get the selected cell value
                          GetTableCellVal(TablePanel,
                                               PNL_TAB,SearchCell,SearchString);
                  else
                          //Get the selected cell value
                          GetTableCellVal(TablePanel,
                                               PNL_TAB,SearchCell,&SearchValue);

                  //Get number of rows in table
                  GetNumTableRows(TablePanel, PNL_TAB,&RowsInTable);
```

Figure 4–21
SearchCB Function Listing *(continued)*

```
53
54                  //Search on all the values in that column
55                  for (row=1; row <= RowsInTable; row++)
56                  {
57                          SetActiveTableCell(TablePanel, PNL_TAB,
58                                                  MakePoint(SearchCell.x,row));
59                          GetActiveTableCell (TablePanel, PNL_TAB, &NewCell);
60                          if (SearchCell.x != RESULTS_COL)
61                                  //Get the selected cell value
62                                  GetTableCellVal(TablePanel,
63                                                          PNL_TAB,NewCell,&NewValue);
64                          else    //Searching on string value
65                                  //Get the selected cell value
66                                  GetTableCellVal(TablePanel,
67                                                  PNL_TAB,NewCell,NewStringValue);
68
69                          //If value found and is not string
70                          if ( (SearchCell.x != RESULTS_COL) && (SearchValue ==
71                                                                  NewValue))
72                          {
73                                  //Highlight the row and put on clipboard
74                                  HighlightRow(NewCell, SearchNumber);
75                                  SearchFlag=1;
76
77                          }//if
78                          else if ( (SearchCell.x == RESULTS_COL) &&
79                                    (strcmp (SearchString,NewStringValue) ==0 ))
80                          {
81                                  //Highlight the row and put on clipboard
82                                  HighlightRow(NewCell, SearchNumber);
83                                  SearchFlag=1;
84                          }//else if
85
86                  }//for
87
88                  //Enable the DISPLAY command button if there is data on clipboard
89                  if (SearchFlag)
90                       SetCtrlAttribute (TablePanel, PNL_DISPLAY, ATTR_DIMMED, 0);
91
92                  SearchNumber++;  //increment the search color index
93                  if (SearchNumber== RESULTS_COL) SearchNumber=0; /*Reset SearchColor
94                                                                       number*/
95
96                  break;
97     }
98     return 0;
99  } //SearchCB
```

Figure 4–21
SearchCB Function Listing *(continued)*

line 66, depending on whether the active cell contains a number or a string. The value of each cell searched is compared against the value selected at line 70 if the cell value is numeric and on line 78 if it is a string. In either case the user-defined function *HighlightRow* is called.

You should be aware that the library function *GetTableCellFromValue* returns the first cell matching the specified value in the table cell range selected. With a few modifications to the code, you can use this function instead of the *for* loop between lines 55 and 86. At line 89 the value for the variable `SearchFlag` is used to verify that at least one value of the search cell was found. If the search was successful, line 90 undims the **DISPLAY** command button on the GUI to enable the display of the **SEARCHED DATA** GUI displaying the searched data.

Highlighting/Pasting Rows to Clipboard

The *HighlightRow* function is called from the *SearchCB* function when the search selection is found. This function highlights the selected row(s) and places them on the system clipboard. The listing for this function is shown in Figure 4–22.

```
//Highlight the selections and place on system clipboard
int HighlightRow(Point NewCell, int SearchNumber)
{
//Hex values for colors
int SearchColor[5]=
  {0xCCCCCC,           0x00FF00,        0x00FFFF,       0xFF00FF,             0xFFFF00};
//                    VAL_LT_GRAY     VAL_GREEN     VAL_CYAN    VAL_MAGENTA  VAL_YELLOW


  int *ColorPtr;
  ColorPtr=SearchColor;

  //Color the row containing the found cell
  SetTableCellRangeAttribute (TablePanel, PNL_TAB,
          VAL_TABLE_ROW_RANGE(NewCell.y), ATTR_TEXT_BGCOLOR,
                                                  *(ColorPtr+SearchNumber));

  SearchedRows++; //increment number of rows with searched values

  //Paste selection to the clipboard
  ClipboardPutTableVals (TablePanel, PNL_TAB,
                                         VAL_TABLE_ROW_RANGE(NewCell.y));

  //Create rows on the Searched Table control
  InsertTableRows (SearchedPanel, SRCH_PNL_TAB, -1, 1,
                                              VAL_USE_MASTER_CELL_TYPE);

  //Paste selection from the clipboard
  status=ClipboardGetTableVals (SearchedPanel, SRCH_PNL_TAB,
                            VAL_TABLE_ROW_RANGE(SearchedRows), &exists);
  return 0;

} // HighlightRow
```

Figure 4–22
HighlightRow Function Listing

The row containing the cell value selected is enhanced using a color from among five different colors in the `SearchColor` array at line 5. Every time a new search value is selected, the searched rows are highlighted with the next color in the array. The colors are selected round robin, so after the fifth selection the first color is used to paint the selected row. The *SetTableCellRangeAttribute* library function at line 14 is used to color the row with the color selected. At line 39 in the *main* function of this project (Figure 4–13), a panel handle `SearchedPanel` is created to display the search results in a table control. This GUI is shown in Figure 4–23. The *ClipboardPutTableVals* function at line 21 of the *HighlightRow* function puts onto the system clipboard the cell range selected for the table control. Every time the cell value is searched, a new row is created on the **SEARCHED DATA** GUI using the *InsertTableRows* library function at line 25. Line 29 pastes the data from the system clipboard onto the table control in the **SEARCHED DATA** GUI (Figure 4–23) using the library function *ClipboardGetTableVals.*

When selected, the **RESET** command button in Figure 4–12 calls the *ResetTableCB* function (not listed in text) and removes the colors from the rows searched and removes unbolds from the cell(s) selected. The data searched, however, is not removed from the system clipboard.

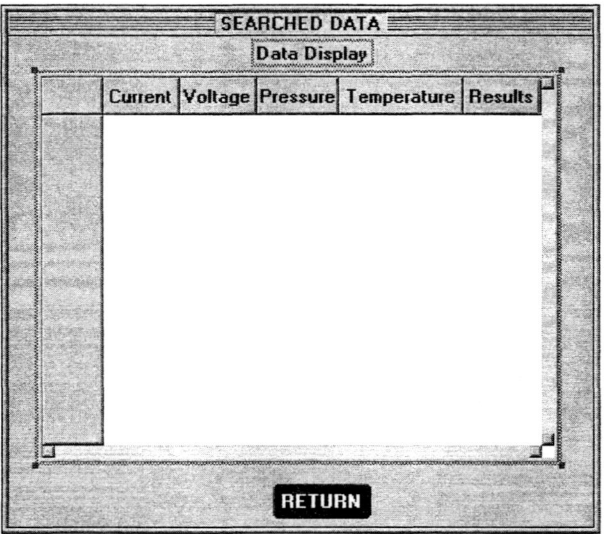

Figure 4–23
SEARCHED DATA GUI

Summary

The table control basics were explained in this chapter. You learned how to set the attributes of the cells, rows, and columns of table control. You were shown how to control the attributes of table control through use of the **Edit** window in the **User Interface** editor and through the use of library functions. You were introduced to the table control pop-up menu. You were shown how to hide, display, and create menu items on this pop-up menu. You were also shown how this pop-up menu could be used to sort a row or a column on a table control. The project demonstrated the various table control library functions to search and sort on the cells and how to paste and retrieve data from the system clipboard.

Library Function Prototypes and Definitions

This section lists alphabetically the *CVI* library functions that were introduced in this chapter.

ClipboardGetTableVals Function

The *ClipboardGetTableVals* function retrieves the text data from the system clipboard and formats it into the cell range specified for the table cells. It also indicates if the text data is available on the system clipboard. Its prototype is shown here and its arguments explained in Table 4–1.

```
int status = ClipboardGetTableVals (int panelHandle,
                              int ControlID, Rect cellRange,
                                        int *available);
```

ClipboardPutTableVals Function

The *ClipboardPutTableVals* function copies the data from the cell range specified for a table control to the system clipboard. Its prototype is shown here and its arguments explained in Table 4–2.

```
int status = ClipboardPutTableVals (int panelHandle,
                              int ControlID, Rect cellRange);
```

Table 4–1 ClipboardGetTableVals *Function*

Input/ Output	Name	Type	Description
Input	*panel Handle*	integer	panel handle loaded in memory
	controlID	integer	control identifier
	cellRange	Rect	*Rect* structure of the cell range of the table control to which to copy the clipboard data
Output	*available*	integer*	indicates whether data is available on the clipboard: `1` data is available `0` there is no data on the clipboard
	status	integer	refer to User Interface Library error codes; negative value represents error

Table 4–2 ClipboardPutTableVals *Function*

Input/ Output	Name	Type	Description
Input	*panel Handle*	integer	panel handle loaded in memory
	controlID	integer	control identifier
	cellRange	Rect	*Rect* structure of the cell range of the table control that is copied to the clipboard
Output	*status*	integer	refer to User Interface Library error codes; negative value represents error

DeleteTableRows Function

The *DeleteTableRows* function deletes from the table control the number of rows specified. Its prototype is shown here and its arguments explained in Table 4–3.

Table 4–3 DeleteTableRows *Function*

Input/ Output	Name	Type	Description
Input	*panel Handle*	integer	panel handle loaded in memory
	controlID	integer	control identifier
	rowIndex	integer	one-based index of the first row to delete
	number_ofRows	integer	number of rows to delete; enter `-1` to delete all the rows in the table control
Output	*status*	integer	refer to User Interface Library error codes; negative value represents error

```
int status = DeleteTableRows (int panelHandle,
                                   int ControlID, int rowIndex,
                                            int number_ofRows);
```

FillTableCellRange Function

The *FillTableCellRange* function sets to the same value the value of all the cells in the cell range specified. All cells must be of the same cell type in the cell range. Its prototype is shown here and its arguments explained in Table 4–4.

```
int status = FillTableCellRange (int panelHandle,
                                   int ControlID, rect cellRange,
                                                  void *value);
```

FileToArray Function

The *FileToArray* function converts a data file to an array. Its prototype is shown here and its arguments explained in Table 4–5.

```
int status = FileToArray (char *fileName, void *array,
                          int DataType, int NumberOfElements,
                          int NumberOfGroups,int arrayDataOrder,
                                     int fileLayout,int fileType);
```

Table 4–4 FillTableCellRange *Function*

Input/ Output	Name	Type	Description
Input	*panel Handle*	integer	panel handle loaded in memory
	controlID	integer	control identifier
	cellRange	Rect	*Rect* structure specifying the cell range for which you want to set the values
	value	any type*	new value of each cell in the range
Output	*status*	integer	refer to User Interface Library error codes; negative value represents error

Table 4–5 FileToArray *Function*

Input/ Output	Name	Type	Description
Input	*fileName*	string*	file pathname
	DataType	integer	data type of the array elements from among the following: ■ `VAL_CHAR`—character ■ `VAL_SHORT_INTEGER`—short integer ■ `VAL_INTEGER`—integer ■ `VAL_DOUBLE`—double ■ `VAL_UNSIGNED_SHORT_INTEGER`—unsigned short integer ■ `VAL_UNSIGNED_INTEGER`—unsigned integer ■ `VAL_UNSIGNED_CHAR`—unsigned character
	NumberOf Elements	integer	number of elements in the array
	NumberOf Groups	integer	number of groups into which the data in the file is divided; rows or columns can be read as groups; pass `1` if the data is not divided into groups

(continued)

Table 4–5 FileToArray *Function (continued)*

Input/ Output	Name	Type	Description
	arrayData Order	integer	specifies how the data is grouped into an array if you have specified the `NumberOfGroups` as other than 1; you have the following two choices: ■ `VAL_GROUPS_TOGETHER`. All elements of first data group are followed by all elements from the second data group, and so on. ■ `VAL_DATA_MULTIPLEXED`. First elements of all data groups are stored together, followed by second elements of all data groups, and so on.
	fileLayout	integer	applies to ASCII files only; this indicates if the groups in the file are arranged as columns or rows; the following are valid values: ■ `VAL_GROUPS_AS_COLUMNS` ■ `VAL_GROUPS_AS_ROWS` if there is only one group, use `VAL_GROUPS_AS_COLUMNS` to indicate that each element of the file is on a separate line
	fileType	integer	specifies whether the file is in ASCII or binary format; the two choices are: ■ `VAL_ASCII` ■ `VAL_BINARY`
Output	*array*	void*	array containing the data in the format specified
	status	integer	success is indicated by a 0 return value; negative value represents error code; for an explanation of error codes, see On-line Help for this function

GetActiveTableCell Function

The *GetActiveTableCell* function returns the *point* structure of the active (selected) cell of the table control. Its prototype is shown here and its arguments explained in Table 4–6.

```
int status = GetActiveTableCell (int panelHandle,
                                 int ControlID, Point *cell);
```

GetBitmapFromFile Function

The *GetBitmapFromFile* function reads the bitmap file and converts it to a bitmap image. Its prototype is shown here and its arguments explained in Table 4–7.

```
int status = GetBitmapFromFile (char filename[],
                                           int *bitmapID);
```

GetNumTableRows Function

The *GetNumTableRows* function obtains the number of rows in the table control. Its prototype is shown here and its arguments explained in Table 4–8.

```
int status = GetNumTableRows(int panelHandle, int  ControlID,
                                         int *numberOfRows);
```

Table 4–6 GetActiveTableCell *Function*

Input/ Output	Name	Type	Description
Input	*panel Handle*	integer	panel handle loaded in memory
	controlID	integer	control identifier
Output	*cell*	Point*	*Point* structure of the active cell
	status	integer	refer to User Interface Library error codes; negative value represents error

Table 4–7 GetBitmapFromFile *Function*

Input/ Output	Name	Type	Description
Input	*filename*	string	image file pathname; the following image types are supported: `.pcx`, `.bmp`, `.dib`, `.rle`, `.ico`, `.wmf`, and `.emf`.
Output	*bitmapID*	integer*	handle to the bitmap object
	status	integer	refer to User Interface Library error codes; negative value represents error

Table 4–8 GetNumTableRows *Function*

Input/ Output	Name	Type	Description
Input	*panel Handle*	integer	panel handle loaded in memory
	controlID	integer	control identifier
Output	*numberOf Rows*	integer*	number of rows in the table
	status	integer	refer to User Interface Library error codes; negative value represents error

GetTableCellFromVal Function

The *GetTableCellFromVal* function returns the first cell in the cell range of a table control with a value matching the value specified. Its prototype is shown here and its arguments explained in Table 4–9.

```
int status = GetTableCellFromVal (int panelHandle,
                     int ControlID, Point beginningCell,
                     Rect cellRange, Point *cell,
                     int searchDirection, int cellType,
                     int dataType,void *value);
```

Table 4–9 GetTableCellFromVal *Function*

Input/ Output	Name	Type	Description
Input	*panel Handle*	integer	panel handle loaded in memory
	controlID	integer	control identifier
	beginning Cell	Point	cell from where to start searching
	cellRange	Rect	cell range to search specified as a rectangle structure
	search Direction	integer	search order; specify either of the following: ■ VAL_ROW_MAJOR. Values are searched in a row from left to right, before traversing to next rows immediately below this row till a match is found. ■ VAL_COLUMN_MAJOR. Values are searched in a column from top to bottom before traversing to the next columns immediately to its right until a match is found.
	cellType	integer	type of cell to conduct search; the following choices are available: ■ VAL_CELL_NUMERIC 0 ■ VAL_CELL_STRING 1
	dataType	integer	data type of value to be searched
	value	any type*	value to search; must match *cellType* and *dataType*
Output	*cell*	Point*	*Point* structure of the cell matching the value specified
	status	integer	refer to User Interface Library error codes; negative value represents error

GetTableCellVal Function

The *GetTableCellVal* function obtains the current value of the cell selected. Its prototype is shown here and its arguments explained in Table 4–10.

```
int status = GetTableValCell (int panelHandle,
                    int ControlID, Point cell,void *value);
```

HideBuiltInCtrlMenuItem Function

The *HideBuiltInCtrlMenuItem* function hides the built-in menu item from the table control menu. Its prototype is shown here and its arguments explained in Table 4–11.

```
int status =  HideBuiltInCtrlMenuItem (int panelHandle,
                    int ControlID, int builtInMenuItemID);
```

InsertTableColumns Function

The *InsertTableColumns* function inserts a new column on the table control at the column location specified. Its prototype is shown here and its arguments explained in Table 4–12.

```
int status = InsertTableColumns (int panelHandle,
                    int ControlID, int columnIndex,
                    int number_ofColumns, int cellTypse);
```

InsertTableRows Function

The *InsertTableRows* function inserts a new row on the table control at the row location specified. Its prototype is shown here and its arguments explained in Table 4–13.

```
int status = InsertTableRows (int panelHandle,
                    int ControlID, int rowIndex,
                    int number_ofRows,int cellType);
```

Table 4–10 GetTableCellVal *Function*

Input/ Output	Name	Type	Description
Input	*panel Handle*	integer	panel handle loaded in memory
	controlID	integer	control identifier
	cell	Point	*Point* structure of the cell for which you want to get the value
Output	*value*	any type*	value of the cell selected
	status	integer	refer to User Interface Library error codes; negative value represents error

Table 4–11 HideBuiltInCtrlMenuItem *Function*

Input/ Output	Name	Type	Description
Input	*panel Handle*	integer	panel handle loaded in memory
	controlID	integer	control identifier
	builtIn MenuItemID	integer	menu item to hide; the following are available choices: ■ `VAL_GOTO`. Go to specified cell by the row and column number. ■ `VAL_SEARCH`. Find a certain cell with the range specified. ■ `VAL_SORT`. Sort by ascending or descending value for the row or column key selected.
Output	*status*	integer	refer to User Interface Library error codes; negative value represents error

Table 4–12 InsertTableColumns *Function*

Input/ Output	Name	Type	Description
Input	*panel Handle*	integer	panel handle loaded in memory
	controlID	integer	control identifier
	column Index	integer	one-based index of the column where you want to insert a new column; enter `-1` to insert columns at the end of the table
	number_of Columns	integer	number of new columns to be inserted
	cellType	integer	cell type of the cell to insert in the new row; the following are valid cell types: ■ `VAL_CELL_NUMERIC`—for cell to hold numeric values ■ `VAL_CELL_STRING`—for cell to hold text ■ `VAL_CELL_PICTURE`—for cell to hold images ■ `VAL_USE_MASTER_CELL_TYPE`—to use the default cell types for each column
Output	*status*	integer	refer to User Interface Library error codes; negative value represents error

MakePoint Function

The *MakePoint* function creates a *Point* structure using the x and y coordinates specified for the point. For a table control the x and y coordinates of the cell are used. Its prototype is shown here and its arguments explained in Table 4–14.

```
int point = MakePoint (int xCoordinate, int yCoordinate);
```

Table 4–13 InsertTableRows *Function*

Input/ Output	Name	Type	Description
Input	*panel Handle*	integer	panel handle loaded in memory
	controlID	integer	control identifier
	row Index	integer	one-based index of the row where you want to insert a new row. –1 to insert rows at the end of the table
	number_of Rows	integer	number of new rows to be inserted
	cellType	integer	cell type of the cell to insert in the new row; the following are valid cell types: ■ `VAL_CELL_NUMERIC`—for cell to hold numeric values ■ `VAL_CELL_STRING`—for cell to hold text ■ `VAL_CELL_PICTURE`—for cell to hold images ■ `VAL_USE_MASTER_CELL_TYPE`—to use the default cell types for each column
Output	*status*	integer	refer to User Interface Library error codes; negative value represents error

Table 4–14 MakePoint *Function*

Input/ Output	Name	Type	Description
Input	*xCoordinate*	integer	horizontal location of point
	yCoordinate	integer	vertical location of point
Output	*point*	Point	*Point* structure containing the point (cell) coordinates

MakeRect Function

The *MakeRect* function creates a *Rect* structure using the x and y coordinates of one corner of the rectangle and the width of the rectangle. Its prototype is shown here and its arguments explained in Table 4–15.

```
Rect rect = MakeRect (int top, int left, int height,
                                            int width);
```

MakeRect function returns the *Rect* structure defined as follows:

```
typedef struct
{
    int top;
    int left;
    int height;
    int width;
} Rect;
```

For instance, if the cell range selected were as shown in Figure 4–24, the rectangle coordinates would be

```
Rect range;  //Declare the variable

            range.top =    2;
            range.left =   3;
            range.height = 3;
            range.width =  2;
```

Table 4–15 MakeRect *Function*

Input/ Output	Name	Type	Description
Input	*top*	integer	location of the top edge of the rectangle
	left	integer	location of the left edge of the rectangle
	height	integer	height of the rectangle
	width	integer	width of the rectangle
Output	*rect*	Rect	*Rect* structure containing the coordinate values specified

	1	2	3	4	5
1			left		
2		top			↑
3					height
4					↓
5			← width	→	
6					

Figure 4–24
Example Cell Range

The *MakeRect* function is of the form `MakeRect(top, left, height, width)`, which for this example can represent the cell range by the function `MakeRect(2,3,3,2)`.

The user interface header `userint.h` has built-in macros that facilitate selecting the range in the *MakeRect* function. These macros are defined below and can be viewed by opening the `userint.h` header file.

- To select all the rows and columns in the table control, use the macro `VAL_TABLE_ENTIRE_RANGE`. This is equivalent to calling the function MakeRect`(1,1,,-1,-1)`.
- To select the entire row of a table control, use `VAL_TABLE_ROW_RANGE(r)`, where `r` is the one-based row number of the row selected.
- To select all the cells in the column, select `VAL_TABLE_COLUMN_RANGE(c)`, where `c` is the one-based column number of the column selected.

NewCtrlMenuItem Function

The *NewCtrlMenuItem* function creates a new menu item on menu for the table control. Its prototype is shown here and its arguments explained in Table 4–16.

```
int status =  NewCtrlMenuItem (int panelHandle,
                   int ControlID, char itemLabel[],
                   int beforeMenuItemID,
                   CtrlMenuCallbackPtr eventFunction,
                           void eventCallbackData);
```

Table 4–16 NewCtrlMenuItem *Function*

Input/ Output	Name	Type	Description
Input	*panel Handle*	integer	panel handle loaded in memory
	controlID	integer	control identifier
	itemLabel	string	name of new menu item you want to add
	beforeMenu ItemID	integer	menu item ID before (above) where you want to place the new menu item; you can specify the built-in menu items constants before placing the menu item using the following constants: ■ `VAL_GOTO -2` ■ `VAL_SEARCH -3` ■ `VAL_SORT -4`
	event Function	CtrlMenu Callback-Ptr	name of the callback function that processes the menu callback; the *event-Function* takes the following form: `void CVICALLBACK FunctionName (int panelHandle, int controlID, int MenuItemID, void * callbackData);` this function is called when the menu item generates a commit event; the *panel-Handle* and the *controlID* are from the control that owns the menu generating the event; the *MenuItemID* is the identifier of the menu item selected; this function also receives the value you pass to *eventCallbackData*
	eventCall backData	void*	specify a pointer to the data that you want to pass to your callback function in the *eventFunction* argument, otherwise pass a zero value
Output	*status*	integer	refer to User Interface Library error codes; negative value represents error

SetActiveTableCell Function

The *SetActiveTableCell* function sets the cell specified by the *Point* structure as the active cell. Its prototype is shown here and its arguments explained in Table 4–17.

```
int status = SetActiveTableCell (int panelHandle,
                                        int ControlID, Point cell);
```

SetTableCellAttribute Function

The *SetTableCellAttribute* function sets the attribute of the cell specified. Its prototype is shown here and its arguments explained in Table 4–18.

```
int status = SetTableCellAttribute (int panelHandle,
                    int ControlID, Point cell,
                    int cellAttribute, int attributeValue);
```

SetTableCellVal Function

The *SetTableCellVal* function sets the value specified on the table control cell selected. Its prototype is shown here and its arguments explained in Table 4–19.

```
int status = SetTableValCell (int panelHandle,
                    int ControlID, Point cell, void  *value);
```

Table 4–17 SetActiveTableCell *Function*

Input/ Output	Name	Type	Description
Input	*panel Handle*	integer	panel handle loaded in memory
	controlID	integer	control identifier
	cell	Point	*Point* structure of the cell for which you want to set the value
Output	*status*	integer	refer to User Interface Library error codes; negative value represents error

Table 4–18 SetTableCellAttribute *Function*

Input/ Output	Name	Type	Description
Input	*panel Handle*	integer	panel handle loaded in memory
	controlID	integer	control identifier
	cell	Point	*Point* structure for the cell for which you want to set the attribute
	cell Attribute	integer	specifies the attribute you want to set
	attribute Value	integer	value you want to set of the attribute specified
Output	*status*	integer	refer to User Interface Library error codes; negative value represents error

Table 4–19 SetTableCellVal *Function*

Input/ Output	Name	Type	Description
Input	*panel Handle*	integer	panel handle loaded in memory
	controlID	integer	control identifier
	cell	Point	*Point* structure of the cell for which you want to set the value
	value	any type	specify the cell value you want to set
Output	*status*	integer	refer to User Interface Library error codes; negative value represents error

SetTableCellRangeVals Function

The *SetTableCellRangeVals* function replaces the values of a table control cell range by the values in a specified array. All cells must be of the same cell type in the cell range. Its prototype is shown here and its arguments explained in Table 4–20.

```
int status = SetTableCellRangeVals (int panelHandle,
                                    int ControlID, rect cellRange,
                                                   void *valueArray,
                                                      int direction);
```

Table 4–20 SetTableCellRangeVals *Function*

Input/ Output	Name	Type	Description
Input	*panel Handle*	integer	panel handle loaded in memory
	ControlID	integer	control identifier
	cellRange	Rect	*Rect* structure specifying the cell range for which you want to set the values
	valueArray	any array*	array with the new cell values
	direction	integer	order in which to store each cell's value in *valueArray;* you have the following choices: ■ VAL_ROW_MAJOR. Values are placed from the array in a given row from left to right, before traversing to the next row. ■ VAL_COLUMN_MAJOR. Values are placed from the array in a given column from top to bottom, before traversing to the next column.
Output	*status*	integer	refer to User Interface Library error codes; negative value represents error

SetTableCellRangeAttribute Function

The *SetTableCellRangeAttribute* function sets the attribute for a range of specified cells. Its prototype is shown here and its arguments explained in Table 4–21.

```
int status = SetTableCellRangeAttribute (int panelHandle,
                                   int ControlID, Rect cellRange,
                                   int cellAttribute,int attributeValue);
```

SetTableColumnAttribute Function

The *SetTableColumnAttribute* function sets the attribute of a table column. Its prototype is shown here and its arguments explained in Table 4–22.

```
int status = SetTableColumnAttribute (int panelHandle,
                        int ControlID, int columnIndex,
                        int columnAttribute, int attributeValue);
```

Table 4–21 SetTableCellRangeAttribute *Function*

Input/ Output	Name	Type	Description
Input	*panel Handle*	integer	panel handle loaded in memory
	controlID	integer	control identifier
	cellRange	Rect	*Rect* structure of the cell range for which you want to set the attribute
	cell Attribute	integer	specify attribute to set
	attribute Value	integer	value you want to set of the attribute specified
Output	*status*	integer	refer to User Interface Library error codes; negative value represents error

Table 4–22 SetTableColumnAttribute *Function*

Input/ Output	Name	Type	Description
Input	*panel Handle*	integer	panel handle loaded in memory
	controlID	integer	control identifier
	column Index	integer	one-based index of the column for which you want to set the attribute
	column Attribute	integer	specifies the attribute you want to set
	attribute Value	integer	value you want to set of the attribute specified
Output	*status*	integer	refer to User Interface Library error codes; negative value represents error

SetTableRowAttribute Function

The *SetTableRowAttribute* function sets the attribute of a table row. Its prototype is shown here and its arguments explained in Table 4–23.

```
int status = SetTableRowAttribute (int panelHandle,
                            int ControlID, int rowIndex,
                            int rowAttribute,int attributeValue);
```

ShowBuiltInCtrlMenuItem Function

The *ShowBuiltInCtrlMenuItem* function makes a built-in pop-up menu item visible on the table control. The menu item can be made visible only if you previously hid the menu item by calling *HideBuiltInCtrlMenuItem*. Its prototype is shown here and its arguments explained in Table 4–24.

```
int status = ShowBuiltInCtrlMenuItem(int panelHandle,
             int ControlID, int builtInMenuItemID,
                                 int beforeMenuItemID);
```

Table 4–23 SetTableRowAttribute *Function*

Input/ Output	Name	Type	Description
Input	*panel Handle*	integer	panel handle loaded in memory
	controlID	integer	control identifier
	row Index	integer	one-based index of the row for which you want to set the attribute
	row Attribute	integer	specifies the attribute you want to set
	attribute Value	integer	value you want to set of the attribute specified
Output	*status*	integer	refer to User Interface Library error codes; negative value represents error

Table 4–24 ShowBuiltInCtrlMenuItem *Function*

Input/ Output	Name	Type	Description
Input	*panel Handle*	integer	panel handle loaded in memory
	controlID	integer	control identifier
	builtIn MenuItemID	integer	built-in menu item ID that you want to make visible selecting from among the following constants: ■ `VAL_GOTO` -2 ■ `VAL_SEARCH` -3 ■ `VAL_SORT` -4
	beforeMenu ItemID	integer	menu item ID before which you want to place the new menu item; you can specify the built-in menu items constants using the following constants:

(continued)

Table 4–24 ShowBuiltInCtrlMenuItem *Function (continued)*

Input/ Output	Name	Type	Description
			■ VAL_GOTO -2 ■ VAL_SEARCH -3 ■ VAL_SORT -4 to place the menu item at the end of the menu item list, use -1.
Output	*status*	integer	returns the ID of the new menu item; refer to User Interface Library error codes; negative value represents error

SortTableCells Function

The *SortTableCells* function sorts a group of specified cells in a table control. Its prototype is shown here and its arguments explained in Table 4–25.

```
int status = SortTableCells (int panelHandle,
                             int ControlID, Rect cellRange,
                             int sortingDirection, int keyIndex,
                             int direction,
                             CellCompareCallbackPtr sortingFunction,
                                                  void *callbackData);
```

Table 4–25 SortTableCells *Function*

Input/ Output	Name	Type	Description
Input	*panel Handle*	integer	panel handle loaded in memory
	controlID	integer	control identifier
	cellRange	Rect	*Rect* structure of the cell range used in the sort

(continued)

Table 4–25 SortTableCells *Function (continued)*

Input/ Output	Name	Type	Description
	sorting Direction	integer	select to sort by row or column by specifying either ■ `VAL_ROW_MAJOR` ■ `VAL_COLUMN_MAJOR`
	keyIndex	integer	row or column containing the sort key
	direction	integer	direction of sort; pass 0 for ascending and 1 for descending sort
	sorting Function	Cell Compare-Callback-Ptr	name of an optional user-defined function to perform a customized cell comparison; see On-line Help for this function for more details; pass `0` if you do not want to use *SortTableCells* function to perform the cell comparisons
	callback Data	void*	pointer to a user-defined data passed to the callback function in *sortingFunction;* pass `0` if you do not want to use this argument
Output	*status*	integer	refer to User Interface Library error codes; negative value represents error

5

VXI COMMUNICATION USING VISA

Chapter Highlights

- Introduction
- Short History
- VXI Chassis, Modules, and Connectors
- Controlling the VXI System
- VXI Address Space and Configuration Registers
- VXI Device Classes
- Communicating with Message-Based Devices
- Resource Manager
- Basics of Programming with VISA
- VISA Project
- Summary
- Library Function Prototypes and Definitions

The basic features of a VME (Versa-Modular Eurocard) Extensions for Instrumentation (VXI) system and how to communicate with the VXI devices are introduced here. You will learn about the VXI hardware and the various address spaces that the VXI devices can map to. You will be shown the different VXI device classes and their configuration registers, and how they are used and configured. You will learn about the functions performed by the VXI Resource Manager and the Virtual Instrumentation Software Architecture (VISA) Default Resource Manager. You will learn how to use VISA functions to set up communication and "talk" to devices that is demonstrated by means of a project.

Introduction

The VXIbus (VMEbus eXtensions for Instrumentation) standard for instruments-on-a-card systems was developed to accommodate the test equipment industry's need to reduce the physical size and cost of rack-mounted equipment, increase performance, and use more precise timing and synchronization between instruments. The instrumentation industry wanted an open architecture for all instruments, enabling them to use the best techniques of both General-Purpose Interface Bus (GPIB) and plug-in Data Acquisition (DAQ) boards.

VXIbus consists of a mainframe chassis with modular instruments on a board (also called *modules*) that plug into chassis slots. VXIbus defines a standard communication protocol to devices using American Standard Code for Information Interchange (ASCII) commands to control instruments, similar to GPIB.

The VXIbus standard is an extension of the VMEbus (VERSAbus Modular Eurocard), standard, which uses the same backplane connectors as the VME. VMEbus is a backplane standard (IEEE 1014) for many electronic platforms. This standard defines the electrical and mechanical backplane characteristics to work with equipment developed by different manufacturers. VXIbus includes additional signals from the VMEbus specification. This standard adds two more board sizes and additional connectors on the board using different board widths. These are explained in the section *VXI Chassis, Modules, and Connectors* below.

Short History

In 1970, Motorola, along with a number of other companies, developed the VMEbus as a backplane standard to communicate with instruments using different electronic platforms. At that time the instrument manufacturers wrote the drivers for their instruments, which did not conform to any established industry standards.

In 1987, the VXIbus Systems Alliance (or VXIbus Consortium, for short) was formed to create a standard protocol for instrumentation encompassing both hardware and software. The VXIbus standard was adopted by IEEE in March 1993 and became known as IEEE 1155. The VXIbus Consortium was founded in September 1993 and comprised of companies such as National

Instruments, GenRad, Hewlett-Packard, Racal Instruments, Tektronix, and Wavetek. Any instrument that is compliant with this software standard is termed VXI*plug&play*-compliant. The VXIbus Consortium meets annually to discuss any new or outstanding issues. The Consortium activities include:

- Issuing new manufacturer ID codes to new VXIbus product developers
- Clarifying specification ambiguities
- Sponsoring vendor interoperability meetings
- Addressing additional standardization issues not covered in the VXIbus specification

To standardize the input/output (I/O) software across the different manufacturers, different hardware controllers, and a wide range of operating systems, National Instruments developed the Virtual Instrumentation Software Architecture (VISA) language. This specification, VPP-4.3, defines a standard for communicating with GPIB devices, VXI devices, or serial interfaces using a common software interface. You will learn more about VISA later in the section *Basics of Programming with VISA*.

VXI Chassis, Modules, and Connectors

The VXI *chassis,* also called *backplane, cage, mainframe,* or *crate,* contains a power supply, cooling system, backplane connections, and physical mounting for VXIbus modules. The VXIbus *modules* (also called *boards* or *cards*) are modular instruments-on-a-board that plug into the VXI mainframe. The mainframe can accommodate boards of four different sizes: A, B, C, and D (see Figure 5–2). The mainframe size is referred to by the largest board size that it can accommodate. You are able to use smaller boards in a larger mainframe through the use of special adapters.

A VXI mainframe has a maximum of 13 slots numbered 0 to 12, from left to right when the mainframe is vertical. Figure 5–1 shows the size C mainframe in the vertical position. This size mainframe is the most commonly used. It is important to use the specified numbering scheme for the slots, since slot 0 of the mainframe has a different backplane signal configuration than the rest of the slots and uses a special board called the *Slot Zero Controller*. The *Slot Zero Controller* is responsible for providing the system clock signals, bus arbitration, and for identifying the slot location of other devices on the system.

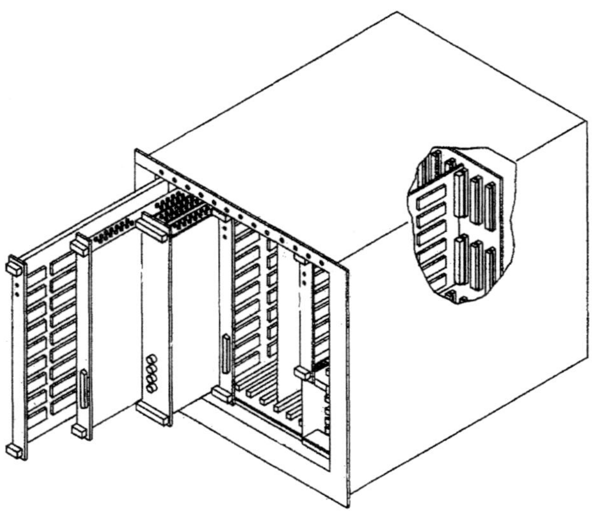

Figure 5–1
13-Slot VXI Chassis with Plug-in Modules
(Copyright National Instruments)

The VMEbus and VXIbus boards can have one to three connectors, depending on the size of the boards. The board sizes and the number of connectors are shown in Figure 5–2. Each connector has three rows of 32 sockets. The board connector is a male connector known as the *P (plug) connector*. The board connector is inserted into the female backplane connector on the mainframe, called the *J (jack) connector*.

The VMEbus mainframe usually houses 20 slots using a 0.8-inch board width for each module. To reduce radiation between the boards, VXIbus increased the board width to 1.2 inches to allow for additional space to enclose the boards in a metal case and connecting the metal case to the backplane ground for better shielding. The increased board width resulted in the VXIbus mainframe having only 13 slots for the boards.

Controlling the VXI System

A VXI system can be controlled using several different configurations:

- **VXI-Based Embedded Computer.** The embedded computer is on an VXI board located inside the VXI mainframe that plugs into the VXI

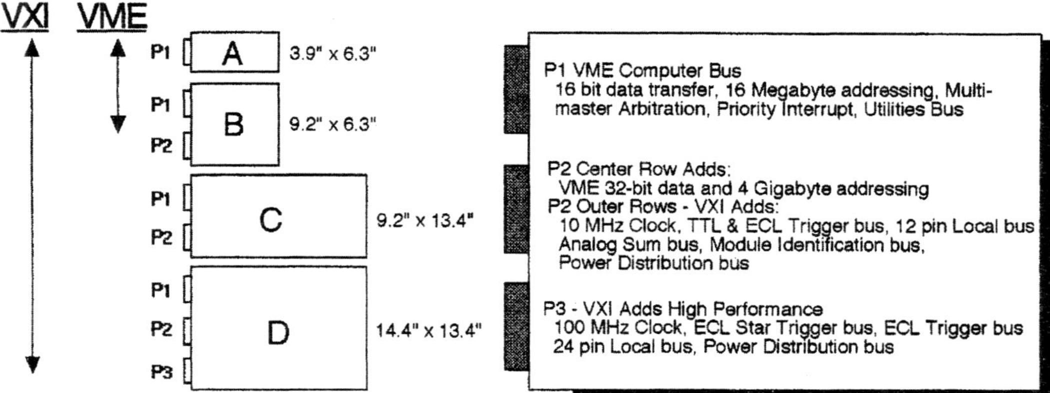

Figure 5–2
VXI Module Sizes and Connectors
(Copyright National Instruments)

backplane. This configuration offers the smallest size and the highest performance.

- **VXI Mainframe Linked Through GPIB.** The GPIB-VXI interface module is installed inside the VXI mainframe and the GPIB interface card on the PC. These are linked together by a GPIB interface cable. The GPIB-VXI interface board translates the GPIB messages and data.
- **High-Speed MXIbus Link.** The VXI backplane is controlled using a highspeed MXIbus, (Multisystem eXtension Interface bus), linked from an external computer to a VXI-MXI extender in Slot 0 of the VXI mainframe. The system performance is equivalent to using the embedded computer connected to the VXI platform. This configuration gives you the flexibility of choosing your own external computer (see Figure 5–3).

If there are not enough slots available on one mainframe, you can add additional mainframes, since in a VXI system there are no limitations on the number of mainframes that you can add. Figure 5–4 shows one of the ways in which you can connect multiple VXI chassis using the MXIbus and VXI-MXI extender boards.

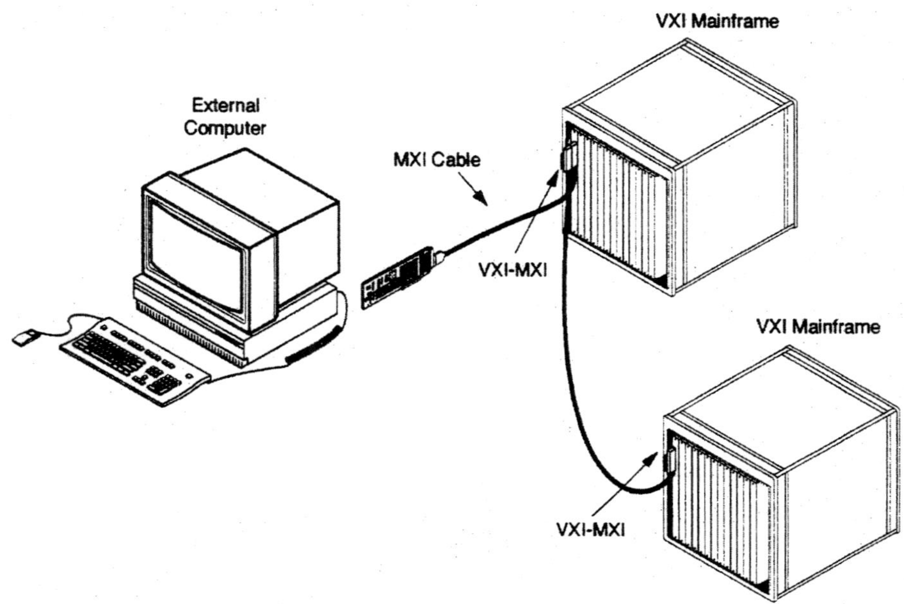

Figure 5–3
PC Using MXI to Control Two-VXIbus System
(Copyright National Instruments)

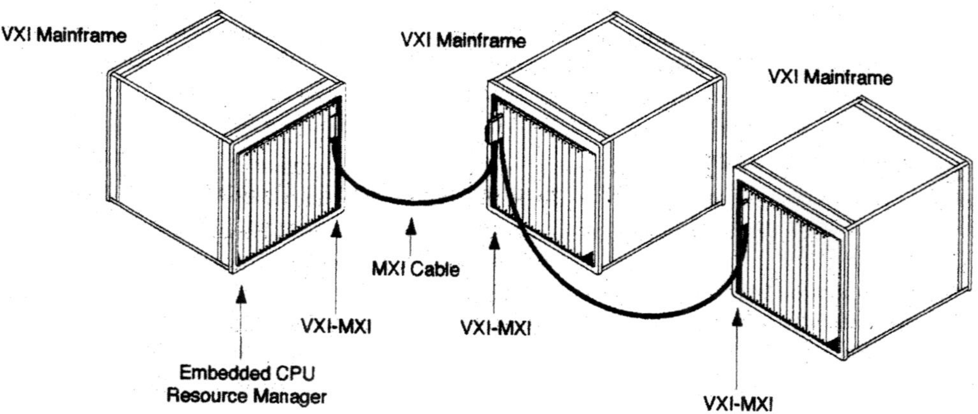

Figure 5–4
VXI-MXI Used for Connecting Multiple VXI Mainframes
(Copyright National Instruments)

VXI Address Space and Configuration Registers

The backplane of the VXIbus has three separate address buses. The first bus is 16 bits wide, the second is 24 bits wide, and the third is 32 bits wide. There is also a data bus on the backplane that is 32 bits wide. The purpose of the address bus is to point to the address (location) of the data. The data is written to or read from the specified address using the data bus. The VXI devices are programmed using the configuration registers. You will need to know the address of the configuration registers in order to configure the device. You can retrieve vital information about the device using these registers, such as the manufacturer ID, model code of the instrument, or the amount of memory requested for normal operation. These are discussed later in this chapter.

The width of the address bus determines what address range can be accessed. The 16-bit address bus can decode addresses from 0000 to FFFF, for a total of 2^{16}, or 64 kB, addresses. These addresses are referred to as the *A16 space*. In the A16 space the upper addresses in the range C000 to FFFF hex are reserved for the configuration registers. The 24-bit address bus can decode addresses from 000000 to FFFFFF, for a total of 2^{24}, or 16 MB, addresses. These addresses are referred to as the *A24 space*. Similarly, the 32-bit address bus can decode addresses from 00000000 to FFFFFFFF, for a total of 2^{32}, or 4 GB, addresses. These addresses are referred to as the *A32 space*. All these address buses are physically separate from each other, and therefore there is no overlapping of any of the address space.

The A16 space address space is broken into 256 windows of 64 decimal registers, enabling each VXI system to have up to 256 VXI devices, using 0 to 254-window space. The 255-window space is reserved for dynamically configured devices that are explained later. The first four configuration registers are the same for all the devices. The remaining registers are device-dependent. The configuration space mapping in the A16 space and to the various registers in the configuration space is shown in Figure 5–5.

The configuration address space consists of 256 windows numbered 0 through 255. These window numbers represent the device's *Logical Address*. For example, the device with logical address 0 has its configuration registers starting at address C000 hex; the device at logical address 1 starts at address C040 hex (offset by 64 bytes, or 40 hex). Offset means how far a particular address is from the absolute address.

Each configuration window consists of 64 registers that are used for configuring the device located at that logical address. The VXI configuration registers can operate in either read or write mode and have different functionalities depending on whether the data is read from or written to the register. The first four configuration registers for each window are:

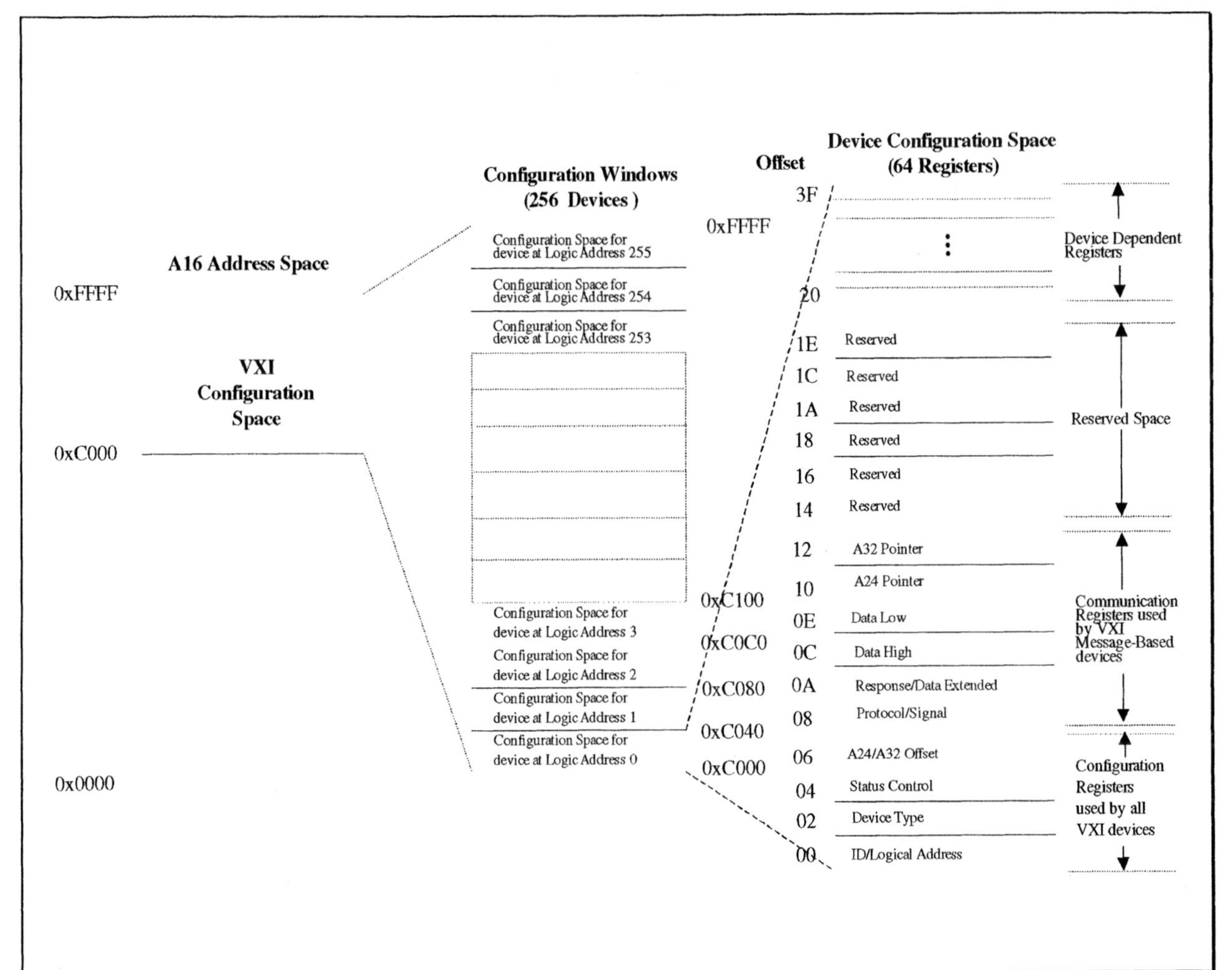

Figure 5–5
VXI A16 Space Address Mapping

1. ID/Logical Address Register
2. Device Type Register
3. Status Control Register
4. A24/A32 Offset Register

The purpose of these registers is explained below with a meaning of the bits used in the registers.

1. **ID/Logical Address Register.** The ID/Logical Address Register is located at offset 0 hex. The register bits used when in the read and write modes are shown in Figure 5–6. In the write mode the lower 8 bits contain the *Logical Address* of the device. The VXI Resource Manager (discussed later in the section *Resource Manager*) writes the logical address in these bits for dynamically configured devices. The device then assumes this new logical address. The upper 8 bits are reserved in this mode. When this register is in the read mode, bits 14 and 15 define the VXI *Device Class,* which can be Register-Based, Message-Based, Extended, or Memory. These device classes are discussed in the section *VXI Device Classes* below. Bits 12 and 13 (*Address Space*) determine whether the device requires addresses in *A16, A24,* and/or *A32 space.* The *Address Space* is decoded as in Table 5–1. The *Manufacturer ID Code* in bits 0–11 holds the unique ID code for the VXI device manufacturer.
2. **Device Type Register.** The Device Type Register is located at offset 02 hex. Figure 5–7 shows the significance of the bits used in this register. In the write mode this register is reserved for future purposes. In the read mode, bits 12–15 (*Required Memory/Model Code* bits) contain the value to be used in computing the memory this device is requesting in the A24 and A32 address space.

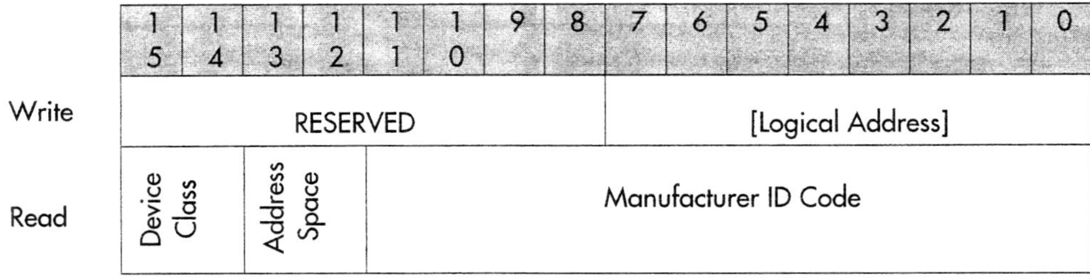

Figure 5–6
ID/Logical Address Register Bit Allocation

Table 5–1 Address Space Bits Decoding

Bit 13	Bit 12	Address Space Used
0	0	A16 and A24
0	1	A16 and A32
1	0	reserved
1	1	A16 only

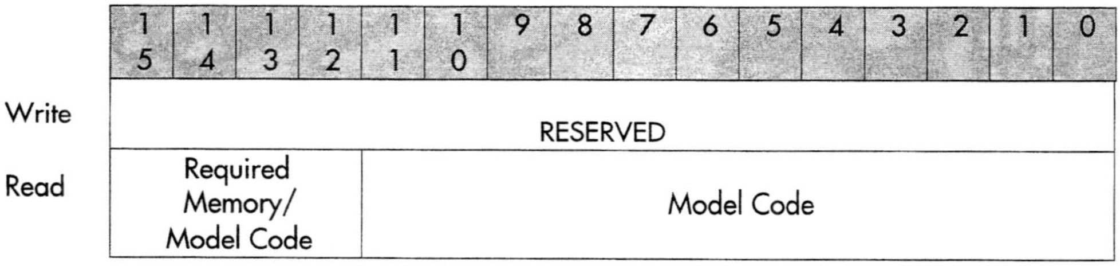

	15	14	13	12	11	10	9	8	7	6	5	4	3	2	1	0
Write	RESERVED															
Read	Required Memory/ Model Code				Model Code											

Figure 5–7
Device Type Register Bit Allocation

The memory requested is computed in the following way:

1. Obtain the decimal value for bits 12–15 in this register (n).
2. If the A24 address space is used, the memory size requested is 2^{23-n}.
3. If the A32 address space is used, the memory size requested is 2^{31-n}.

You can determine the address space by decoding bits 12 and 13 in the ID/Logical Address Register (see Figure 5–6 and Table 5–1).

The lower 12 bits of the Device Type Register in the read mode describe the *Model Code* of the VXI device. If the device does not require any memory in A24 and A32 address space, indicated by bits 12 and 13, both set in the ID/Logical Address Register, all 16 bits in the Device Type Register are used for defining the *Model Code* of the device. Note that the Device Type Register is different from the *Device Class* obtained from the ID/Logical Address Register (Figure 5–6). The *Device Class* is explained in the section *VXI Device Classes* below.

3. **Status/Control Register.** The Status/Control Register is located at offset 04 hex, and its bit allocations are shown in Figure 5–8. When you write to this register, it is used as a control register, and when you read from it, it gives you status information. The bit 0 in the control register (write mode) is used for resetting the device. The bit 1 (*Sysfail Inhibit*) register is used for placing the device in an offline state if it does not pass the self-test. The *A24/A32 Enable* bit (bit 15) is used to activate sharing of the device's onboard memory if it is set in the ID/Logical Address Register (Figure 5–6) by the *Address Space* (bits 12 and 13). The remainder of the bits in this register are device dependent and are not discussed here. In the read mode, this register acts as the status register; bits 0,1, and bits 4–13 are device dependent. Bit 2 (*Pass* bit) is used to indicate if the device passed the power-on self-test, and bit 3 (*Ready* bit) in conjunction with the *Pass* bit indicates that the device is operational. Bit 14 (*MODID* bit) indicates the state of the MODID line. The MODID lines are used by the Slot Zero device to determine in which slot a device with a particular logical address resides. The Slot Zero device needs this information for configuring the VXI during startup. The *A24/A32 Active* bit (bit 15) indicates whether the onboard memory (if any requested) in the A24 or A32 address space is ready to be accessed.
4. **A24/A32 Offset Register.** The A24/A32 Offset Register is located at offset 06 hex, and its bit allocations are shown in Figure 5–9. This register is applicable only to devices that use the A24/A32 address space specified in the ID/Logical Address Register (Figure 5–6) by the *Address*

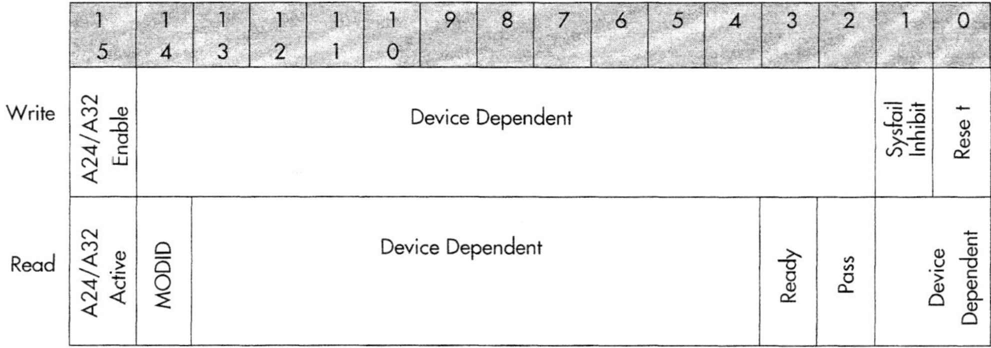

Figure 5–8
Status/Control Register Bit Allocation

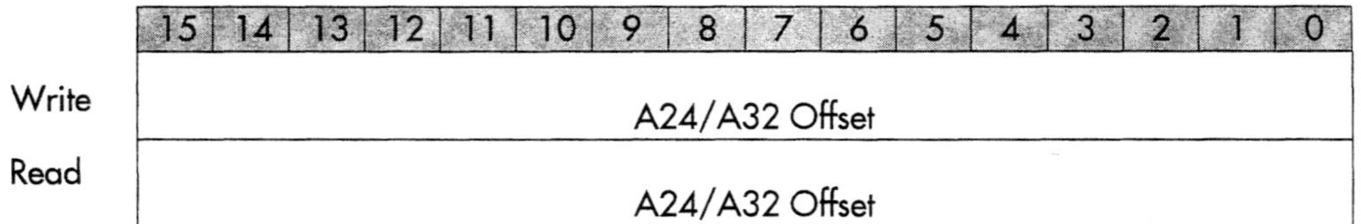

Figure 5–9
A24/A32 Offset Register Bit Allocation

Space (bits 12 and 13). The Resource Manager system startup assigns and writes the base address of the onboard memory of the device in the A24 or A32 address space. In the read mode this register returns the base address of either the A24 or A32 address space of the device's on-board memory.

VXI Device Classes

Here we look at the various classes of VXI devices, their purpose, and how these devices are decoded and configured. All VXIbus devices use the four configuration registers defined above, but some devices use the additional registers, depending on their device class. As we saw above for the ID/Logical Address Register (Figure 5–6), bits 14 and 15 are used for determining the device class. The device class is decoded from these two bits and its representation is shown in Table 5–2. The purpose and functionality of each of these device classes is explained below.

Table 5–2 *Device Class Bits Decoding (ID/Logical Address Register)*

Bit 15	Bit 14	Device Class
1	1	Register-Based
1	0	Message-Based
0	0	Memory
0	1	Extended

- **Register-Based Devices.** Register-Based devices are required only to implement the four VXIbus configuration registers mentioned above, but most of these devices implement additional instrument-specific registers. These devices are controlled through register reads and writes specified by the manufacturer. Each device manufacturer specifies the pattern for register accesses to control the specific functions of the device. There is no standard protocol for controlling Register-Based devices.
- **Message-Based Devices.** Message-Based devices use the basic four configuration registers plus the VXIbus communication registers at offsets 08 through 12 hex (see Figure 5–5). All VXI Message-Based devices must follow the same protocol independent of the device manufacturer. For understanding how the message-based devices communicate and a discussion of the communication registers, refer to the section *Communicating with Message-Based Devices* below.
- **Memory Devices.** Memory devices use the basic four configuration registers similar to the other classes, but also have memory accessible through other addresses in the VXI/VMEbus address space. These devices usually serve the purpose of being used for scratch-pad memory. These memory addresses are accessed explicitly through software.
- **Extended Devices.** Extended devices are used when you want to expand the VXIbus by adding additional devices such as the National Instruments VXI-MXI mainframe extender. Extended devices have a *Subclass* register located at offset 1E hex that defines the subclass to which the extended device belongs. This enables the Resource Manager to identify and configure the device.

Communicating with Message-Based Devices

Some of the basic concepts of Message-Based device communication and the functionalities of the communication registers are discussed in this section. The VXIbus specification defines a *commander/servant hierarchy* to create a communication command and control structure. VXIbus has a multimaster backplane in which multiple processors try to control the same device. The VXIbus specification allows for creating a commander/servant hierarchy to establish a communication order. A commander is any device that has one or more servant devices and uses commands to talk to the Message-Based servant devices or uses device-specific registers to communicate with Register-Based servant devices. A *commander* device can communicate with various *servant* devices. A commander can "speak" to a servant at any time; a servant can respond only when requested. The servant cannot initiate commands to its commander. There can be only one commander in a VXIbus system. The commander commands and controls the configuration and communication registers of the servant devices exclusively. The servant responds to the commander using the word serial protocol (explained below) if the servant is a Message-Based device, or through device-specific registers if the servant is a Register-Based device.

To communicate with the Message-Based devices, you can use one of the three VXI-specified serial communication protocols: *word serial*, *longword serial*, and *extended longword serial*. These protocols are very similar to the GPIB protocol, transferring data to and from the device using one word at a time. Message-Based devices recognize a minimum set of *word serial* commands called base-level commands. The Resource Manager uses some base-level commands during configuration of the devices.

Other capabilities of the Message-Based device are explained below.

- **Master Capability.** A Message-Based servant device with *master* capability can communicate with its commander using the commander's logical address. Such a servant communicates with its commander using VXIbus signals rather than by generating interrupts. A master must have control of the bus by being a bus master to send a signal to its commander signal register. A *bus master* is a device that has requested and been granted control of the VXI bus from the Slot Zero Controller, enabling this device to control the VXI backplane to do VXI cycles (reads/writes, command/query).
- **Instrument Capability.** A Message-Based device can have *instrument* capability to send and receive word serial messages. Servant devices that use the GPIB protocol can send the status byte to its commander.
- **Asynchronous Communication Capability.** Message-Based devices that can be programmed to interrupt are categorized as *programmable interrupters*. The devices can also be programmed to act as interrupt handlers called *programmable handlers*.

The communication registers for Message-Based devices consist of the Protocol/Signal, Response/Data Extended, Data High, and Data Low Registers.

These registers are shown in Figure 5–5 and are located at offset 08 hex through 0E hex. The functionalities of these registers are explained below.

- **Protocol/Signal Register.** The Protocol/Signal Register is located at offset 08 hex. The device's communication capabilities are set up in this register during the configuration setup. The device can be set up as a Commander, Master, Interrupter, or whether the device has a Signal Register.
- **Response Register.** The Response Register is located at offset 0A hex. This register is used for the *extended word serial* protocol and is used to establish handshaking for all three protocols. This register gives the status of the device's communication registers, errors, whether the device is ready to receive or send data, whether the device is locked from word serial access, and other related information.
- **Data High Register.** The Data High Register is located at offset 0C hex. This register is used for the *long word serial* and *extended word serial* protocol to pass messages between Message-Based devices.
- **Data Low Register.** The Data Low Register is located at offset 0E hex. This register is used for the *word serial* protocol to send the data out of the device. This register contains the handshake byte and the ASCII character that are loaded into the register one at a time until the complete string has been transferred.

Resource Manager

The Resource Manager resides on the VXIbus embedded controller or an external computer controlling the VXIbus. The Resource Manager runs the *Resman algorithm* using the `resman.exe` file. `Resman.exe` is installed automatically on a computer when the NI-VXI software is installed and sets up the configuration registers of the devices and the system resources at startup and initiates normal operations.

The Resource Manager can reside physically in any slot in the VXI chassis but must have a logical address of 0. At power-on the VXIbus system first operates in the configuration state in which the Resource Manager configures the devices, then transitions to an operational state in which the normal operation of the device can begin. The Resource Manager is used only during startup and is not used for controlling the VXIbus system after the startup operation is complete.

The Resource Manager performs the following functions at startup:

- The Resource Manager identifies system devices by attempting to read the ID/Logical Address configuration registers in the configuration space for each of the 256 devices. If a value is returned from the register, the device is presumed to exist at that logical address. If there was a bus error in reading the device at that logical address, it would indicate that there is no device present at that address.
- The MODID lines report the slot number of the device to the Resource Manager as explained above for the *Status/Control Register*.
- The Resource Manager writes the logical address of the dynamically configured device in the lower 8 bits of the ID/Logical Address Register, as explained above.
- The *Status/Control Register* for the device keeps a record of whether the device passed the self-test on startup and reports the status to the Resource Manager.
- The Resource Manager configures the A24 and A32 address map by setting the A24/A32 *Enable* bit in the Control Register, and A24/A32 *Active* bit in the Status Register.
- The Resource Manager sets up the commander/servant hierarchy for message-based devices.
- The Resource Manager allocates interrupt lines to programmable interrupters and programmable interrupt handlers for message-based devices.
- After the configuration is complete, the Resource Manager sends the *word serial* command *Begin Normal Operation* to the message-based devices to begin normal operation.

There are a few caveats about running *Resman*. The VXI controller must be initialized before running *Resman*. This is done automatically when National Instruments VXI driver software is installed. You must run *Resman* every time you cycle power on the mainframe to reinitialize the system. This is because the Resource Manager runs only once on power-up. Note that if you need to run the Resource Manager again, the message-based devices do not respond to most of the *word serial* configuration commands after receiving the *Begin Normal Operation* command. To run the Resource Manager again you would need to power cycle on the VXI mainframe(s) or send the system reset (`SYSRESET*`) command to set the devices to their configuration state. This is accomplished through the VISA function *viAssertUtilSignal* (if you are using *CVI* 6.0 or later versions) or by using the NI-VXI function *AssertSysReset* (for

CVI versions 5.5 and earlier). Sending the system reset from either of these functions is equivalent to cycling power on the VXI system, as it asserts the *Sysreset* line on the controller specified.

Basics of Programming with VISA

In this section you are introduced to the basics of communicating with instruments using VISA. Before we start to program using VISA, you should understand that VISA is a standard I/O language for communication with instruments using a high-level API that calls the low-level drivers. As mentioned above, VISA provides interface independence from communicating with VXI, GPIB, and serial instruments by using the same software interface.

Programs written using VISA are portable from one platform to another. This is accomplished by VISA using its own data types to ensure that the data size for variables remain the same when the software is compiled on different platforms. The VISA communication functions are also designed to be portable across different platforms and operating systems.

VISA is a simple language to learn and use and was developed to enable end-users to communicate effortlessly with instruments. VISA uses a few and common set of operations that are required to control and communicate with different interface devices. For example, you can use the same communication function to write to a VXI, GPIB, PXI, or serial device. Similarly, you can use a read function with the similar interface to obtain data from a device irrespective of its interface bus.

VISA is an object-oriented language and some object-oriented terminology that is used in communication with the *resource* is explained here. *Resource* is defined as a complete set of capabilities of a device. This set of capabilities can consist of writing to and reading from the device, requesting service, performing a self-test, resetting the device, and performing other device controls. Object-oriented languages refer to the *functions* of these resources as *operations* and the properties associated with the object containing information about the objects as *attributes*.

Let us look at the basic steps used to set up communication with a device using VISA:

1. Open a session to the VISA Default Resource Manager to initialize the VISA system.
2. Open a communication channel (session) to the device.

3. Write to the device.
4. Read from the device.
5. Optionally, process the received data.
6. Close the communication channel (session) to the device.
7. Close the VISA Default Resource Manager session.

Let us now see how each of these steps is performed using the VISA library functions.

Step 1 calls for establishing a session with the VISA Default Resource Manager using the VISA library function *viOpenDefaultRM(&DefaultRM)*. The VISA Default Resource Manager initializes the VISA system and can search for available resources or open sessions to them. Opening a *session* means establishing a communication link to the VISA Default Resource Manager or the device(s). Note that the VISA Default Resource Manager is different from the VXI Resource Manager. The VXI Resource Manager executes *Resman* when the VXI system is turned on and therefore must be run before VISA. In VXI systems, the VISA Default Resource Manager looks at the `resman.tbl` created by the Resman utility to obtain the instruments available on the system. The *viOpenDefaultRM(&DefaultRM)* function returns a resource manager handle, `DefaultRM`, that identifies this session uniquely. This argument will be used in the *viOpen* and *viClose* library functions when closing the session with the VISA Default Resource Manager.

Step 2 is to open a communication channel to the device using the function *viOpen*. This function has a form similar to

```
status = viOpen(DefaultRM, Instrument_Descriptor,
                  Access Mode,Timeout, &InstrumentHandle);
```

The first argument, `DefaultRM`, of *viOpen* is the resource manager handle obtained from the *viOpenDefaultRM* function. The second argument, `Instrument_Descriptor`, is the instrument descriptor that contains the string with the exact name and location of the VISA resource to identify the instrument uniquely. The `Instrument Descriptor` has the format

```
Interface Type[board]::Address::Resource
```

The Instrument Descriptor syntax is shown in Table 5–3. *Interface Type* in the table shows the keywords that are used to communicate with the various interfaces. The items shown in parentheses in Table 5–3 are optional parameters. The `board index` is the board number to which the device is connected. The board index is used if more than one interface board is present.

Table 5–3 Instrument Descriptor Syntax

Interface Type	Syntax
VXI INSTR	`VXI[board index]::VXI logical address[::INSTR]`
VXI MEMACC	`VXI[board index]::MEMACC`
VXI BACKPLANE	`VXI[board index][::mainframe logical address]::BACKPLANE`
VXI SERVANT	`VXI[board index]::SERVANT`
GPIB-VXI INSTR	`GPIB-VXI[board index]::VXI logical address[::INSTR]`
GPIB-VXI MEMACC	`GPIB-VXI[board index]::MEMACC`
GPIB-VXI BACKPLANE	`GPIB-VXI[board index][::mainframe logical address]::BACKPLANE`
GPIB INSTR	`GPIB[board index]::primary address[::secondary address][::INSTR]`
GPIB INTFC	`GPIB[board index]::INTFC`
GPIB SERVANT	`GPIB[board index]:: SERVANT`
PXI INSTR	`PXI[board index]::device[::function][::INSTR]`
Serial INSTR	`ASRL[board index][::INSTR]`
TCPIP INSTR	`TCPIP[board index]::host address [::LAN device name][::INSTR]`
TCPIP SOCKET	`TCPIP[board index]::host address::port::SOCKET`

In the `Instrument Descriptor` format given above, `Address` is the logical address of the device with which you want to establish communication, and `Resource` refers to a grouping used to encapsulate the VISA operations.

As shown in Table 5–3, the following VISA resource types are supported: `INSTR`, `MEMACC`, `INTFC`, `BACKPLANE`, `SERVANT`, and `SOCKET`. Two of the most commonly used resource types are INSTR and MEMACC.

A VISA Instrument Control (INSTR) Resource lets the controller interact with the device to write and read data from the device, triggering and handling service requests, resetting and initializing the device, obtaining the device status, transferring large blocks of data, accessing the device registers, and other functionalities to communicate with the device.

A VISA Memory Access (MEMACC) Resource is used for accessing the entire memory range for the address space specified. This is used for accessing device registers for multiple devices. This resource allows for moving large blocks of data between addresses using the absolute address within the address space specified.

For a description and functionality of other VISA resource types, see *VISA Resource Types* in Chapter 2 of the *NI-VISA Programmer's Reference Manual.*

As an example, to communicate with a device located at logical address `10`, connected to the VXI interface `VXI0`, use the following as the Instrument Descriptor:

```
VXI0::10::INSTR
```

The next argument in the *viOpen* function is the `Access Mode`. The `Access Mode` argument can consist of `VI_EXCLUSIVE_LOCK`, `VI_LOAD_CONFIG`, or `VI_NULL`. `VI_EXCLUSIVE_LOCK` is used for locking the resource before the operation returns. If the resource cannot be locked, the operation waits up to the time specified in the *Timeout* argument (the next argument in the *viOpen* function) before returning an error. When `VI_EXCLUSIVE_LOCK` is used in this argument, any external user-defined settings will be used rather than the VISA-specified defaults. `VI_LOAD_CONFIG` is used for any external user-defined settings rather than the VISA-specified defaults. If there are no external settings, the operation will use the VISA defaults and return `VI_WARN_CONFIG_NLOADED`. Use `VI_NULL` if you are not using these features.

The last argument of this function is `InstrumentHandle`, which returns the communication channel to the device. This argument is used to communicate with the device selected and is used in all subsequent function calls.

There are a few differences between communicating with message-based devices and a register-based device. Message-based devices communicate using the *word serial* protocol, as mentioned above, whereas register-based devices write and read 16-bit values in the A16, A24, or A32 address space. These address spaces were explained in the section *VXI Address Space and Configuration Registers* above.

Step 3 is to write to the message-based device using the function *viWrite*. The generic form of this function is as follows:

```
status = viWrite (InstrumentHandle, WriteString,
                  NumberOfBytesToWrite, BytesTransferred);
```

In this function, the first argument, `InstrumentHandle,` is the handle you obtained from the *viOpen* function above to open a communication channel to the device. `WriteString` is the buffer containing the string you want to send to the device. `NumberOfBytesToWrite` is the count of the number of bytes in the `WriteString` buffer. The last argument in this function is the number of bytes that were actually transferred. You can receive the count in this argument if you want to be certain that all the bytes were written to the device; otherwise, you can use `VI_NULL` if you do not care to obtain this information.

To read from a message-based device, use the *viRead* function. The generic form of this function is

```
status = viRead(InstrumentHandle, ReadString,
                NumberOfBytesToRead, BytesTransferred);
```

As in *viWrite* function, the first argument is the handle you obtained from the *viOpen* function to open a communication channel to the device. You specify the number of bytes to read in the argument `NumberOfBytes-ToRead.` `BytesTransferred` tells you the actual number of bytes that were read. Again, you can use `VI_NULL` for this argument if you do not need this value.

To write to or read from a register-based device or the configuration registers of message-based devices, the register access functions are used. VISA allows two ways in which you can access register-based devices: high-level access (HLA) operations and low-level access (LLA) operations. The HLA operations use the arguments in the VISA functions to perform the necessary steps to set up the hardware, access the address space, perform the write or read to the register, detect errors, and perform error handling. All these activities are performed through a single HLA function. When using LLA operations, you have to perform the same activities as HLA through multiple function calls. The HLA functions are easier to implement in an application. The LLA operations take longer to create an application, but result in faster communication than when the HLA functions are used. In this chapter we discuss the HLA functions only. Refer to the *NI-VISA User Manual* for an explanation of LLA functions and implementation.

Using the HLA functions to write to a register-based device, you can use the *viOut8, viOut16,* or *viOut32* function, depending on whether you are writing to an 8-, 16-, or 32-bit register. The generic form of the *viOut16* function is shown here; the 8- and 32-bit versions use similar function arguments.

```
status = viOut16(InstrumentHandle, AddressSpace,
                                  Offset, ValueToWrite);
```

As in the *viWrite* function, the first argument is the handle you obtain from the *viOpen* function to open a communication channel to the device. The `AddressSpace` is the VXI address space you want to access. It can be A16, A24, or A32 space. The next argument is `Offset`, which refers to the memory offset from the base address of the device. The use of this argument requires some explanation. Logical addresses in A16 space start at offset 0xC000 (see Figure 5–5). The base address of a device at logical address 2, for example, is computed as follows:

```
0xC000 + (0x40  * 2) = 0xC080
```

since there are 0x40 (64 decimal) address registers in each configuration space of a device. This address calculation is performed automatically by the function since it knows which device you are talking to from the `InstrumentHandle` argument. Suppose you want to refer to the Status/Control Register device at offset 4; its address would be

```
0xC080 + 0x04 = 0xC084
```

All you need to do is to supply the `Offset` for the Status/Control Register in the *viOut* function and the correct register is referenced. The last argument in the *viOut* function is `ValueToWrite` and is the data you want to write to this register.

Step 4 is to read from a register-based device. You can use the *viIn8, viIn16,* or *viIn32* function, depending on whether you are reading to an 8-, 16-, or 32-bit register. The generic form of the *viIn16* function is shown here; the 8- and 32-bit versions use similar arguments.

```
status = viIn16(InstrumentHandle, AddressSpace,
                                  Offset, &ReadBuffer);
```

The arguments used in this function are similar to those defined for the *viOut16* function, except that the data received from the device is read in the buffer `ReadBuffer`.

Step 5 is optional and can be used to analyze or display the data as needed.

Step 6 is to close the session to the device using the *viClose* function. This function uses only one argument—the instrument handle that was created by the *viOpen* function.

Finally, at step 7, you must close the session with the VISA Default Resource Manager using the *viClose* function. The `DefaultRM` handle obtained from the function *viOpenDefaultRM* above is used as an argument in *viClose*.

With these steps in mind, let us now look at an example project using the HLA functions.

VISA Project

Some of the basics of communicating using VISA were shown in the section above. `Project5.prj` will show you how to set up a communication and how to write and read data to/from a message-based VXI device.

This project demonstrates communicating with a Function/Arbitrary Waveform Generator VXI device. The code is written to work with the Hewlett-Packard (now part of Agilent Technologies) E1441A Arbitrary Waveform Generator VXI module. For simplicity, in this book I refer to the Arbitrary Waveform Generator as the function generator. Let us examine the GUI for this project, shown in Figure 5–10. Using this GUI you will be able to configure the function generator to output the selected type of waveform and the frequency from the **Waveform Shape** box, with the peak-to-peak voltage specified set from the **Amplitude** box. You can set the output impedance to either 50 ohms or open-circuit by selecting from the **Impedance** switch, and also apply an offset DC voltage. After you have made the selection, you can click on the **CONFIGURE** command button to send the selections to the function generator and configure the function generator. Echoing the string back from the device and displaying it on the text box on the GUI verifies that the string was sent correctly to the function generator. If you want to find all the resources available on your VXI system, click on the **FIND RESOURCES** command button. All the devices with their logical addresses will be displayed on the text box.

Let us look at the code and the VISA functions used in this project.

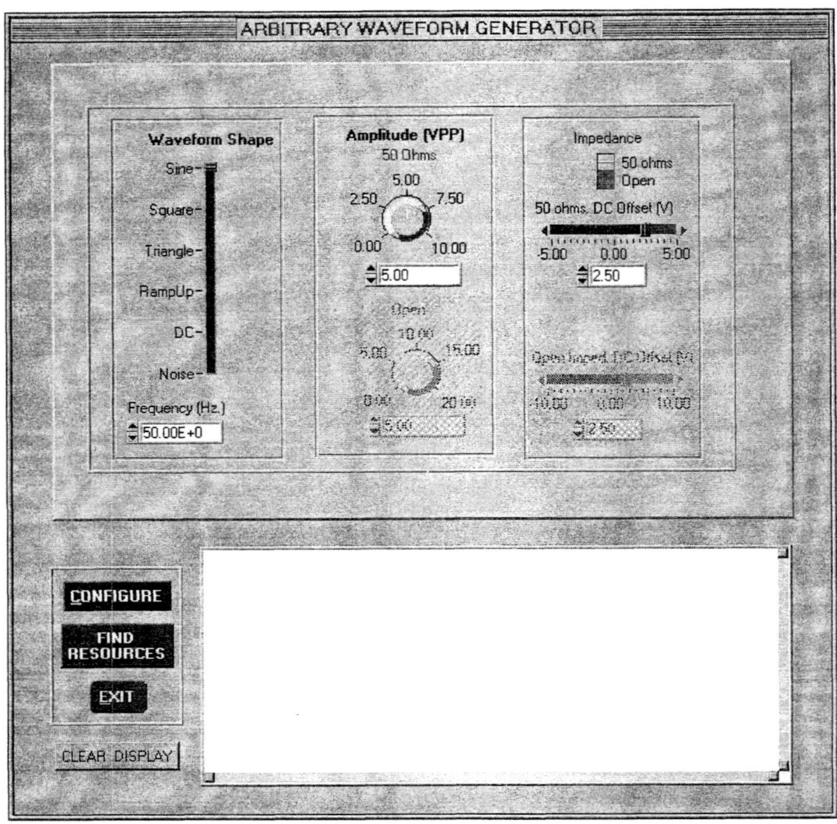

Figure 5–10
Arbitrary Waveform Generator GUI

Header and *main* Function

The header and the *main* function code are listed in Figure 5–11.

The `visa.h` header contains the VISA library functions. At lines 17–21 the global variables using the VISA data types are defined. As mentioned above, the VISA data types are used to keep the data size the same when porting an application to a different platform. The VISA data types are listed in Appendix A of the *NI-VISA Programmer Reference Manual*. You will notice that all VISA data types start with "Vi." At line 17 `ViStatus` is a data type used for obtaining the status of the VISA function. For successful completion of VISA

```
1   #include <utility.h>
2   #include <formatio.h>
3   #include <userint.h>
4   #include <ansi_c.h>
5   #include <stdio.h>
6   #include <stdlib.h>
7   #include <string.h>
8   #include <visa.h>
9   #include "project5.h"
10
11  #define BUFFER_LEN 256      //Buffer Length
12
13  static int AWGPanelHandle;
14  int  DriverIndex;
15  short int  ImpedanceIndex;
16
17  ViStatus    status = VI_SUCCESS;
18  ViSession   InstrHandle, ResourceManager;
19  ViChar buffer[BUFFER_LEN], bytes[BUFFER_LEN], errStr[BUFFER_LEN];
20
21  unsigned char read_buf[BUFFER_LEN], cmdString[BUFFER_LEN];
22  ViReal64 Impedance;
23
24  int SetupCommunication(void);
25
26  int main()
27  {
28    AWGPanelHandle =LoadPanel(0, "project5.uir", AWG_PNL);
29    DisplayPanel(AWGPanelHandle);
30    //Open VISA Default Resource Manager
31    status = viOpenDefaultRM(&ResourceManager);
32    if (status < VI_SUCCESS)
33    {
34            sprintf (buffer, "Error Opening Resource Manager...Exiting Program");
35            InsertTextBoxLine (AWGPanelHandle, AWG_PNL_DISP, -1, buffer);
36
37            return -1;
38    }
39    RunUserInterface();
40  } //main
```

Figure 5–11
project5 Header and *main* Function Listing

functions, the status value is always 0, whereas negative values indicate an error. The value of the status variable is checked against the reserved word VI_SUCCESS, whose value is defined as 0 in visatype.h, which is called from visa.h. If the status value is less than VI_SUCCESS, there is an error in the function call, and error handling is performed by the code. At line

18 the `ViSession` data type is defined for sessions opened with the VISA Default Resource Manager or with a VISA device. The `ResourceManager` handle is used at line 31 to identify the session uniquely. Recall that the first call when opening a session with VISA is to the *viOpenDefaultRM* function and is therefore included in the `main` function. It cannot be emphasized enough that whenever you make a call to any VISA function, you must check the function's return status for possible errors. In case there is a problem with the function call, you will be notified of the error immediately. VISA status codes are 32-bit integers in hexadecimal that can indicate possible errors or warnings or indicate if the operation completed successfully. The status codes are listed in Appendix B of the *NI-VISA Programmer Reference Manual.* Between lines 32 and 37, the value returned from the resource manager is checked and the error (if any) displayed in the text box. If an error is found, the Default Resource Manager is closed and the program terminated.

Finding System Resources

Before we configure any device, we would like to know all the resources that exist on the system and their logical addresses. To do so, we click on the **FIND RESOURCES** command button on the GUI. The callback function *FindResourcesCB* listed in Figure 5–12 is invoked.

The VISA library function *viFindRsrc* shown on line 14 is used to find all the VXI resources on the system. The first argument of this function is the resource manager handle that is obtained from *viOpenDefaultRM.* The second argument is a string consisting of the search expression specifying the type of resource that you want to search. You can use wild-card characters in the string as shown in Figure 5–13. These choices are displayed when you right-click in the **Expression** box on the function panel for the *viFindRsrc* function. The `Matches` variable in the *viFindRsrc* function returns the number of resources found.

In this project we are searching for all instruments of `INSTR` *Resource*. The `FindHandle` argument is a pointer to a list of character strings containing a description of the resources found. These strings are the instrument descriptors and are returned in the `InstrumentDesc` buffer. The *viFindRsrc* function returns only the first instrument descriptor string that matches "?*INSTR". Additional instrument descriptor strings are obtained using the VISA library function *viFindNext* at line 29. The *viFindNext* function uses the `FindHandle` obtained from the *viFindRsrc* function to return the next instrument descriptor in `InstrumentDesc`. To find all the resources, the *viFindNext* function has to be called each time inside a *while* loop to find the string matches found in the *viFindRsrc* function, as shown at lines 25–45.

Line 49 is used to deallocate the memory used by `FindHandle` when you are done searching all the resources.

```
1   //Find VISA Resources
2   int CVICALLBACK FindResourcesCB (int panel, int control, int event,
3             void *callbackData, int eventData1, int eventData2)
4   {
5     int counter=0;
6     char InstrumentDesc[BUFFER_LEN];
7     ViFindList FindHandle;
8     viUInt32 Matches;
9     switch (event)
10    {
11     case EVENT_COMMIT:
12
13    //Find all VISA "INSTR" resources on the system
14    status = viFindRsrc (ResourceManager, "?*INSTR", &FindHandle, &Matches,
15                                                               InstrumentDesc);
16    if (status < VI_SUCCESS)
17    {
18            sprintf (buffer, "Error in finding resources.... Exiting Program");
19            InsertTextBoxLine (AWGPanelHandle, AWG_PNL_DISP, -1, buffer);
20
21            return -1;
22    }
23
24    //Find all instruments on the system
25    while (counter < Matches)
26    {
27            if (counter != 0)
28            { //Find next device
29                    status = viFindNext (FindHandle, InstrumentDesc);
30                    if (status < VI_SUCCESS)
31                    {
32                        sprintf (buffer, "An error occurred finding the next resource");
33                            InsertTextBoxLine (AWGPanelHandle, AWG_PNL_DISP, -1,
34                                                                   buffer);
35
36                            return -1;
37                    }
38            }//end if
39
40            //Display instrument found
41            sprintf(buffer, "Found device:%s", InstrumentDesc);
42            InsertTextBoxLine (AWGPanelHandle, AWG_PNL_DISP, -1, buffer);
43
44            counter++;   //instruments found counter
45    }// end while
46
47    break;
48    }
49    viClose (FindHandle); //Free space allocated by system to the list reference
50   return 0;
51  } //FindResourcesCB
```

Figure 5–12
FindResourcesCB Function Listing

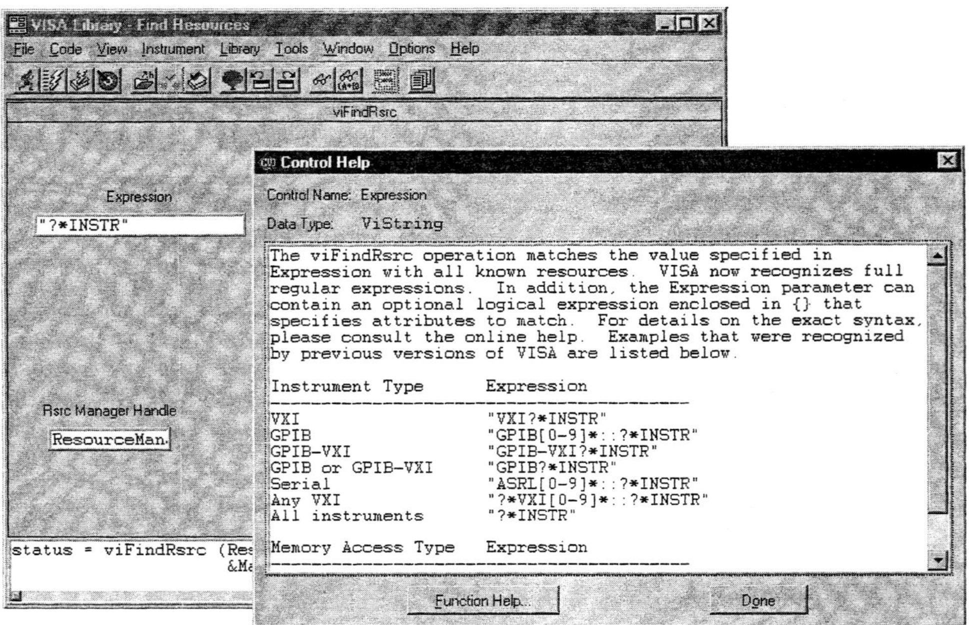

Figure 5–13
Expression List in *viFindRsrc* Function

Setting Up Communication with the Function Generator

Once the resources on your system have been found, you are ready to open a communication channel to the device for communication. In the *Setup-Communication* function the device is opened for communication and the timeout limit established, initialized, and identified. Let us now look at the code for the *SetupCommunication* function and see how these are achieved.

You use the VISA library function *viOpen* shown in the *SetupCommunication* function (Figure 5–14) at line 5. The first argument of the *viOpen* function is a handle returned by *viOpenDefaultRM* from the *main* function. This handle identifies the communication channel with VISA. The next argument in the *viOpen* function is the string consisting of the instrument descriptor of the device with which you want to open communication. Here we are communicating with a VXI device at interface `0` located at logical address `10`. The next two arguments are not used in this function and are therefore set to `VI_NULL`. The last argument of the *viOpen* function is `InstrHandle`, a unique identifier that is used to communicate with this device and is used in subsequent function calls.

```
1    //Open device and setup the device
2    int SetupCommunication(void)
3    {
4       //Open a session to the device
5       status = viOpen(ResourceManager,"VXI0::10::INSTR",VI_NULL,  VI_NULL, &InstrHandle);
6
7       if (status < VI_SUCCESS)
8       {
9
10         viStatusDesc (ResourceManager, status, errStr);
11         MessagePopup ("Error!", errStr);
12      }
13
14      //Set timeout to 5 seconds
15      status = viSetAttribute (InstrHandle, VI_ATTR_TMO_VALUE, 5000);
16      if (status < VI_SUCCESS)
17      {
18         viStatusDesc (InstrHandle, status, errStr);
19         MessagePopup ("Error!", errStr);
20      }
21
22      //Initialize the device
23      status = viWrite (InstrHandle, (ViBuf) "*RST", 5, VI_NULL);
24      status = viWrite (InstrHandle, (ViBuf) "*IDN?", 6, VI_NULL);
25      status = viRead (InstrHandle, read_buf, BUFFER_LEN-1, VI_NULL);
26
27      if (status < VI_SUCCESS)
28      {
29         viStatusDesc (InstrHandle, status, errStr);
30         MessagePopup ("Error!", errStr);
31      }
32      InsertTextBoxLine (AWGPanelHandle, AWG_PNL_DISP, -1, (char*) read_buf;
33
34     return 0;
35   } //SetupCommunication
```

Figure 5–14
SetupCommunication Function Listing

The VISA library function *viSetAttribute* at line 15 is used for setting the attributes of the communication channel. There are numerous attributes you can set by bringing up the function panel and clicking on the **Attribute Name** field (see Figure 5–15). Here we are selecting the timeout attribute and setting the timeout value to 5 seconds. The timeout value is entered in milliseconds as the last argument of this function. It is necessary to set this attribute for a message-based device to prompt you in case the device "hangs up."

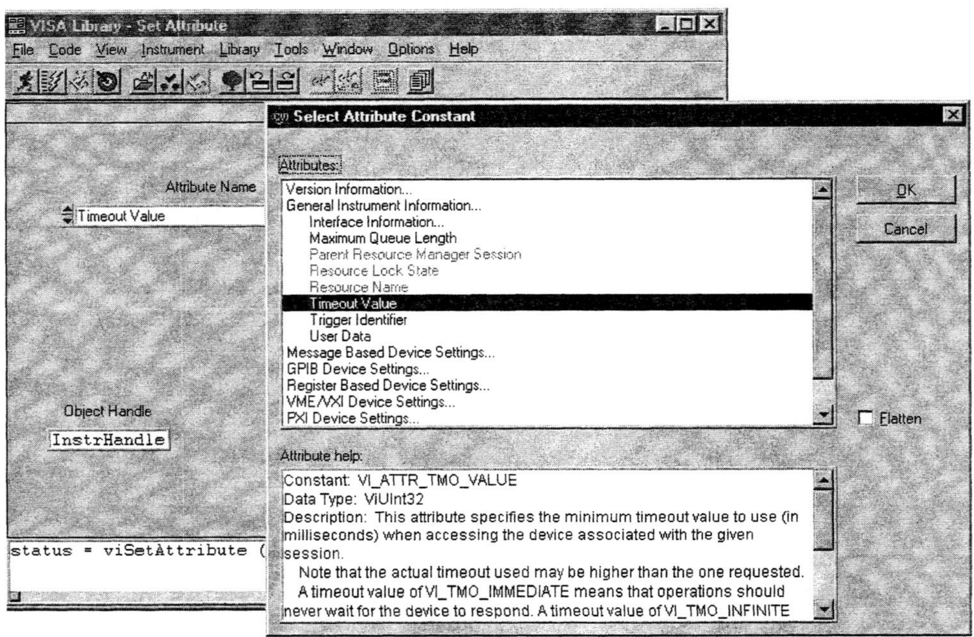

Figure 5–15
Attributes in *viSetAttribute* Function Panel

The VISA library function *viStatusDesc* at line 18 takes the VISA status code and converts it to a meaningful error message. In this program, the error string is displayed on a message pop-up panel whenever there is a communication error.

To send a string to a message-based device, you use the *viWrite* library function. The reply from a message-based device is obtained in the string returned by the *viRead* function. These functions are shown at lines 23–25.

The first argument of *viWrite* function is `InstrHandle`, the instrument handle that was obtained from *viOpen* function. This identifies the device to which you want to send the string message. At line 23 the string containing the reset command ("`*RST`") is sent to the device and the device identification ("`*IDN?`") is requested at line 24 in the second argument of this function. These strings are standard IEEE 488.2 commands that are acceptable to all VXI message-based devices.

The *viRead* function receives the string returned in `read_buf` and contains the identification information for this device.

As mentioned above in the section *Basics of Programming with VISA*, to write and read to a register-based device you use the *viIn* and *viOut* library functions. This program is communicating with a message-based device, and therefore functions relevant to such a device will be explained only for this project.

Configuring the Function Generator

The *ConfigAWGCB* function, listed in Figure 5–16, is invoked from the **CONFIGURE** command button and configures the function generator based on the selections made from the GUI in Figure 5–10. Some of the main tasks performed by this function are mentioned here. For simplicity, the function status is not tested in this callback function, although it is highly recommended that you do so when running code with the hardware.

After establishing communication using the *SetupCommunication* function called at line 12, the frequency and impedance values selected are obtained. Depending on the **Impedance** switch selection (**50 ohms** or **Open**), the appropriate **Amplitude** knob and the **DC Offset** slider controls are enabled. The enabled controls allow you to select the appropriate amplitude and DC offset values for either 50 ohms or for open-circuit impedance. This is

```
1   //Invoked from the CONFIGURE command button
2   int CVICALLBACK ConfigAWGCB (int panel, int control, int event,
3            void *callbackData, int eventData1, int eventData2)
4   {
5   int WaveIndex;
6   ViReal64 frequency, Amp, DC_Offset;
7   ViChar impedance[BUFFER_LEN] Waveform[BUFFER_LEN];
8     unsigned char imp_buf[BUFFER_LEN];
9     switch (event)
10    {
11      case EVENT_COMMIT:
12           SetupCommunication();  //Open resources
13           //Get selected frequency
14           GetCtrlVal(AWGPanelHandle,AWG_PNL_FREQ, &frequency);
15           //Get impedance
16           GetCtrlVal(AWGPanelHandle,AWG_PNL_IMP, &ImpedanceIndex);
17           if (ImpedanceIndex)   //50 ohms impedance
```

Figure 5–16
ConfigAWGCB Function Listing *(continued)*

```
        {
                GetCtrlVal(AWGPanelHandle,AWG_PNL_AMP_50_OHMS, &Amp);
                GetCtrlVal(AWGPanelHandle,AWG_PNL_DC_50HMS, &DC_Offset);

                Fmt(cmdString, "%s<OUTP:LOAD 50"); //Set impedance to 50 ohms
                status = viWrite (InstrHandle, cmdString, NumFmtdBytes(), VI_NULL);
                Fmt(cmdString, "%s<OUTP:LOAD?");  //create command string
                status = viWrite (InstrHandle, cmdString, NumFmtdBytes(), VI_NULL);
                status = viRead (InstrHandle, imp_buf,BUFFER_LEN, VI_NULL);
                Fmt(impedance, "%s<Impedance = %s\n", imp_buf);
                InsertTextBoxLine (AWGPanelHandle, AWG_PNL_DISP, -1, impedance);

        }
        else    //Open circuit
        {
                GetCtrlVal(AWGPanelHandle,AWG_PNL_AMP_OPEN_OHMS, &Amp);
                GetCtrlVal(AWGPanelHandle,AWG_PNL_OPEN_DC, &DC_Offset);

                Fmt(cmdString, "%s<OUTP:LOAD INF"); //Set impedance to OPEN
                status = viWrite (InstrHandle, cmdString, NumFmtdBytes(), VI_NULL);
                Fmt(cmdString, "%s<OUTP:LOAD?");  //read back configured impedance
                status = viWrite (InstrHandle, cmdString, NumFmtdBytes(), VI_NULL);
                status = viRead (InstrHandle, imp_buf,BUFFER_LEN, VI_NULL);
                Fmt(impedance, "%s<Impedance = %s\n", imp_buf);
                InsertTextBoxLine (AWGPanelHandle, AWG_PNL_DISP, -1, impedance);

        }

    //Select waveform
    GetCtrlIndex(AWGPanelHandle,AWG_PNL_WAVE_SHAPE, &WaveIndex);
    switch(WaveIndex)
    {
            case 0:
                Fmt(cmdString, "%s<APPL:SIN %f, %f VPP, %f V",
                                                    frequency, Amp, DC_Offset);
                status = viWrite (InstrHandle, cmdString, NumFmtdBytes(), VI_NULL);

                Fmt(cmdString, "%s<APPL?");  //read back configuration
                status = viWrite (InstrHandle, cmdString, NumFmtdBytes(), VI_NULL);

                status = viRead (InstrHandle, read_buf,BUFFER_LEN, VI_NULL);
                CopyString(Waveform, 0, "Sine",0, -1);
                break;
            case 1:
                Fmt(cmdString, "%s<APPL:SQU %f, %f, %f ",
                                                    frequency, Amp, DC_Offset);
                status = viWrite (InstrHandle, cmdString, NumFmtdBytes(), VI_NULL);

                Fmt(cmdString, "%s<APPL?");  //read back configuration
                status = viWrite (InstrHandle, cmdString, NumFmtdBytes(), VI_NULL);

                status = viRead (InstrHandle, read_buf,BUFFER_LEN, VI_NULL);
```

Figure 5–16
ConfigAWGCB Function Listing *(continued)*

```
                                                CopyString(Waveform, 0, "Square",0, -1);
                    break;
                case 2:
                    //Set Waveform shape, amplitude, offset and frequency
                    Fmt(cmdString, "%s<APPL:TRI %f, %f VPP, %f V",
                                                        frequency, Amp, DC_Offset);
                    status = viWrite (InstrHandle, cmdString, NumFmtdBytes(), VI_NULL);

                    Fmt(cmdString, "%s<APPL?");  //read back configuration
                    status = viWrite (InstrHandle, cmdString, NumFmtdBytes(), VI_NULL);

                    status = viRead (InstrHandle, read_buf,BUFFER_LEN, VI_NULL);

                    CopyString(Waveform, 0, "Triangle",0, -1);
            break;

                case 3:
                    //Set Waveform shape, amplitude, offset and frequency
                    Fmt(cmdString, "%s<APPL:RAMP %f, %f VPP, %f V",
                                                frequency, Amp, DC_Offset );
                    status = viWrite (InstrHandle, cmdString, NumFmtdBytes(), VI_NULL);

                    Fmt(cmdString, "%s<APPL?");  //read back configuration
                    status = viWrite (InstrHandle, cmdString, NumFmtdBytes(), VI_NULL);

                    status = viRead (InstrHandle, read_buf,BUFFER_LEN, VI_NULL);
                    CopyString(Waveform, 0, "Ramp",0, -1);
                    break;

                case 4:
                    //Set Waveform shape, amplitude, offset and frequency
                    Fmt(cmdString, "%s<APPL:DC %f, %f VPP, %f V",
                                                frequency, Amp, DC_Offset );
                    status = viWrite (InstrHandle, cmdString, NumFmtdBytes(), VI_NULL);

                    Fmt(cmdString, "%s<APPL?");  //read back configuration
                    status = viWrite (InstrHandle, cmdString, NumFmtdBytes(), VI_NULL);

                    status = viRead (InstrHandle, read_buf,BUFFER_LEN, VI_NULL);

                    CopyString(Waveform, 0, "DC",0, -1);
                    break;

                case 5:
                    //Set Waveform shape, amplitude, offset and frequency
                    Fmt(cmdString, "%s<APPL:NOIS %f, %f VPP, %f V",
                                                      frequency, Amp, DC_Offset );
                    status = viWrite (InstrHandle, cmdString, NumFmtdBytes(), VI_NULL);

                    Fmt(cmdString, "%s<APPL?");  //read back configuration
                    status = viWrite (InstrHandle, cmdString, NumFmtdBytes(),VI_NULL);

```

Figure 5–16
ConfigAWGCB Function Listing *(continued)*

```
122                     status = viRead (InstrHandle, read_buf,BUFFER_LEN, VI_NULL);
123                     CopyString(Waveform, 0, "Noise",0, -1);
124                     break;
125                 default:
126                     Fmt(buffer, "%s<Error in Wavetype selection");
127                     InsertTextBoxLine (AWGPanelHandle, AWG_PNL_DISP, -1, buffer);
128                     break;
129
130             }
131
132                     Fmt(read_buf, "%s<WAVE = %s FREQ= %f  AMPLITUDE= %f  OFFSET
133                                        = %f", Waveform,frequency,Amp,DC_Offset);
134                     InsertTextBoxLine (AWGPanelHandle, AWG_PNL_DISP, -1, read_buf);
135
136         break;
137     }
138
139     return 0;
140 } //ConfigAWGCB
```

Figure 5–16
ConfigAWGCB Function Listing *(continued)*

shown between lines 14 and 42. Notice that throughout this function we are communicating with the device using the *Standard Commands for Programmable Instrumentation* (SCPI). SCPI are basically strings in a hierarchical tree structure using a standard vocabulary to command and control the devices. An introduction to SCPI is given in the book *LabWindows/CVI Programming for Beginners,* but for instrument-specific communications, refer to the instrument's *User's Manual.* In the *ConfigAWGCB* function, the commands specific to the function generator are used and are explained in the source code listing.

At line 47 the waveform index selected is obtained and the SCPI command(s) is/are used to send the frequency, amplitude, and DC offset to the function generator. For verification purposes the command(s) sent to the function generator are read back from the function generator using the "`APPL?`" command and displayed in the text box on the GUI.

When you exit the program, be sure first to close the communication channel to the device and then the channel to the VISA Default Resource Manager using the respective handles in the library function *viClose.* This is shown in the source code listing of the *ExitAWGCB* callback function in Figure 5–17.

```
//Exit device
int CVICALLBACK ExitAWGCB (int panel, int control, int event,
        void *callbackData, int eventData1, int eventData2)
{
      switch (event)
        {
        case EVENT_COMMIT:
                DiscardPanel(AWGPanelHandle);
                viClose(InstrHandle);
                viClose(ResourceManager);
                QuitUserInterface(0);
                break;
        }
      return 0;
}//ExitAWGCB
```

Figure 5–17
ExitAWGCB Function Listing

Summary

The aim of this chapter was to explain the hardware and software features of the VXI system. This chapter gave you the basic features of VXI devices. You were shown the differences between the various types of VXI devices and how to communicate with them using VISA functions. By means of the project and with the appropriate hardware connected to your computer, you were able to send data to and receive data from the device. A short introduction to VISA was given here. Due to the vastness of VISA functionalities, all the features could not be mentioned in this chapter. If you are interested in learning its advanced features, refer to the On-line manuals.

Library Function Prototypes and Definitions

This section lists alphabetically the *CVI* library functions that were introduced in this chapter.

AssertSysReset Function

The *AssertSysReset* function asserts the *Sysreset* line to the controller specified on the VXI system. Its prototype is shown here and its arguments explained in Table 5–4.

```
NIVXI_STATUS status= AssertSysreset (INT16 Controller,
                                     UINT16 Mode);
```

viAssertUtilSignal Function

The *viAssertUtilSignal* function asserts or deasserts the utility bus signal specified. Its prototype is shown here and its arguments explained in Table 5–5.

```
viStatus status= viAssertUtilSignal (ViSession vi,
                                     ViUInt16 line);
```

Table 5–4 AssertSysReset *Function*

Input/ Output	Name	Type	Description
Input	*Controller*	INT16	logical address of the controller to use; a value of -1 denotes the local controller; range: 0–254
	Mode	UINT16	select from the following reset modes: 0 reinitialize to original configuration 1 reset the VXI system 2 reset the VXI system and reinitialize local controller
Output	*status*	NIVXI_ STATUS	NI-VXI status codes consist of the following value: 0 successful -1 unsupported function -2 invalid controller

Table 5–5 viAssertUtilSignal *Function*

Input/ Output	Name	Type	Description
Input	*vi*	ViSession	session handle identifier
	line	ViUInt16	utility bus signal to assert; select from the following: ■ VI_UTIL_ASSERT_SYSRESET ■ VI_UTIL_ASSERT_SYSFAIL ■ VI_UTIL_DEASSERT_SYSFAIL
Output	*status*	viStatus	return status can be the following: ■ VI_SUCCESS—operation completed successfully ■ VI_ERROR_INV_OBJECT—session reference identifier invalid ■ VI_ERROR_NSUP_OPER— *vi* given does not support this operation ■ VI_ERROR_RSRC_LOCKED—operation cannot be performed; resource identified by *vi* is locked for this kind of access

viClose Function

The *viClose* function closes a communication channel to a resource. Its prototype is shown here and its arguments explained in Table 5–6.

```
ViStatus status = viClose (ViSession Session_Handle);
```

Table 5–6 viClose *Function*

Input/ Output	Name	Type	Description
Input	*Session_ Handle*	ViSession	session handle to close session
Output	*status*	ViStatus	VISA status codes; negative value indicates an error

viFindNext Function

The *viFindNext* function finds the instrument descriptor of the next resource. Its prototype is shown here and its arguments explained in Table 5–7.

```
ViStatus status = viFindNext (ViFindList Find_Handle,
                        char Instrument_Descriptor[]);
```

viFindRsrc Function

The *viFindRsrc* function finds a list of matching instrument descriptors available on the system. Its prototype is shown here and its arguments explained in Table 5–8.

```
ViStatus status = viFindRsrc (ViSession ResourceManagerHandle,
                              ViString Expression,
                              ViFindList *Find_Handle,
                              ViUInt32 *Return_Count,
                              char Instrument_Descriptor);
```

viIn16 Function

The *viIn16* function reads a 16-bit value from the A16, A24, or A32 address space for register-based devices. Its prototype is shown here and its arguments explained in Table 5–9.

```
ViStatus status = viIn16 (ViSession Instrument_Handle,
                              ViUint16 AddressSpace,
                              ViBusAddress Offset,
                              ViUint16 *ReadValue);
```

viOpen Function

The *viOpen* function opens a communication channel to a device. Its prototype is shown here and its arguments explained in Table 5–10.

```
ViStatus status = viOpen (ViSession ResourceManagerHandle,
                    ViRsrc Instrument_Descriptor,
                    ViAccessMode Access_Mode,
                    ViAccessMode Timeout,
                    ViSession *Instrument_Handle);
```

Table 5–7 viFindNext *Function*

Input/ Output	Name	Type	Description
Input	*Find_Handle*	ViFindList	pointer to a list of character strings containing the list of resources found
Output	*Instrument_ Descriptor*	char[ ]	next instrument descriptor from the list of resources found
	status	ViStatus	VISA status codes; negative value indicates an error

Table 5–8 viFindRsrc *Function*

Input/ Output	Name	Type	Description
Input	*Resource Manager Handle*	ViSession	handle returned from *viOpenDefaultRM* function
	Expression	ViString	search expression to find the type of resources
	Find_Handle	ViFind List*	pointer to a list of character strings containing the list of resources found
Output	*Return_ Count*	ViUInt32*	number of resources found with the matching *Expression* string
	Instrument_ Descriptor	char	first instrument descriptor from the list of resources found
	status	ViStatus	VISA status codes; negative value indicates an error

Table 5–9 viIn16 *Function*

Input/ Output	Name	Type	Description
Input	*Instrument_ Handle*	ViSession	handle used to establish session with a device
	AddressSpace	ViUint16	the address space to read from; select from either of the following address spaces: ■ VI_A16_SPACE ■ VI_A24_SPACE ■ VI_A32_SPACE
	Offset	ViBus Address	offset in bytes to access
Output	*ReadValue*	ViUint16*	value read back
	status	ViStatus	VISA status codes; negative value indicates an error

Table 5–10 viOpen *Function*

Input/ Output	Name	Type	Description
Input	*Resource Manager Handle*	ViSession	handle returned from *viOpenDefaultRM* function
	Instrument_ Descriptor	ViRsrc	VISA instrument descriptor to which the session is opened
	Access_Mode	ViAccess Mode	select from the following: ■ `VI_EXCLUSIVE_LOCK`—locks the resource before operation returns ■ `VI_LOAD_CONFIG`—used for external user-defined settings ■ `VI_NULL`—used when *Access_Mode* not required

(continued)

Table 5–10 viOpen *Function (continued)*

Input/ Output	Name	Type	Description
	Timeout	ViAccess Mode	maximum time to wait for opening a resource
Output	*Instrument_ Handle*	ViSession*	handle to uniquely define a session with a device
	status	ViStatus	VISA status codes; negative value indicates an error

viOpenDefaultRM Function

The *viOpenDefaultRM* function opens a communication channel to the VISA Default Resource Manager. Its prototype is shown here and its arguments explained in Table 5–11.

```
ViStatus status = viOpenDefaultRM (ViSession
                                   ResourceManagerHandle);
```

viOut16 Function

The *viOut16* function writes a 16-bit value to the A16, A24, or A32 address space for register-based devices. Its prototype is shown here and its arguments explained in Table 5–12.

```
ViStatus status = viOut16 (ViSession Instrument_Handle,
                           ViUint16 AddressSpace,
                           ViBusAddress Offset,
                           ViUint16 *WriteValue);
```

viRead Function

The *viRead* function reads the data from the device specified. Its prototype is shown here and its arguments explained in Table 5–13.

```
ViStatus status = viRead (ViSession Instrument_Handle,
                           unsigned char buffer[ ],
                              ViUInt32 BufferLength,
                                 ViUInt32 ReturnCount);
```

Table 5–11 viOpenDefaultRM *Function*

Input/ Output	Name	Type	Description
Output	*Resource Manager Handle*	ViSession	handle used to establish the session with the Default Resource Manager
	status	ViStatus	VISA status codes; negative value indicates an error

Table 5–12 viOut16 *Function*

Input/ Output	Name	Type	Description
Input	*Instrument_ Handle*	ViSession	handle used to establish session with a device
	AddressSpace	ViUint16	address space to write value; select from either of the following address spaces: ■ VI_A16_SPACE ■ VI_A2 4_SPACE ■ VI_A32_SPACE
	Offset	ViBus Address	offset in bytes to access
Output	*WriteValue*	ViUint16*	value to write
	status	ViStatus	VISA status codes; negative value indicates an error

viSetAttribute Function

The *viSetAttribute* function sets a specified attribute for the object given. Its prototype is shown here and its arguments explained in Table 5–14.

```
ViStatus viSetAttribute (ViObject Object_Handle,
                         ViAttr Attribute_Name,
                         ViAttrState Attribute_Value);
```

Table 5–13 viRead *Function*

Input/ Output	Name	Type	Description
Input	*Instrument_ Handle*	ViSession	handle used to establish session with a device
	Buffer Length	ViUInt32	number of bytes to read
Output	*buffer*	unsigned char[ ]	buffer to read the data from the device
	ReturnCount	ViUInt32	number of bytes actually transferred; use `VI_NULL` if not interested in obtaining this value
	status	ViStatus	VISA status codes; negative value indicates an error

Table 5–14 viSetAttribute *Function*

Input/ Output	Name	Type	Description
Input	*Object_ Handle*	ViObject	handle used to establish session with a device
	Attribute_ Name	ViAttr	attribute you want to set
	Attribute_ Value	ViAttr State	value to set the specified attribute
Output	*status*	ViStatus	VISA status codes; negative value indicates an error

viStatusDesc Function

The *viStatusDesc* function converts the status code to a string message. Its prototype is shown here and its arguments explained in Table 5–15.

```
ViStatus viStatusDesc (ViObject Object_Handle,
                       ViStatus Status,
                       char Description[ ]);
```

Table 5–15 viStatusDesc *Function*

Input/ Output	Name	Type	Description
Input	*Object_ Handle*	ViObject	handle used to establish session with a device
	Status	ViStatus	status code you want to convert
Output	*Description*	char[ ]	status code value converted to string message
	status	ViStatus	VISA status codes; negative value indicates an error

viWrite Function

The *viWrite* function writes the data to the device specified. Its prototype is shown here and its arguments explained in Table 5–16.

```
ViStatus status = viWrite (ViSession Instrument_Handle,
                           ViBuf buffer,
                           ViUInt32 BufferLength,
                           ViUInt32 ReturnCount);
```

Table 5–16 viWrite *Function*

Input/ Output	Name	Type	Description
Input	*Instrument_ Handle*	ViSession	handle used to establish session with a device
	Buffer Length	ViUInt32	number of bytes to write
	buffer	ViBuf	buffer to write the data to the device
Output	*ReturnCount*	ViUInt32	number of bytes actually transferred; use `VI_NULL` if this value is not required
	status	ViStatus	VISA status codes; negative value indicates an error

6

DATA ACQUISITION

Chapter Highlights

- Introduction
- Data Acquisition Board Architecture
- Signal Conditioning
- Analog Input/Output Parameters
- DAQ Designer Tool
- Installing and Setting Up the DAQ Board
- Using the DAQ Channel Wizard
- Hardware Configurations
- Using DAQ Library Functions
- Summary
- Library Function Prototypes and Definitions

This chapter introduces you to some of the basic hardware and software features of data acquisition. You are introduced to the different configurations in which the data acquisition can be performed by means of components of a multifunction input/output DAQ (data acquisition) board. Signal conditioning is discussed and the various analog input/output parameters explained. You are shown how to select the appropriate DAQ board using the DAQ Designer tool and to install and set up the DAQ board. The use of the DAQ Channel Wizard to configure and verify operations of DAQ channels is explained. Finally, the DAQ library functions for analog input/output, digital input/output, and counter applications are explained by means of code fragments.

Introduction

DAQ is a process by which a real-world physical phenomenon is converted into an electrical signal by means of a transducer/sensor using data acquisition hardware and software. A transducer/sensor is a device that converts physical phenomena (such as temperature, pressure, strain, light, motion, flow, or any other measurable entity) into a proportional electrical stimulus. DAQ hardware and software are used to acquire raw data from transducers/sensors connected to the DAQ system. The DAQ software acquires the data and sends it on your computer. The DAQ system consists of transducers, a DAQ board, and optional signal conditioning devices. Signal conditioning devices are discussed in the section *Signal Conditioning*.

To acquire data, you can use the DAQ board in various configurations. Figure 6–1 shows the plug-in DAQ board residing inside the computer. A single DAQ board receives the multiplexed and conditioned signals from the Signal Conditioning eXtensions for Instrumentation (SCXI) modules.

If you are using a laptop computer, you can use a PCMCIA DAQ card, as shown in Figure 6–2.

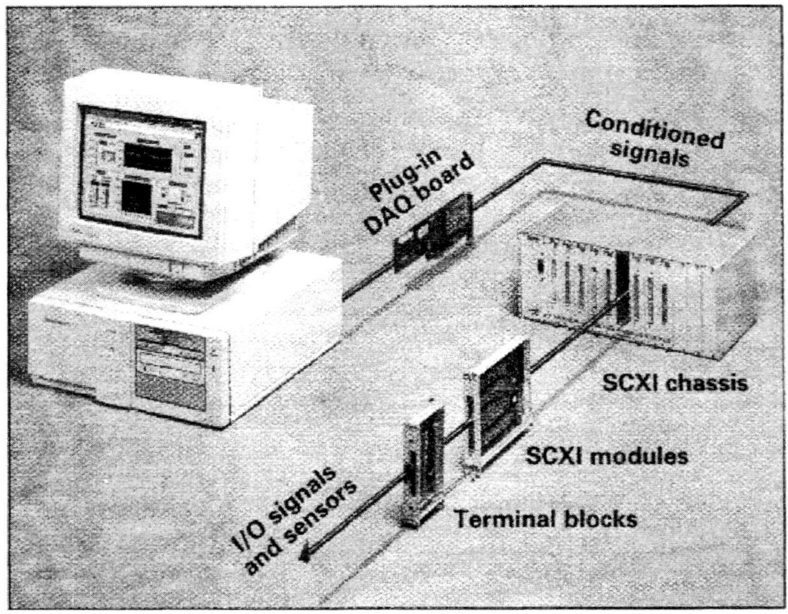

Figure 6–1
Plug-in DAQ Board Configuration (Copyright National Instruments)

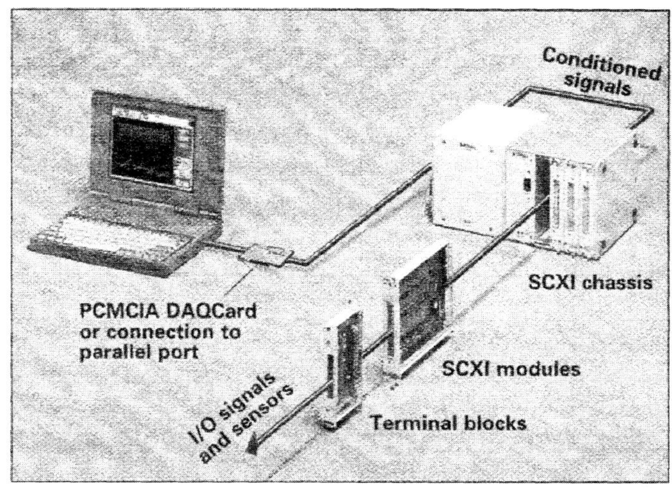

Figure 6–2
PCMCIA DAQ Board Configuration (Copyright National Instruments)

To acquire data from a device located a distance from your computer, you can use the serial, parallel, USB, or Ethernet interface to connect to a remote DAQ board. The setup you use can be similar to that shown in Figure 6–3 using a remote SCXI chassis and SCXI modules. One serial bus, RS-485, allows you to use up to 31 SCXI chassis per serial port at distances of up to 1200 meters from your computer.

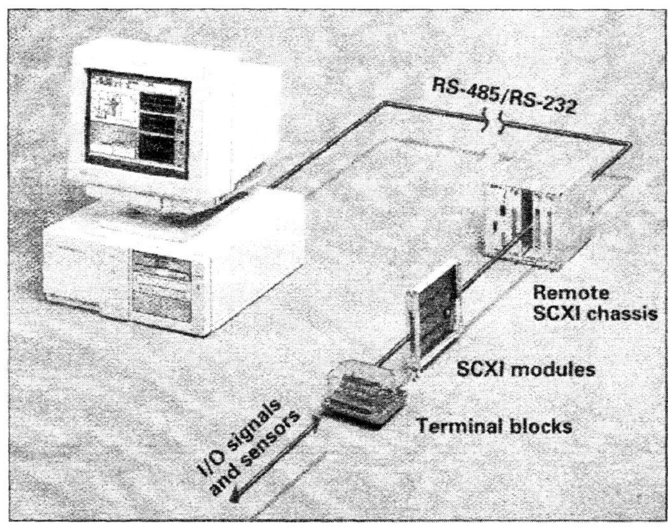

Figure 6–3
Remote DAQ Board Configuration (Copyright National Instruments)

Using an external DAQ board configuration, the data throughput to the computer is reduced, with the serial interface being the slowest.

Data Acquisition Board Architecture

A DAQ board can read analog input signals and convert them to digital signals using an analog-to-digital (A/D) converter. The DAQ board can generate analog output signals using digital-to-analog (D/A) conversion and can send and receive digital signals.

The DAQ board typically consists of one or more analog-to-digital converters (ADCs), digital-to-analog converters (DACs), digital input/output (I/O) ports, and counters/timers for frequency measurement and pulse generation. These components interface with transducers and a Real-Time System Integration (RTSI) connector. A multifunction input/output (MIO) DAQ board is the most versatile of the plug-in DAQ boards. Its block diagram is shown in Figure 6–4. Some of the components of the MIO DAQ board are discussed below.

Analog Input Circuitry receives the analog input signals from the transducer (or signal condition device) and converts them to digital values using the ADCs. The Analog Input Circuitry contains the analog multiplexer, the instrumentation amplifier, and sample-and-hold circuitry.

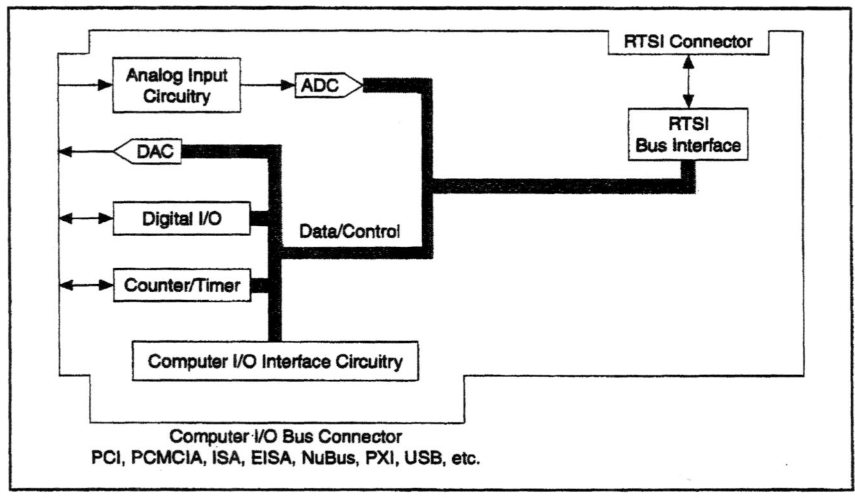

Figure 6–4
Multifunction DAQ Board Block Diagram (Copyright National Instruments)

The *analog multiplexer* is a switch that connects one or more of the input channels to the instrumentation amplifier. During data acquisition from several channels the multiplexer switches the channels in sequence to connect to the instrumentation amplifier.

The *instrumentation amplifier* applies gain to the input voltage to keep it within the ADC range. The gain is determined by the *CVI* function and is applied according to the input signal limits.

The *sample-and-hold circuitry* ensures that the input voltage does not fluctuate during analog-to-digital conversion. The sample-and-hold circuitry accomplishes this by holding the analog voltage from the instrumentation amplifier and inputting it to the ADC as a steady unchanging input.

The *DAC* is used to convert the digital values into analog signals that can be output from the DAQ board.

Digital I/O is discussed in the section *Digital Input/Output*.

The *Counter/Timer circuitry* on the DAQ board is used for counting pulses, measuring pulse widths, pulse transition frequency, and for generating timing signals and pulse trains.

The *RTSI* bus interface on the DAQ card is a National Instruments innovation used to route timing and trigger signals between multiple functions on one DAQ board or between two or more boards. The RTSI is used to synchronize A/D conversions, D/A conversions, digital inputs, digital outputs, and counter/timer operations on multiple boards. This is useful when you want to use two boards to capture data simultaneously or use the third board to generate an output pattern synchronized with the sampling rate of the inputs.

Signal Conditioning

At times, signals produced by the transducers/sensors may need to be conditioned before being input to the DAQ board. Some form of signal condition device(s) is used to clean up the signal. Signal conditioning is most appropriate under the following situations:

- When you want to multiplex a large number of channels (more than are supported by your DAQ board)
- When you are using a mix of input/output (I/O) or transducer types
- When operating in a noisy environment

- When you need to amplify low-level signals
- When you need to isolate transducer signals from a computer
- When you need to filter unwanted noise from signals

Signal conditioning also provides external voltage and current excitation and linearizes the response to some of the transducers that require this feature. Here are some common types of signal conditioning, with an explanation of their purpose:

- **Amplification.** Amplification maximizes use of the digitizer range to increase the accuracy of the digitized signal and increases the signal-to-noise ratio (SNR). It is best to amplify the low-level signals close to a signal source instead of these closer to the DAQ board, to increase the SNR. SNR is defined as the ratio of the overall root mean square (rms) signal level to the overall rms noise level. Amplifying the signal close to the signal source eliminates any noise that may be picked up by the lead wires along the signal path. Amplification is most commonly used for thermocouples and low-voltage transducers.
- **Transducer Excitation.** Some transducers require external voltage or current excitation as input to their circuitry. Examples of transducers requiring such excitations are strain gauges and resistance temperature detectors (RTDs).
- **Linearization.** Linearization is a method of conditioning a signal to determine a linear relationship with respect to measured phenomena. Some transducers, such as strain gauges, themistors, RTDs, and thermocouples, do not generate voltages linearly with respect to their measured phenomena.
- **Isolation.** To avoid damage to a computer from large voltage spikes in transducer signals, some form of isolation must be used. The two common methods are optical isolation and analog isolation. Optical isolation refers to transferring data without the use of electrical continuity using an optoelectrical transmitter and receiver. Analog isolation blocks voltage spikes to the DAQ board using an isolation instrumentation amplifier.
- **Filtering.** Filtering refers to removing unwanted noise in signals to avoid acquiring erroneous data. This is accomplished by means of lowpass and highpass filters, which are built into the SCXI devices and are configurable through jumpers or software configuration.

Analog Input/Output Parameters

When measuring analog signals with a DAQ board, you must consider a number of factors that will allow you to use DAQ products effectively in your application. The DAQ board can be placed in either the differential, referenced single-ended (RSE), or nonreferenced single-ended (NRSE) category. These categories are explained below.

- **Differential.** In differential mode the transducer voltage is determined by using two analog input channels for each transducer, connected to the positive and negative terminals of the transducer and measuring the difference in voltage between the two channels. This is the most accurate method of reading a signal in all cases; however, it does require two channels per signal.
- **Referenced Single-Ended (RSE).** In RSE mode the transducer voltage is determined by reading one channel with respect to the onboard ground signal of the board.
- **Nonreferenced Single-Ended (NRSE).** In NRSE mode the transducer voltage is determined by reading one channel with respect to an external ground signal provided to the board.

The basic specifications for DAQ products consist of the sampling rate, resolution, and input range. *Sampling rate* refers to the maximum rate at which analog-to-digital (A/D) conversions take place. To obtain an accurate representation of the input signal, a faster scan rate must be used, allowing you to acquire more points in a given time frame. The Nyquist theorem states that you must sample at twice the frequency component of the signal to prevent *aliasing*. Aliasing refers to the distortion that appears in the digitized signal that makes it appear to have an incorrect frequency. Depending on your application, sampling at a rate five to 10 times the highest frequency of a signal is recommended for an accurate representation of the waveform.

Resolution is the number of bits that an analog-to-digital converter (ADC) uses to convert an analog signal. High resolution causes the voltage range to be divided into more divisions, thus making it possible to detect a smaller voltage change. This is made clear by looking at the example in Figure 6–5, which shows a sine wave and its corresponding digital image using a 3-bit converter. As you notice in this figure, the 10-volt range is divided into 2^3 or 8 divisions, represented by the binary code 000 through 111. The digitized waveform from the 3-bit ADC is not an accurate representation of the original signal since the resolution is not high enough. This is shown in the bottom part of the figure. Using a 16-bit converter will divide the voltage range into 2^{16} or 65,536 divisions, causing a higher resolution and more accurate representation of the analog waveform, as shown on the bottom right of Figure 6–5.

It is often required to provide stimuli to the DAQ system from the analog output circuitry. Settling time, slew rate, and resolution of the DAC determine the quality of the output signal produced. *Settling time* is the time required for the output to stabilize to within a specified accuracy. *Slew rate* is the maximum rate of change that the DAC can produce on the output signal. A DAC with a short settling time and a high slew rate will change the output to a new voltage level within a short time, making it possible to generate high-frequency signals. The output *resolution* is similar to the input resolution discussed above.

Range, Gain, and Code Width

The maximum and minimum analog voltage levels that the ADC can digitize are referred to as the *input range*. Typically, you can set the range of the

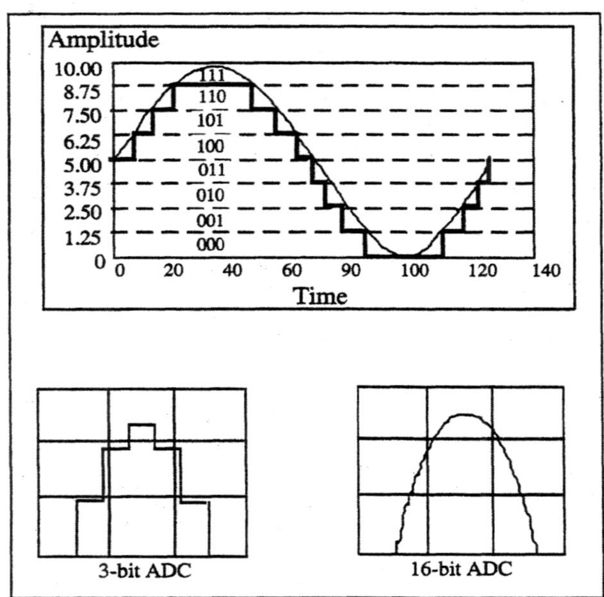

Figure 6–5
Resolution Example Using 3- and 16-bit ADC (Copyright National Instruments)

board to be polar, such as 0 to 10, or bipolar (such as –10 to 10). Selectable ranges enable you to take advantage of the available resolution to measure the signal accurately, as you will see in the examples given below. The NI-DAQ driver software will select a *gain* for you, depending on the input range you specify. Gain is a factor by which the signal is amplified before being digitized. The smallest detectable change is calculated using the code width equation as follows:

$$(\text{voltage range of board})/(\text{gain} * 2^{\text{resolution in bits}})$$

This change in voltage represents 1 LSB (least significant bit) of the digital value and is called the *code width*. If, for example, you are using a 12-bit DAQ board with a 0- to 10-volt input range and a gain of 10, the smallest detectable change (code width) will be calculated as follows using the equation above:

$$10/(10 * 2^{12}) = .244 \text{ mV}$$

Now if the range were changed to –10 to 10 volts using the same DAQ board attributes as above, the smallest detectable change would become

$$20/(10 * 2^{12}) = .488 \text{ mV}$$

since the voltage range is 20 (–10 volts to +10 volts) in this case. By changing the range and the gain, you can adjust the value for the smallest detectable change in your application.

DAQ Designer Tool

There are numerous data acquisition boards available from National Instruments and other manufacturers. To find the correct board to satisfy your application needs is not always an easy task. You can look at the National Instruments *Measurement and Automation* catalog and select the specifications to decide on the right DAQ board for your application or use the *DAQ Designer* CD-ROM available free of charge from National Instruments. DAQ Designer is interactive software that prompts you through various questions regarding your DAQ requirements and enables you to decide the hardware best suited for your application(s). A sample prompt screen is shown in Figure 6–6. If you are not familiar with data acquisition hardware, using the DAQ Designer tool is the best way to find the most suitable DAQ board for your application.

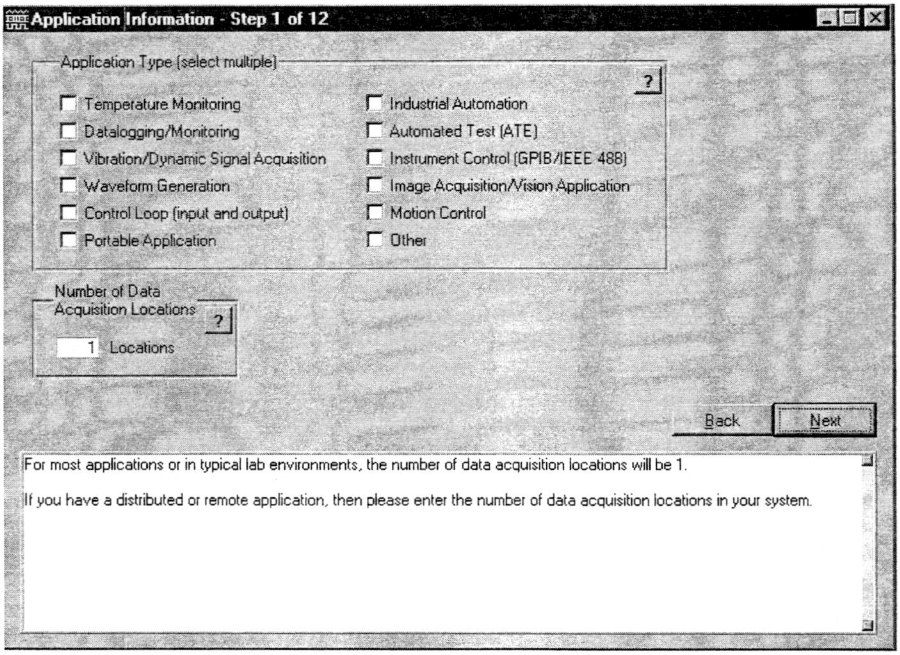

Figure 6–6
DAQ Designer Query Screen

Installing and Setting Up the DAQ Board

The NI-DAQ board is shipped with a NI-DAQ CD that contains the DAQ drivers and utilities to configure and test your board. You can also download the latest version of the DAQ driver from the National Instruments Web site (`www.ni.com`). Install the NI-DAQ software from the CD by selecting **Install NI DAQ** from the menu (Figure 6–7) and follow the instructions on the screen. You can also install the software from the Windows **Start** menu by selecting **Run** and selecting the file `setup.exe` on the CD using the **Browse...** button. When installation is complete, click on **Quit Install** on the window displayed and shut down the computer. Install the DAQ devices, accessories, and cables. You complete the installation by turning on your computer. Installation should start automatically when you turn on your computer. If not, select the **Programs>>National Instruments>>NI-DAQ>>NI-DAQ Setup** from the Windows **Start** menu (Figure 6–32).

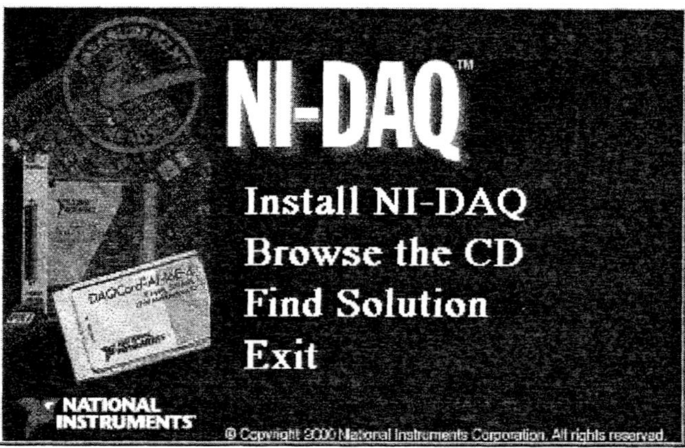

Figure 6–7
NI-DAQ Installation

When the installation is complete, you are prompted to the **Configure Measurement and Automation System**. Accept this option to bring up the **Measurement & Automation Explorer (MAX)** window, as shown in Figure 6–8. Your installation screens may look different, depending on the NI-DAQ version you are using.

To confirm that your device is installed properly, click on the `Devices and Interfaces` folder to see the device name in this folder. If at any time you want to reconfigure or test the device, you can double-click on the **Measurement & Automation** icon on your Windows desktop to run this utility. Depending on the NI-DAQ version installed, your **MAX** window may look different but as a minimum will consist of the following folders: `Data Neighborhood`, `Devices and Interfaces`, `IVI` Instruments, and `Scales`. In earlier versions of NI-DAQ, the `Software` folder is not included.

The `Data Neighborhood` folder includes the DAQ Channel Wizard found in previous NI-DAQ versions. The DAQ Channel Wizard is explained in the section *Using the DAQ Channel Wizard*.

Double-clicking on the `Devices and Interfaces` folder gives a list of National Instruments devices with the device number shown in parentheses (see Figure 6–8). This device number in used in the *CVI* function when referring to the board to perform the DAQ operations. Here the PCI-MIO-16E-4 board is installed as device 1. Devices will not be installed automatically if you are using Windows NT, PCMCIA, and non-PnP (*Legacy* board) AT devices.

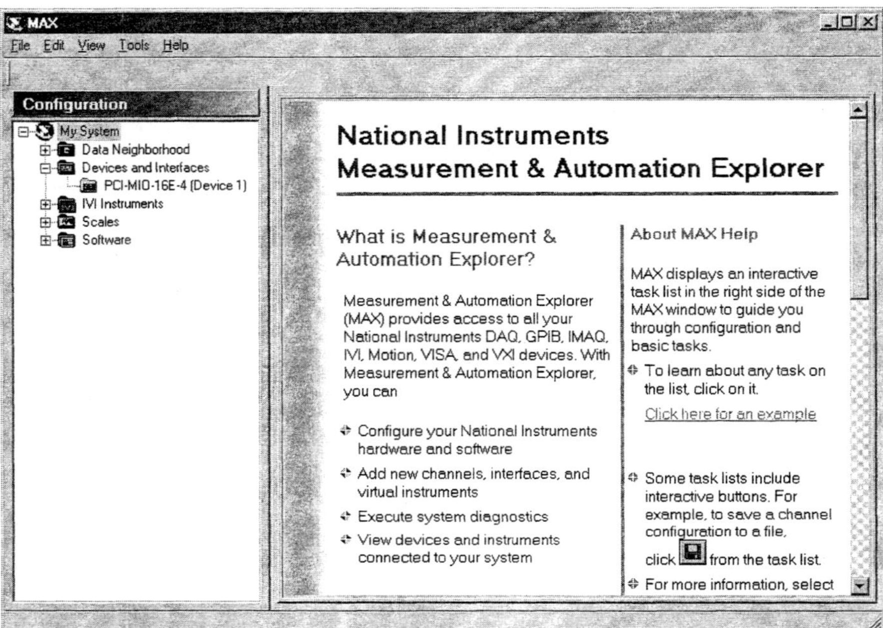

Figure 6–8
Measurement & Automation Explorer (MAX) Window

For the non-PnP (non-Plug&Play) board you must configure the board manually using the **Add New Hardware** option under **Control Panel** in Windows. For more information on installing and configuring DAQ devices in Windows NT, see the NI-DAQ release notes or the DAQ Troubleshooting Wizard on the National Instruments Web site.

Double clicking on the `Software` folder shows a list of installed software from where you can view, launch, or install the latest versions of the software. You are shown how to accomplish this by the help available in the right half of the **MAX** window shown in Figure 6–9. Again, your display screens may look different depending on the DAQ version being used.

To configure and test the board, right-click on the `PCI-MIO-16-E4` device in the `Devices and Interfaces` folder, and select **Properties** from the menu. The **Configuring Device** window is displayed (Figure 6–10).

Under the **System** tab in the **Configuring Device** window, the configuration for **Memory Range**, **Interrupt Request**, and **Direct Memory Access** assigned to the board from the Windows registry are displayed.

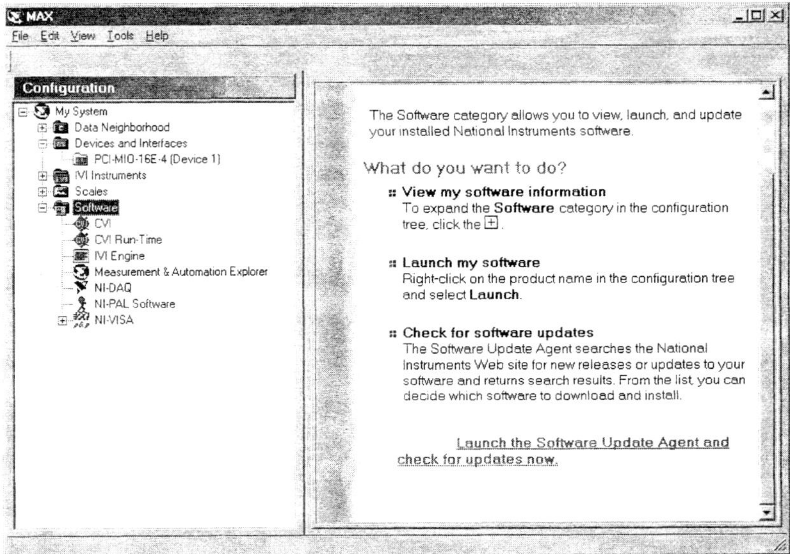

Figure 6–9
MAX Window with Software Installed

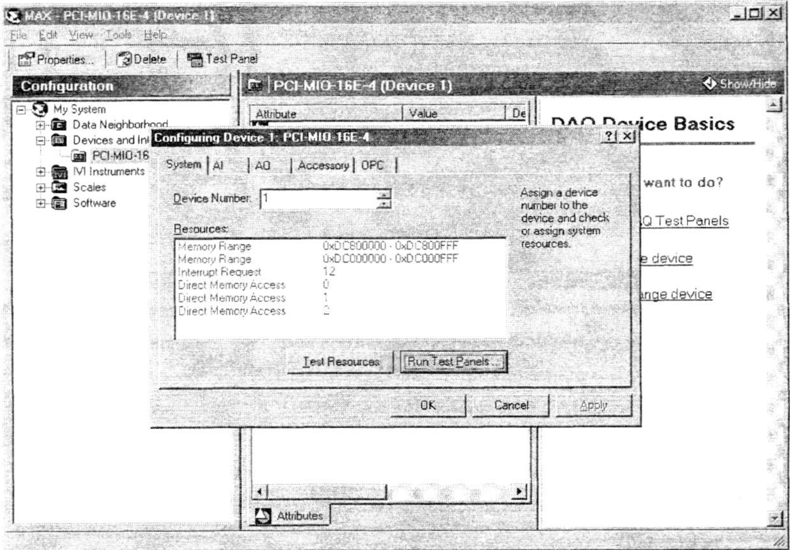

Figure 6–10
Devices and Interfaces Configuration: **System** Tab

Selecting the **AI** (Analog Input) tab, the options shown in Figure 6–11 are displayed. In this window you can set **Polarity/Range** and **Mode** for the board's operation from the pull-down menu. You can select **Polarity/Range** to obtain the desired precision, as explained in the section *Range, Gain, and Code Width* above. The **Modes** for the DAQ board were explained in the section *Analog Input/Output Parameters*.

The **AO** (Analog Output) tab allows you to set the **Polarity** of the analog output channels (see Figure 6–12). You have an option to choose between **Unipolar**, **Bipolar**, **Unipolar External Reference**, or **Bipolar External Reference**.

- *Unipolar* refers to a signal range that is always positive.
- *Bipolar* is a signal range that has both positive and negative values.
- *Bipolar External Reference* is a bipolar range that has a minimum and maximum value equal to the supplied external reference positive and negative values.
- *Unipolar External Reference* is a unipolar signal that has a maximum value equal to the supplied external reference value.

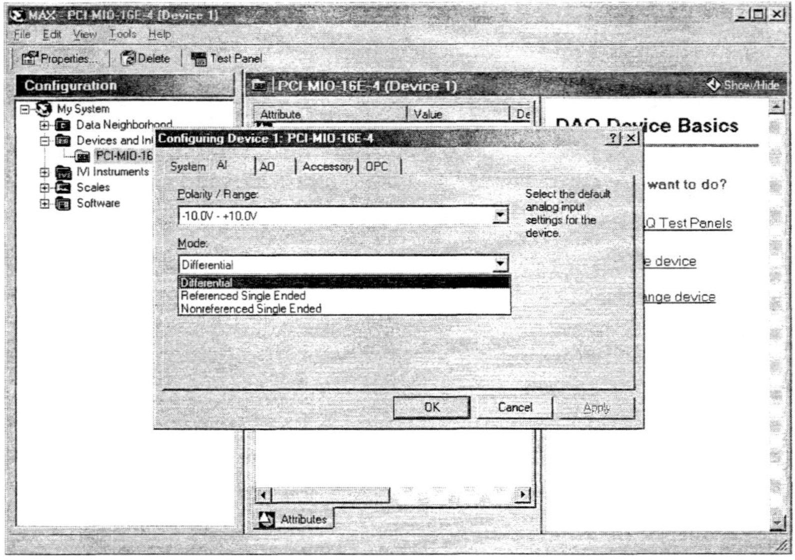

Figure 6–11
Devices and Interfaces Configuration: **AI** Tab

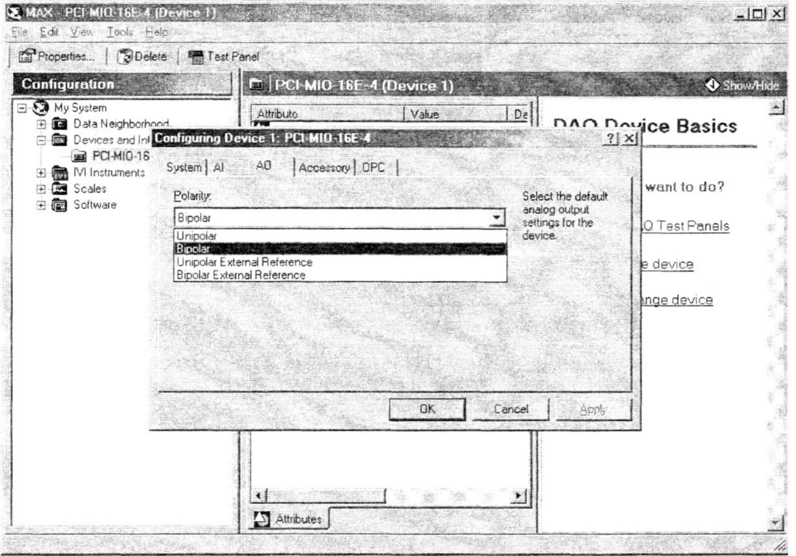

Figure 6–12
Devices and Interfaces Configuration: **AO** Tab

You can select the accessories attached to the DAQ board (if any) by clicking on the **Accessory** tab and selecting from among the accessories shown in the pull-down menu (see Figure 6–13).

The last tab, **OPC**, sets the analog input recalibration period in seconds that is used with the NI-DAQ OPC server. If you are not using the OPC server, you can ignore this setting, since it is disabled by default.

The **Test Resources** button on the **System** window (see Figure 6–10) will test the system resources assigned to the DAQ board. A pop-up window displaying the message "The device has passed the test" will appear if the board is configured properly. Select **OK** to return to the **System** tab.

To test the individual functions of the DAQ board, select **Run Test Panels...** from the **System** window. Figure 6–14 shows the Analog Input test for the transducer connected on channel 0.

When you select the **Analog Output** tab on the **Test Panel** window, Figure 6–15 is displayed. In this window you can either set a single voltage or a sine wave, or set the output voltage on one of the DAQ board's analog output channels in the **Channel Selection:** dialog box. The output selected will be displayed continuously on the **Test Panel** window.

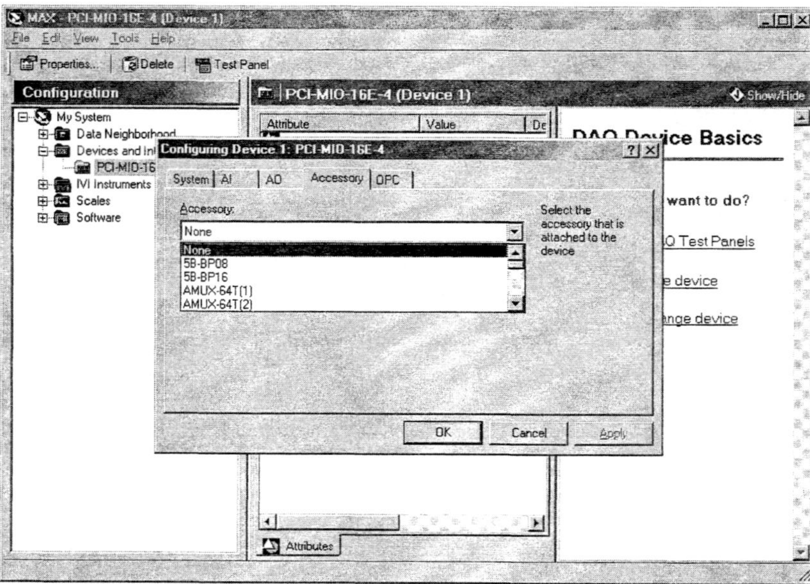

Figure 6–13
Devices and Interfaces Configuration: **Accessory** Tab

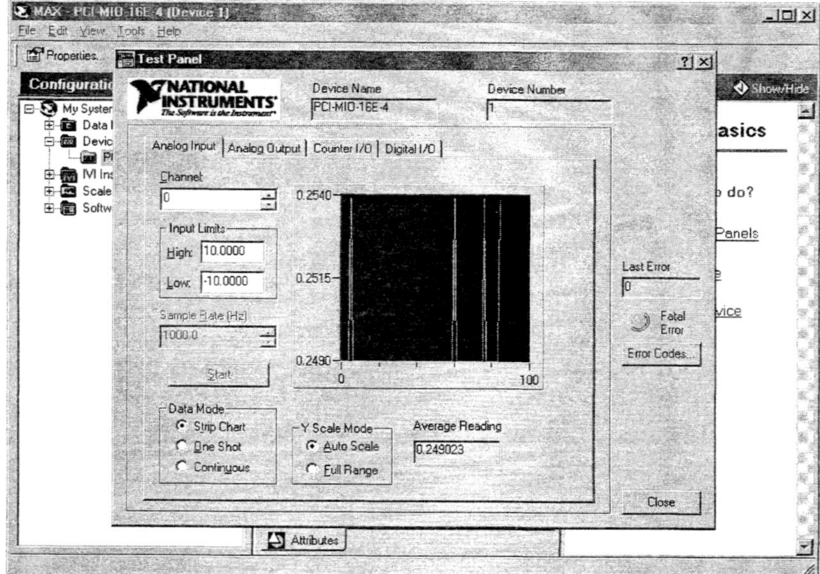

Figure 6–14
Run Test Panel: **Analog Input** Window

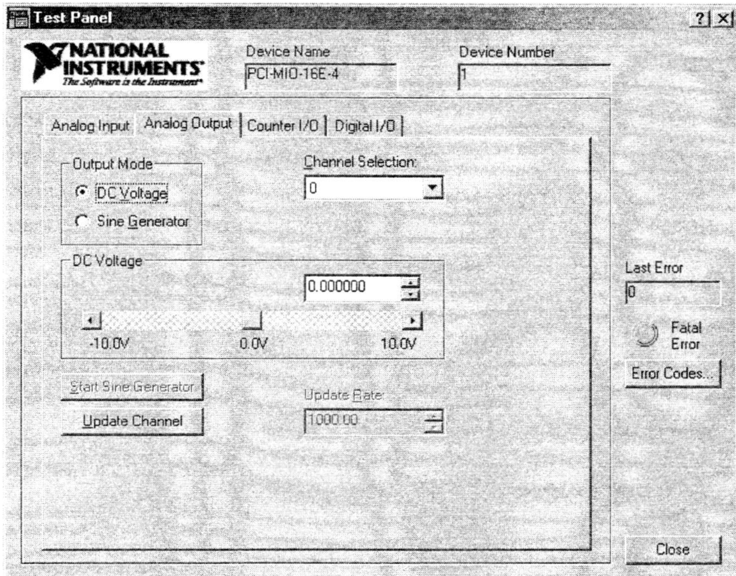

Figure 6–15
Run Test Panel: **Analog Output** Window

When the **Counter I/O** tab on the **Test Panel** window is selected, the **Counter I/O** window is displayed (Figure 6–16) and is used for testing the counter timers on the DAQ board. Select **Simple Event Counting** in the **Counter Mode** dialog box, leave the counter to test in the **Counter:** dialog box to **GPCTR0**, and select **Start**. The **Counter Value:** box will be updated continuously and the **GPCTR0** LED will be enabled in the **Counter Status** subpanel. To stop the counter test, click on the **Reset** command button. Test similarly for other counters by selecting from the **Counter:** dialog box.

The **Digital I/O** tab on the **Test Panel** window is used for testing the digital lines on the DAQ board. To test these lines, check the **Output** radio buttons for the lines selected as output lines (see Figure 6–17, where lines 0–3 are output lines) and notice that the LEDs are turned off on the panel. To test these lines, check on the **Logic Level** boxes for the **Output** lines and read the logic on these lines using a voltmeter, oscilloscope, or LEDs connected to these lines. Select the **Close** button to return to the board's configuration screen.

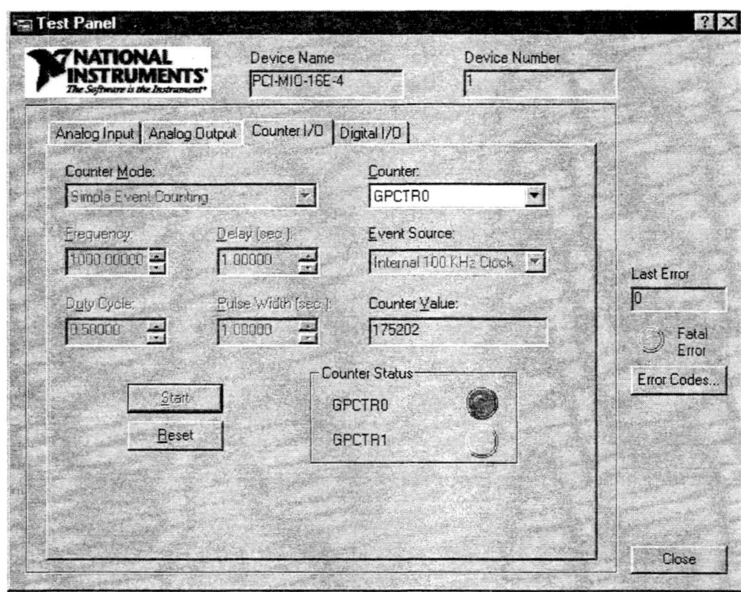

Figure 6–16
Run Test Panel: **Counter I/O** Window

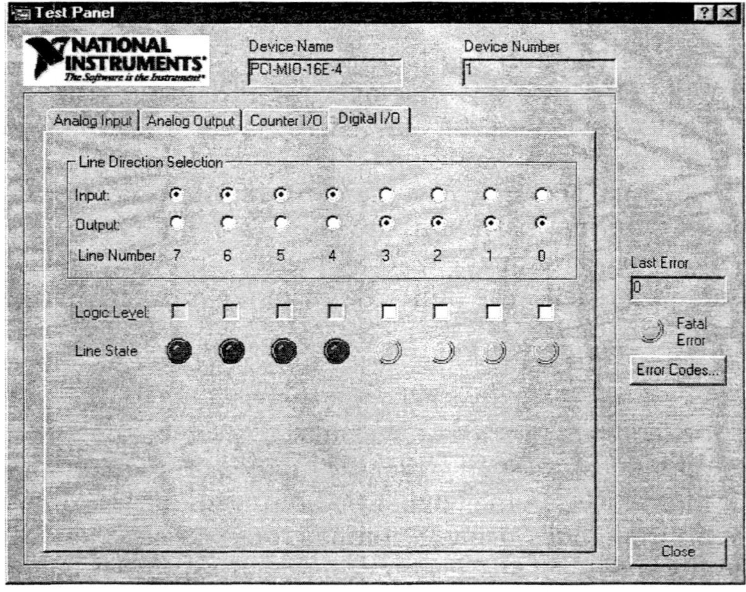

Figure 6–17
Run Test Panel: **Digital I/O** Window

If you have SCXI hardware installed on your system, you would need to configure it also. The SCXI installed will appear as a folder under the `Devices and Interfaces` folder in the **MAX** window (not shown in Figure 6–8). When you click on the SCXI chassis folder, the right half of the window displays the modules in the SCXI chassis with the Chassis Address and Chassis Identification. Right-click on the SCXI Chassis in the left half of the window, and select **Properties** from the pull-down menu. The **Configuring SCXI** window is displayed, from where you can set the **Chassis ID** and the **Chassis Address**. To configure the modules in the SCXI chassis, right-click on each module in the right-half window one at a time and select **Properties**. The window to configure the SCXI module is displayed from where you can set the operating mode, DAQ board to control the SCXI chassis, channel gains, and filters (if any are used). You can also select the accessory connected (if any).

The `Data Neighborhood` folder contains the *DAQ Channel Wizard* to create aliases for your signals connected to the channels on your DAQ board. The `IVI Instruments` folder shown in Figure 6–18 contains the **Logical Name Wizard** and the Interchangeable Virtual Instruments (IVI) logical names. The right half of the window contains a description of the items selected. The `Scales` folder contains a wizard to customize the scales for your channel configurations. The purpose of the `Scales` folder is given in the next section.

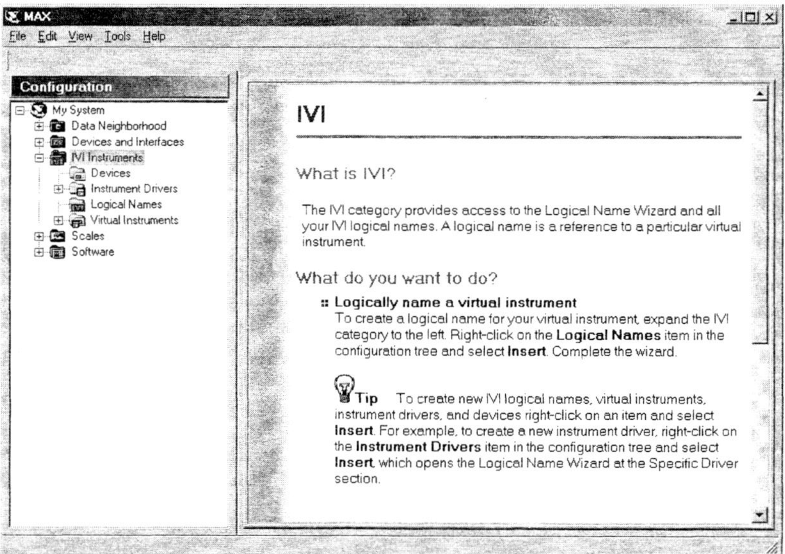

Figure 6–18
MAX Window: `IVI Instruments`

Using the DAQ Channel Wizard

The DAQ Channel Wizard is a utility to create aliases between the signals and the channels connected on the DAQ board, set the configuration, and apply scaling to the channels. In this section you will see how to configure analog input, analog output, and digital I/O channel types. You enter the name of the channels, channel description, units and range of the signal, transducer type, scaling coefficients, and unit conversion factors in the **DAQ Channel Wizard** dialog windows that guide you through the configuration process. Once configured, you can reference the channel by its name in your application; the scaling and conversion will be performed automatically for you in the background. You can create and save multiple configuration files specifying different settings. These files are saved with a `.daq` extension. You can save DAQ files to any location from **MAX** by selecting **NI-DAQ Configuration** under the **Tools** menu. You can set the active DAQ configuration to another DAQ file, set up remote (networked) data acquisition, and create new DAQ configuration files.

Let us now configure a channel. Start the **Measurement & Automation Explorer** by double-clicking on the icon on the Windows desktop. Right-click on the `Data Neighborhood` folder and select **Create New...** from the pull-down menu. A **Create New...** window will be displayed (Figure 6–19).

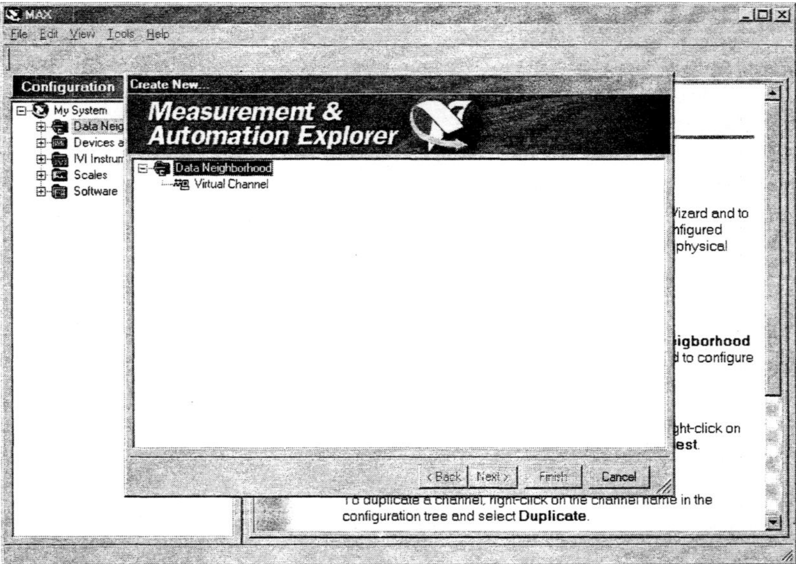

Figure 6–19
DAQ Channel Wizard: **Create New** Window

Highlight the **Virtual Channel** option in the **Create New...** window and select the **Finish** button. This will bring up a **Create New Channel** window (Figure 6–20) from where you can create an alias channel name and configure either an **Analog Input**, **Analog Output**, or **Digital I/O** channel.

An example of setting up an analog input channel is given below. Select **Analog Input** from the pull-down menu in Figure 6–20 and click on **Next>.** This will bring up a dialog window (Figure 6–21) where you can enter the **Channel Name** and **Description**. Enter `Temp_Sensor0` in this box. Type a brief description of the purpose of this channel in the **Channel Description** box. After this channel is configured, you can refer to it in your software by its alias name (`Temp_Sensor0`).

Select **Next>** to display Figure 6–22, from where you can select the type of sensor or measurement from the pull-down menu list. Select **Voltage** from the pull-down list box and a new window will be displayed (figure not shown) with your choice. Also, be sure to check the box next to **This will be a temperature measurement** if you are measuring temperature using this channel, as in this example.

In the next window (not shown) you define the **Units** and **Range** of the temperature sensor. When you select **Next>** from this window, Figure 6–23 is displayed, where you define the signal's scaling.

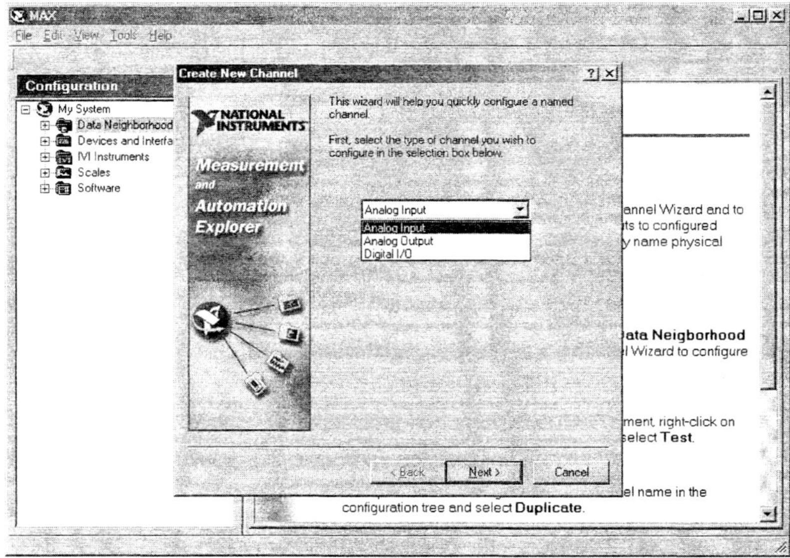

Figure 6–20
DAQ Channel Wizard: **Create New Channel** Window

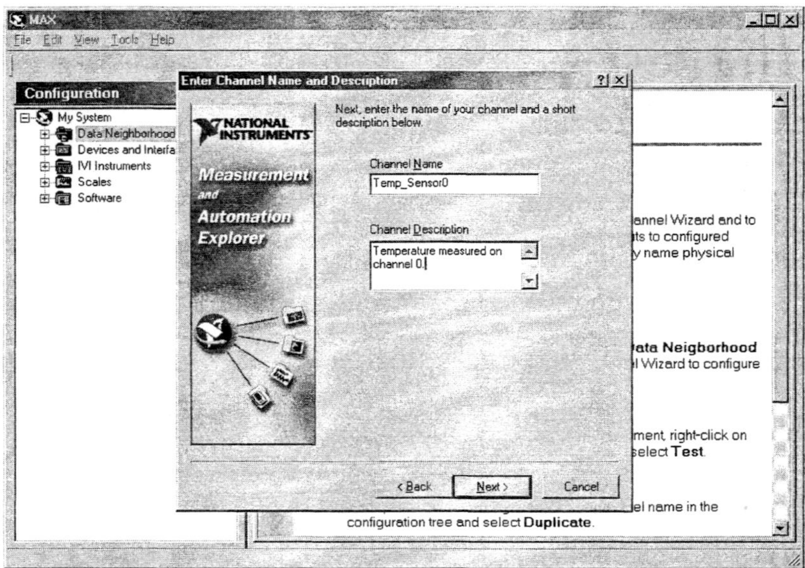

Figure 6–21
DAQ Channel Wizard: **Channel Name and Description** Window

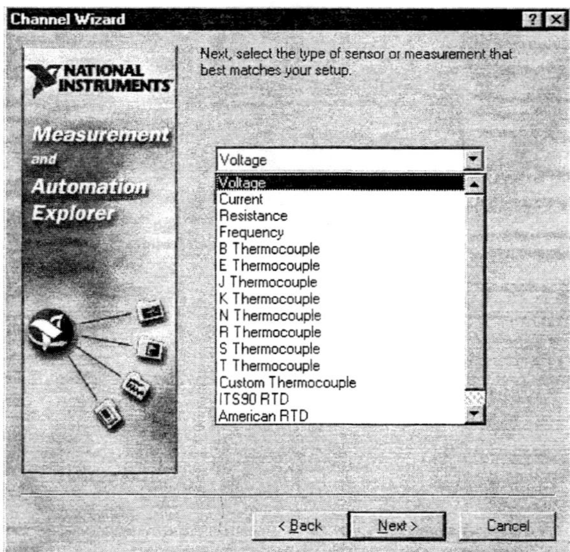

Figure 6–22
DAQ Channel Wizard: Type of **Sensor** or **Measurement** Window

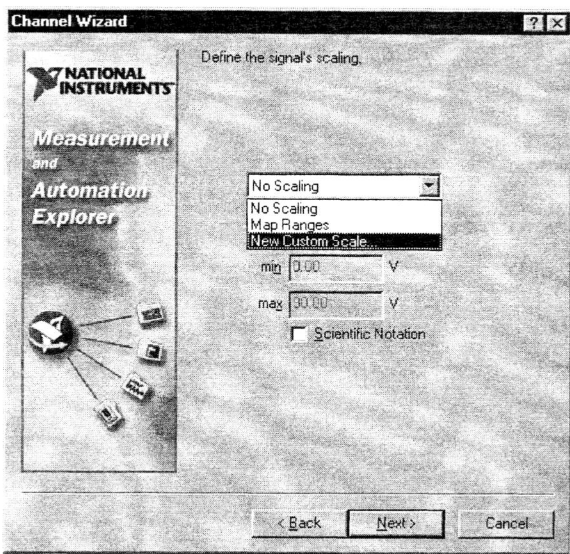

Figure 6–23
DAQ Channel Wizard: **Signal Scaling Choices** Window

If you are using your own scaling, select **New Custom Scale...** and a window will be displayed asking you to input the **Scale Name, Scale Description,** and **Scale Type.** Enter the relevant information as shown in Figure 6–24 and click on **Scale Type**. You have a choice of **Linear**, **Polynomial,** or **Table** scales. In this case, since you are using a temperature sensor, select **Linear Scale**. For a description and meaning of the different types of scales and other information about the **Measurement & Automation Explorer**, select **Help>>Help Topics** in the Measurement & Automation Explorer menu. From there you can navigate to the appropriate topic.

When you select **Next>** in Figure 6–24 you are asked to enter the values for the linear scale values. The scaling factor for the temperature sensor is to multiply the voltage reading by 100 to obtain the degrees. Therefore, enter `100.0` for the **m** value in Figure 6–25. Since there is no offset, leave **b** as `0`. Click **Next>** to verify that the scale is configured correctly. A linear plot with the scaling factors you selected will be shown as in Figure 6–26.

When **Finish** is selected in Figure 6–26, a window showing the scale's name just created (`TempScale` in this example) with scale values selected is displayed. Note that this scale name is also displayed when you select the `Scales` folder in the **Data Neighborhood** window.

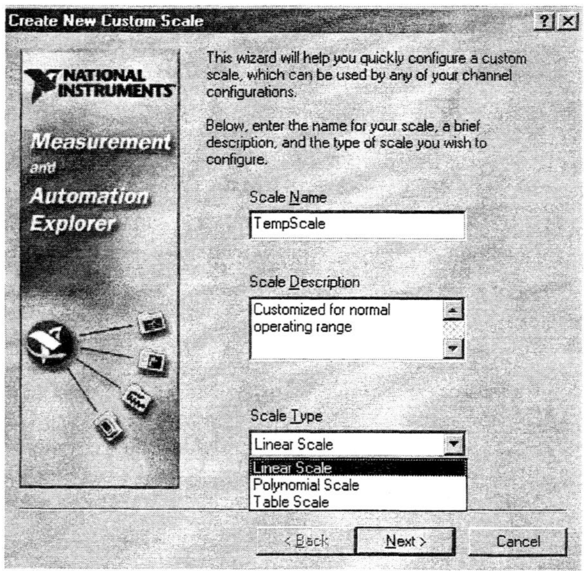

Figure 6–24
DAQ Channel Wizard: **Create New Custom Scale** Window

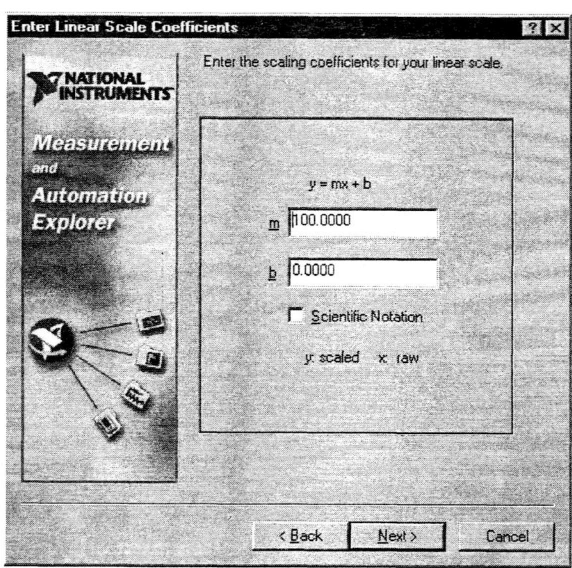

Figure 6–25
DAQ Channel Wizard: **Scale Coefficient** Window

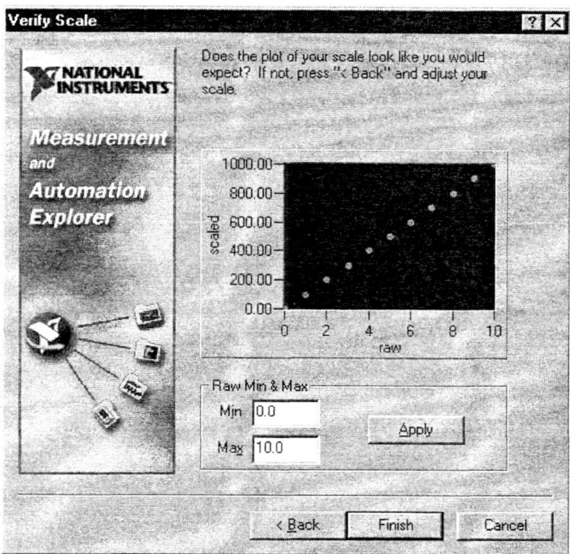

Figure 6–26
DAQ Channel Wizard: **Verify Scale** Window

Select **Next>** from the window showing the scale information; Figure 6–27 is displayed with the DAQ hardware settings. Select the appropriate values in the dialog boxes for this channel.

The **Finish** button configures the channel that is now visible in the **Data Neighborhood** window. Similarly, you can define and configure other channels on your DAQ board. When configured, these will appear in the **Data Neighborhood** window. Figure 6–28 shows the three channels configured with their alias names: `Temp_Sensor0`, `SineWave_Ch1`, and `SquareWaveCh2` for channels 0, 1, and 2, respectively, with their settings in the window on the right.

To view the configuration of an individual channel, right-click on the channel name in the **Data Neighborhood** window (Figure 6–28) and select **Properties** from the pull-down menu. Figure 6–29 gives the related configuration values for the channel selected.

Before you run your application, you may want to test the individual channels with the appropriate signals connected. Connect the signal to that channel, right-click on the channel name in the **Data Neighborhood** window, and select **Test**. A test window will display the signal waveform on the channel selected. Figure 6–30 is the waveform for a square-wave input connected to

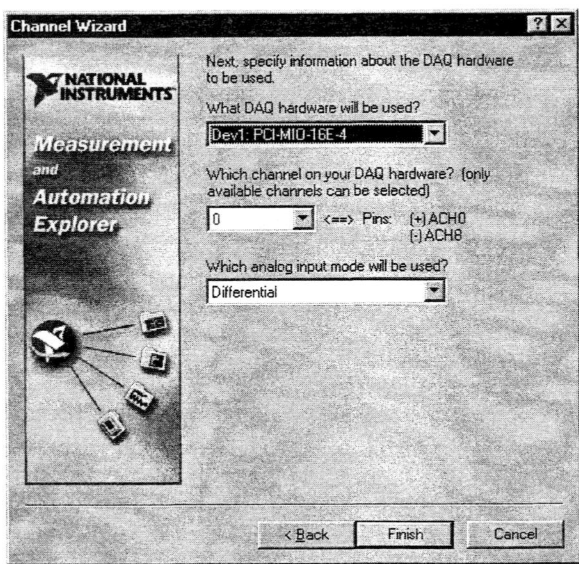

Figure 6–27
DAQ Channel Wizard: Hardware Settings

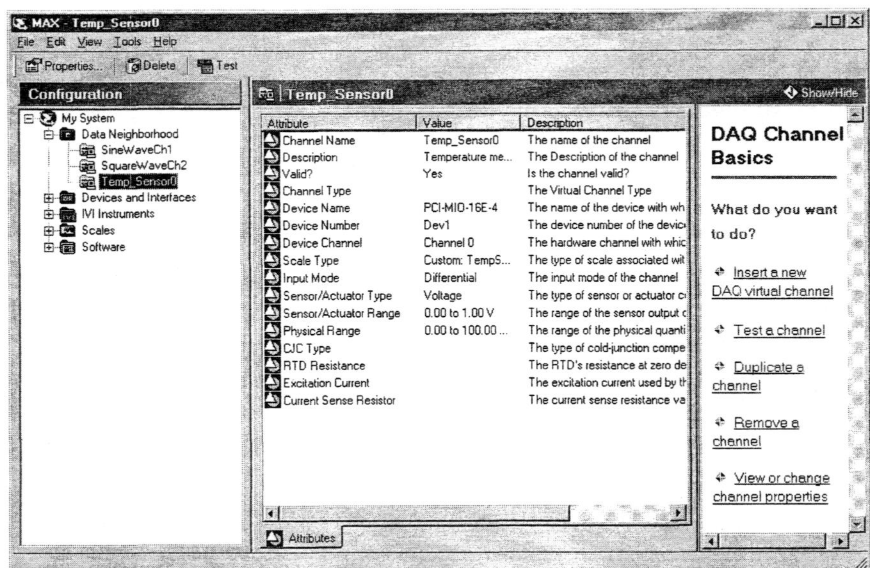

Figure 6–28
Data Neighborhood: Created Channels

analog input channel 2. This test will give you confidence that your channel is configured and you can measure the signal correctly.

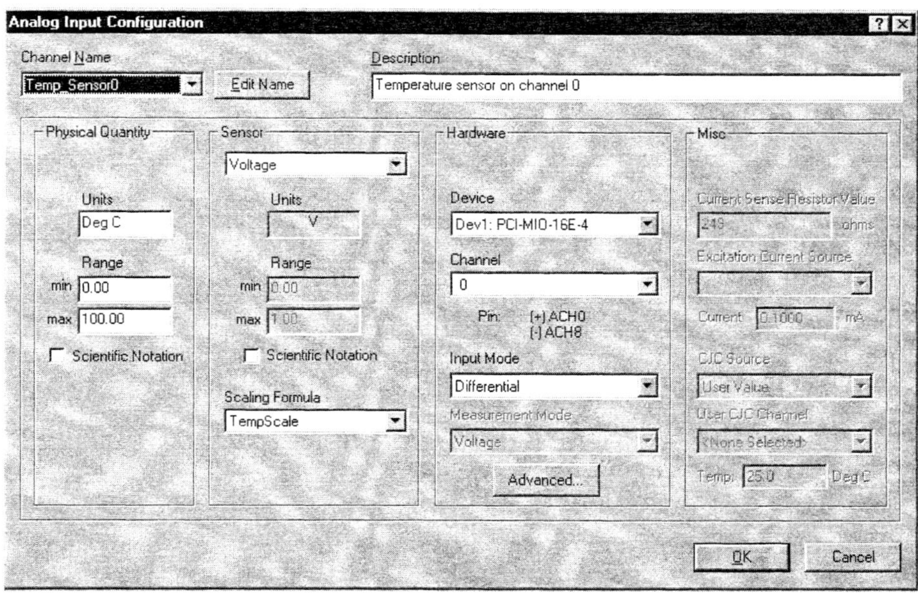

Figure 6–29
Channel Configuration Window

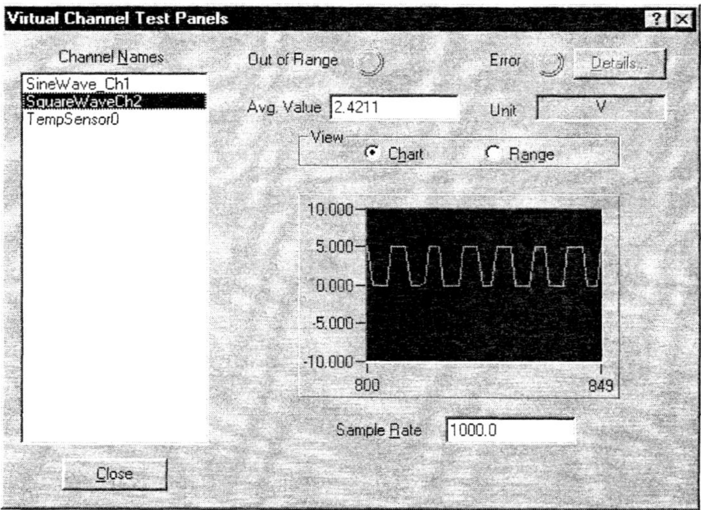

Figure 6–30
DAQ Channel Wizard: Test Panel

The scale names created appear in the `Scales` folder in the **MAX** window (see Figure 6–31). When you right-click on the scale name that you created, you have a choice of **Delete** and **Properties...** from the pull-down menu. Selecting the **Properties...** menu item will display the current scale settings that can be modified from the window displayed.

Hardware Configurations

There are three parameters that must be set up on the DAQ board: Base I/O Address, Interrupt Level, and Direct Memory Access (DMA). On *Plug&Play* DAQ boards, these parameters are configured from the software. On non-Plug&Play DAQ boards, the configuration is done through jumpers and dip-switches provided that there is no conflict with the hardware. If you make a change to the jumper settings on the board, you must change the software configuration accordingly.

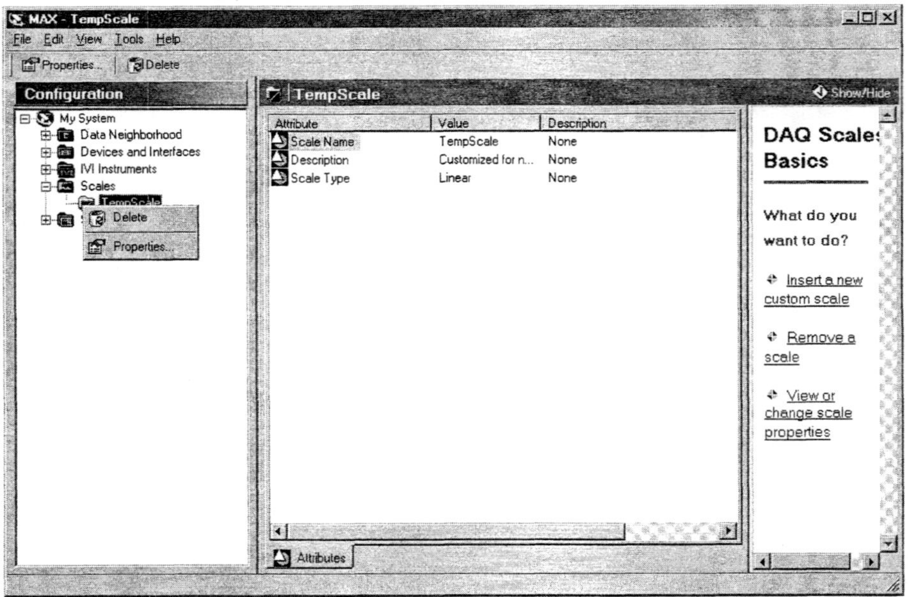

Figure 6–31
DAQ Channel Wizard: **Scales** Window

- **Base I/O Address.** Base I/O Address settings are used to determine where in the computer's I/O space the board's registers reside. The DAQ board uses these registers to communicate with the computer. The DAQ driver software writes to the configuration registers to configure the board and reads from the data registers on board to obtain the board' s status and data.
- **Interrupt Level.** Interrupt Level is a means of informing the processor of the priority with which it must handle the interrupt request from the DAQ board. Each interrupt-generating device is assigned a different interrupt level in order for the processor to decide uniquely when to handle the interrupt in order to avoid conflicts.
- **Direct Memory Access (DMA).** The DAQ board can communicate with the computer through DMA. DMA increases the data throughput by transferring the data directly from the peripheral to the computer memory, bypassing the processor. DMA is genrally used when high-speed data transfer is required.

If at any later time you want to set up NI-DAQ, obtain On-line Help, or access the DAQ documentations, you can do so from the Windows **Start** menu and follow the path shown in Figure 6–32.

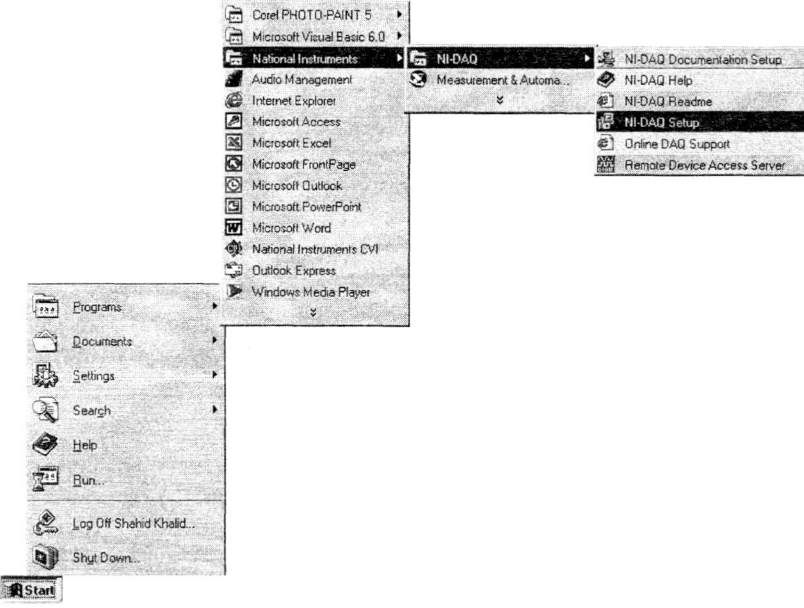

Figure 6–32
NI-DAQ Setup Path

Using DAQ Library Functions

In this section you will see how to acquire data using the DAQ Library functions. There are two types of DAQ library functions: Easy I/O for DAQ and Data Acquisition. To get started with data acquisition, it is simpler to use the Easy I/O for DAQ library functions that will suffice for most of your simple data acquisition applications. The Easy I/O for DAQ library functions currently supports only Analog I/O, Counter/Timers, and simple Digital I/O and does not support multirate scanning. If your data acquisition needs are more sophisticated, you should use the Data Acquisition Library functions. It is recommended that you do not mix calls to the Data Acquisition Library functions with similar types of calls to the Easy I/O for DAQ Library functions in the same program.

Analog Input

To access the Analog Input Easy I/O DAQ library functions, select **Library>> Easy I/O for DAQ>>Analog Input**.

Let us look at the fundamental steps involved in data acquisition using the library functions when sampling analog input channels:

1. Identify the channel(s) you wish to acquire the data from, and the data type to be created. This is accomplished by using the *nidaqAICreateTask* library function. This function returns a unique identifier called the task identifier associated with the data acquisition task. This task identifier is used in subsequent function calls to identify the DAQ board, channel(s) and the type of analog input task.
2. Call the *nidaqAIStart* library function to start acquiring the data from the channel(s) using the task identifier created in the *nidaqAICreateTask* library function.
3. Retrieve analog input data from the channel(s) selected by calling the *nidaqAIRead* library function. In this function you can specify a subset of channels to retrieve data from selected channels if you are scanning multiple channels. In this function you also specify the number of scans to acquire. A scan is one sample from each channel in multichannel data acquisition.

4. Plot or display the data acquired on your user interface.
5. Stop data acquisition by calling the library function *nidaqAIStop* using the task identifier created in the *nidaqAICreateTask* library function.
6. Relinquish the resources created in the *nidaqAICreateTask* library function by calling the *nidaqAIDestroyTask* library function with the task identifier.

You will see how these steps are accomplished by looking at a simple data acquisition project. Load and run `project6-1` to bring up the GUI shown in Figure 6–33. This GUI is shown with the channel selection options in the **Channel String** ring control.

The data acquisition library functions in the source code (Figure 6–34) are annotated with the step numbers shown above.

When you select the **START** command button the *StartAcquisitionCB* function at line 27 is invoked. At line 40 the *nidaqAICreateTask* library function is called and is repeated here for your convenience.

```
daq_error = nidaqAICreateTask (ChannelSelected,
                NidaqPointByPoint, &NumOfChannels, &TaskID);
```

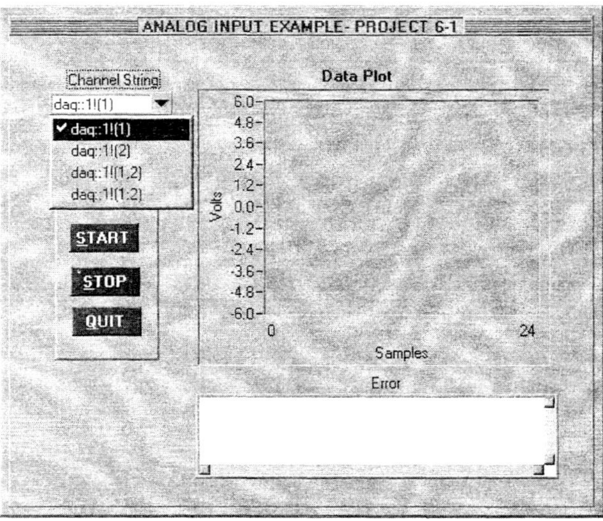

Figure 6–33
Analog Input Example GUI for `project6-1`

```
#include <utility.h>
#include <cvirte.h>
#include <userint.h>
#include <easyio.h>
#include <ansi_c.h>
#include "project6-1.h"

static int DAQHandle;
long TaskID, daq_error, NumOfChannels;
char *ErrorString, Channel_Selected[100];
static double AcquiredData[2000];
int StartFlag=0;

int main (int argc, char *argv[ ])
{
  if (InitCVIRTE (0, argv, 0) == 0)
          return -1;      /* out of memory */
  if ((DAQHandle = LoadPanel (0, "project6-1.uir", PANEL)) < 0)
          return -1;
  DisplayPanel (DAQHandle);
  RunUserInterface ();
  DiscardPanel (DAQHandle);
  return 0;
} //main

//Invoked from START command button
int CVICALLBACK StartAcquisitionCB (int panel, int control, int event,
          void *callbackData, int eventData1, int eventData2)
{


  switch (event)
  {
       case EVENT_COMMIT:
          //Obtain the channel string from the control box
          GetCtrlVal (DAQHandle, PANEL_CHL_STR, Channel_Selected);
          ClearStripChart (DAQHandle, PANEL_STRIP_CHART);  //Clear the strip chart

          //Step 1 - Create data acquisition task
          daq_error = nidaqAICreateTask (Channel_Selected, kNidaqPointByPoint,
                                                      &NumOfChannels, &TaskID);
          if ( daq_error !=0)
          {
                  ErrorString = nidaqGetErrorString (TaskID, daq_error);
                  InsertTextBoxLine (DAQHandle, PANEL_ERROR_BOX, -1, ErrorString);
          }

          //Step 2 - Start data acquisition
          daq_error=nidaqAIStart(TaskID);
          if ( daq_error !=0)
          {
```

Figure 6–34
Source Code Listing for `project6-1` *(continued)*

```
                ErrorString = nidaqGetErrorString (TaskID, daq_error);
                InsertTextBoxLine (DAQHandle, PANEL_ERROR_BOX, -1, ErrorString);
                }
        if (NumOfChannels)
                        SetCtrlAttribute (DAQHandle, PANEL_STRIP_CHART,
                                                ATTR_NUM_TRACES, NumOfChannels);

        if (NumOfChannels > 1)
                        SetTraceAttribute (DAQHandle, PANEL_STRIP_CHART, 2,
                                                ATTR_TRACE_COLOR, VAL_RED);

         if (NumOfChannels > 2)
                        SetTraceAttribute (DAQHandle, PANEL_STRIP_CHART, 3,
                                           ATTR_TRACE_COLOR,  VAL_DK_GREEN);
        StartFlag=1;
        break;
  }
  return 0;
}//StartAcquisitionCB

//Run the timer
int CVICALLBACK TimerTickCallback (int panel, int control, int event,
        void *callbackData, int eventData1, int eventData2)
{

        if  (( event == EVENT_TIMER_TICK) && (StartFlag) ){
        //Step 3 - Read data
        daq_error = nidaqAIRead (TaskID, Channel_Selected, 1, -1.0,
                                                            AcquiredData);
        if ( daq_error !=0)
        {
                ErrorString = nidaqGetErrorString (TaskID, daq_error);
                InsertTextBoxLine (DAQHandle, PANEL_ERROR_BOX, -1, ErrorString);
        }

        //Step 4 - Plot data
        PlotStripChart (DAQHandle, PANEL_STRIP_CHART, AcquiredData,
                                        NumOfChannels, 0, 0,  VAL_DOUBLE);
        }
  return 0;
}//TimerTickCallback
//Invoked from the STOP command button
int CVICALLBACK StopAcquisitionCB (int panel, int control, int event,
        void *callbackData, int eventData1, int eventData2)
{
  switch (event)
  {
        case EVENT_COMMIT:
        //Step 5 - Stop data acquisition
        daq_error =nidaqAIStop(TaskID);
        if ( daq_error !=0)
```

Figure 6–34
Source Code Listing for `project 6-1` *(continued)*

```
            {
                    ErrorString = nidaqGetErrorString (TaskID, daq_error);
                    InsertTextBoxLine (DAQHandle, PANEL_ERROR_BOX, -1, ErrorString);
            }
            daq_error =nidaqAIDestroyTask (TaskID);
            if ( daq_error !=0)
            {
                    ErrorString = nidaqGetErrorString (TaskID, daq_error);
                    InsertTextBoxLine (DAQHandle, PANEL_ERROR_BOX, -1, ErrorString);
            }

            StartFlag=0;  //Stop timer - disable flag
            break;
    }
  return 0;
}

//Invoked from QUIT command button
int CVICALLBACK QuitAcquisitionCB (int panel, int control, int event,
            void *callbackData, int eventData1, int eventData2)
{
  switch (event)
  {
            case EVENT_COMMIT:
            /*Stop data acquisition either in this function
            or in "Stop Acquisition CB" function. If you try
            to stop a task that is already stopped, a
            library error occurs.*/




            //Step 6 - Relinquish the resources
            daq_error =nidaqAIDestroyTask (TaskID);
            if ( daq_error !=0)
            {
                    ErrorString = nidaqGetErrorString (TaskID, daq_error);
                    InsertTextBoxLine (DAQHandle, PANEL_ERROR_BOX, -1, ErrorString);
            }
            StartFlag=0;  //Stop timer - disable flag
            QuitUserInterface (0);
            break;
  }
  return 0;
} //QuitAcquisitionCB
```

Figure 6–34
Source Code Listing for `project6-1` *(continued)*

The arguments for the *nidaqAICreateTask* function are explained. The first argument, `ChannelSelected` of *nidaqAICreateTask,* is a string called the *channel path* that specifies the channel(s) from where you want to acquire the data. There are two ways in which you can specify the channel paths: using the channel name aliases set up by the **Data Channel Wizard** as explained above in the section *Using the DAQ Channel Wizard* or by using the NI-DAQ syntax. The channel path in this example is obtained from the **Channel String** ring control (Figure 6–31). The first list item in the **Channel String** ring control is

```
"daq::1!(1)"
```

Let us examine what this syntax means and how it is structured. The channel path always starts with "`daq::`", followed by the device number, which in this case is "`1,`" followed by "`!`". The character "`!`" is a separator between the strings, followed by the channel number in parentheses, "`(1)`" in this example.

You can also specify the channel path as a range. For example, if you wanted to acquire data from channels 1–3, on device 1 you could use the NI-DAQ syntax as follows:

```
"daq::1!(1:3)"
```

In addition, the channel numbers you may want to acquire data from may not be contiguous. For example, to acquire data from channels 1, 2, 3, and 8, the NI-DAQ syntax will be as follows:

```
"daq::1!(1:3,8)"
```

The NI-DAQ syntax used to acquire data using the SCXI chassis, the module number in the SCXI chassis, and the channel number is of the form

```
"daq::1!(ob#!sc#!md#!ch#)"
```

where

```
# following ob is the plug-in DAQ board channel number
# following sc is the SCXI chassis number
# following md is the module number in the chassis
# following ch is the channel number on the SCXI
```

For example, in the NI-DAQ syntax

```
"daq::1!(ob0!sc1!md2!ch2)"
```

would acquire data from SCXI channel 2 from the module in slot 2 of SCXI chassis 1, that is sent to channel `0` of the onboard DAQ (device `1`).

The second argument of the *nidaqAICreateTask* function is called the *task type* and specifies the type of analog data to be created. There are two choices:

1. **Point by point**, as the name implies, acquires single points or single scans of data so that the data is reported to your application as it is acquired.
2. **Capture Waveform** data is used when you want to acquire a complete set of data at timed intervals and write it to the data buffer. The data is read from the buffer as needed by your application.

In this example we are capturing the data point by point using the reserved word `nidaqPointByPoint`.

The third argument, `NumOfChannels`, returns the number of channels in the channel path argument. You may use this information in your application to set up the number of channels to plot on the strip chart, for example.

The last argument of this function is the *task identifier*. This identifier is used in subsequent function calls when referring to the task created with the configuration in *nidaqAICreateTask*.

Each NI-DAQ library function returns a `status code` to indicate whether the function has performed successfully. In this example the `status code` is returned in the variable `nidaq_error`. As always, you must check for errors whenever making a call to the library functions. When the `status code` returns the value `0`, it implies that there are no errors or warnings. Warnings are indicated by the `status code` being a positive number and errors by a negative value. You can get a detailed listing of status codes from the desktop by selecting **Start>>Programs>> National Instruments>>Ni-DAQ>>NI DAQ Help**, selecting the **Contents** tab, and clicking on **Status Codes**.

A better way to obtain the meaning of the status code(s) is to let your application decipher the status code(s) using the NI-DAQ library function *nidaqGetErrorString*, which gives a text description of the status code. This function is shown on line 44 and is repeated here.

```
ErrorString = nidaqGetErrorString (TaskID, daq_error);
```

This function takes the task identifier as the first argument and the status code is the second argument. This function returns the description of the status code in the string `ErrorString`. In this program the `ErrorString` is displayed in the text box on the GUI.

The library function *nidaqAIStart* at line 49 starts the data acquisition for the channel(s) specified by the task identifier created in the library function *nidaqAICreateTask*. This function consists of the task identifier as the only argument.

Data is obtained from the DAQ card using the *nidaqRead* library function at line 79, called by the *TimerTickCallback* function. The *TimerTickCallback* function reads the data at the rate specified in the timer control. The arguments of the *nidaqRead* library function are as follows:

```
daq_error = nidaqAIRead (TaskID, Channel_Selected ,1, -1.0,
                                                    AcquiredData);
```

The first argument of this function is the task identifier that was created using the function *nidaqAICreateTask*. The second argument is the channel path, as explained above for the *nidaqAICreateTask* function argument. This channel path can be a list of one or more of the channels specified in the *nidaqAICreateTask* function, or an empty string ("") can be used to specify all channels in the task. The third argument is the number of scans to acquire (one in this case). The fourth argument is the maximum time in seconds to wait for the data before the function returns a timeout error. As shown in this example, "`-1.0`" is the default argument value and is used to let the function calculate the timeout based on the acquisition rate and the number of scans. You can disable the time limit by entering a "`-2.0`" for this argument. The last argument of this function contains the data acquired in the array `AcquiredData`. This array contains the data acquired from the channel(s) specified in the channel path. If you had created the channel path using the **DAQ Channel Wizard**, the data is the units of channel that you had specified in the DAQ Channel Wizard. If you are using the NI-DAQ syntax, all the data is in volts. At line 88 the `AcquiredData` is plotted on the strip chart using the *PlotStripChart* library function.

When you select the **QUIT** command button, the *QuitAcquisitionCB* function is called to stop data acquisition, relinquish the resources being used, and stop event processing from the user interface. The *nidaqAIStop* library function at line 101 stops data acquisition for the task specified as the argument in this function. The *nidaqAIDestroyTask* library function at lines 107 and 136 contains the task identifier for which you want to relinquish the resources. After calling this function, the task identifier is no longer valid and cannot be used until you create the task using the *nidaqAICreateTask* library function again. The *StopAcquisitionCB* function stops data acquisition without exiting the program.

An instance of this project's execution with sine wave input on channel 1 and square wave input on channel 2 is shown in Figure 6–35. This is accomplished by selecting `daq::1!(1:2)` from the **ChannelString** ring control. If you prefer, you can run this application for a sine wave or a square wave input individually by selecting the appropriate channels from the **Channel String** ring control.

You can obtain a single value from an analog input channel, a single scan of values from analog input channel, or a finite number of scans from an analog input channel using the DAQ library functions *nidaqAISinglePointOp*, *nidaqAISingleScanOp*, or *nidaqAIScanOp*, respectively. These high-level functions are useful but should not be used in continuous operation for good performance since they create and destroy data acquisition tasks every time they are called. `project6-2` is modified from `project6-1` to obtain and plot on the graph control for a finite **Number of Scans** from the channel selected at the **Scan Rate** selected from the GUI using the function *nidaqAIScanOp*. A sample scan run is shown in Figure 6–36. Try running this program using different values for **Number of Scans** and **Scan Rate** to see how the graph plot changes.

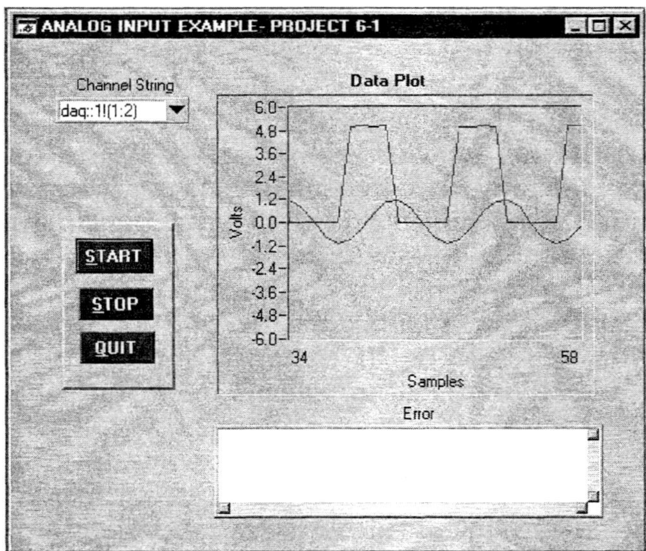

Figure 6–35
`project6-1` Sample Run

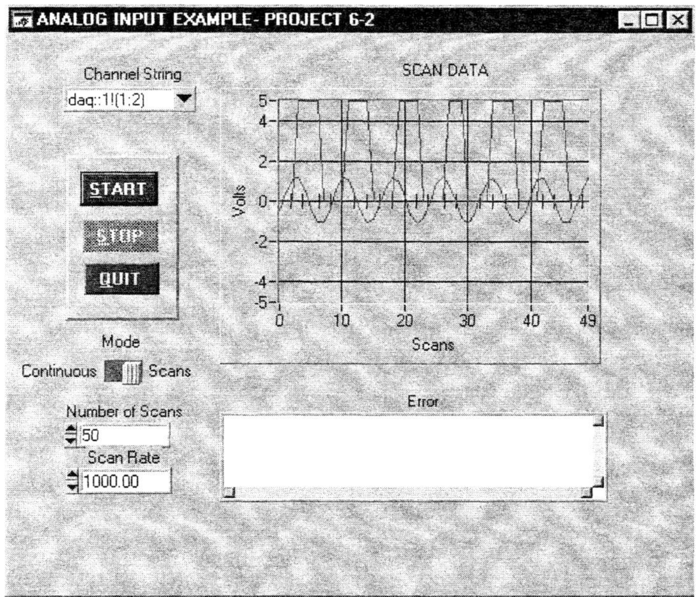

Figure 6–36
`project6-2` Sample Run

The code segment that is executed when **Scans** is selected from the **Mode** binary switch on the GUI is shown in Figure 6–37.

The *nidaqAIScanOp* function, called at line 11, is

```
nidaqAIScanOp (Channel_Selected, NmbrScans, ScanRate,
                    UPPER_LIMIT, LOWER_LIMIT, kNidaqGroupByChannel,
                                             &ActualRate, ScanData);
```

Let us look at the arguments for this function. The `Channel_Selected` is the channel string obtained from the **Channel String** ring control on the GUI. `NmbrScans` and `ScanRate` are obtained from the **Number of Scans** and **Scan Rate** controls on the GUI. The `UPPER_LIMIT` and `LOWER_LIMIT` set the upper and lower limits of the input signal range and are specified in the `#define` statement at the top of the program. The `kNidaqGroupBy-Channel` is the *fillmode* and specifies that the data array will be grouped by channels. These settings actually don't change how data is obtained

```
SetInputMode (DAQHandle, PANEL_START_BUTTON, 1);  //Enable Start Button

SetInputMode (DAQHandle, PANEL_STOP_CMD, 0);  //Disable Stop Button

SetCtrlAttribute (DAQHandle,PANEL_STRIP_CHART, ATTR_VISIBLE, 0);
CreateGraphControl();
GetCtrlVal(DAQHandle,PANEL_SCAN_NMBR,&NmbrScans);
GetCtrlVal(DAQHandle,PANEL_SCAN_RATE, &ScanRate);

ScanData= (double *) malloc(NmbrScans* NumOfChannels * sizeof(double) );
daq_error = nidaqAIScanOp (Channel_Selected, NmbrScans, ScanRate,
   UPPER_LIMIT, LOWER_LIMIT, kNidaqGroupByChannel, &ActualRate,ScanData);

if ( daq_error !=0)
{
        ErrorString = nidaqGetErrorString (TaskID, daq_error);
        InsertTextBoxLine (DAQHandle, PANEL_ERROR_BOX, -1, ErrorString);
}

for (i=0; i < NumOfChannels; i++)
   if (i==0)
         PlotY (DAQHandle, GRAPH_CTRL_ID, &ScanData[NmbrScans*i],
                                   NmbrScans, VAL_DOUBLE, VAL_THIN_LINE,
                                  VAL_EMPTY_SQUARE, VAL_SOLID, 1,VAL_RED);
else
       PlotY (DAQHandle, GRAPH_CTRL_ID, &ScanData[NmbrScans*i],
          NmbrScans, VAL_DOUBLE, VAL_THIN_LINE,
                              VAL_EMPTY_SQUARE, VAL_SOLID, 1, VAL_DK_GREEN);
ScanFlag =1;
```

Figure 6–37
`project6-2` Code Segment

from the DAQ card. It is always obtained in round-robin scanning. It just changes how the data buffer is organized when it is returned. *Grouped by Channels* means that the one-dimensional array has each channel's data grouped together. Another option for *fillmode, Grouped by Scan,* is indicated by using the `kNidaqGroupByScan` constant in this argument. In the Grouped by Scan option, the one-dimensional data array has each scan's data grouped together.

It is important to know whether the data array will be displayed on a graph control or on a strip chart. Grouped by Channel is used for displaying the data array on the graph control, and *Grouped by Scan* is used to display the data array on the strip chart. If you acquire the data Grouped by Scan, you can later change the data to group by channel using the *GroupByChannel* library function.

Looking at the arguments of *nidaqAIScanOp,* the `ActualRate` argument returns the actual scan rate (in seconds per second) for data acquisition and could be different from the `ScanRate`. The waveform data from the selected channel(s) is returned in the data array `ScanData`, which is ordered as specified by the *fillmode*. The values for the channel will be in the units of the channel if the channel string used was referencing a channel created using the **DAQ Channel Wizard**. The values will be in volts if the channel string uses the NI-DAQ syntax, as in this case.

Analog Output

Multifunction input/output DAQ boards contain digital-to-analog converters (DACs) to convert digital values to analog values. This functionality generates outputs that you can send to the analog output channel(s) of the DAQ board and to the devices connected to these channel(s). To do so, you use the DAQ library functions. To access the Analog Output Easy I/O DAQ library functions, select **Library>>Easy I/O for DAQ>>Analog Output**. The four analog output library functions are displayed in the function tree panel: *AOUpdateChannel, AOUpdateChannels, AOGenerateWaveforms,* and *AOClearWaveforms*. Below we look at the purpose of these library functions and how they are used.

- **AOUpdateChannel.** The *AOUpdateChannel* function outputs the voltage specified to one analog output channel. The code fragment below outputs `3.6` volts to channel `2` on device `1`.

```
short deviceNumber=1;
double voltage =3.6:
        .
        .
        .
        .
AOUpdateChannel( deviceNumber, "2", voltage);
```

- **AOUpdateChannels.** The *AOUpdateChannels* function outputs the voltages specified to the analog output channel(s) specified in the second argument of the function. The code fragment below will output `1.1` volts on channel `0`, `2.1` volts on channel `2`, and `0.8` volt on channel `4` on device `1`.

```
short deviceNumber=1;
double voltage[3] ={1.1, 2.1,.8};
        .
        .
        .
AOUpdateChannels( deviceNumber, "0,2,4", voltage);
```

- **AOGenerateWaveforms.** The *AOGenerateWaveforms* function generates a waveform of voltages on the analog output channels specified in the channel string. You specify the number of D/A conversions to comprise a waveform in the argument `updatesChannel` for a particular channel, the number of waveform iterations in the argument `iterations`, and the number of updates performed per second by specifying the value in the argument `updatesSecond`. If the number of iterations is 0, the waveform will be generated continuously. You will then need to call *AOClearWaveforms* to clear waveform generation.

```
short deviceNumber=1;
double updatesSecond = 5.0; //Number of updates per second
int updatesChannel =9;       //Number of D/A conversion that
                             //comprises a waveform
int iterations =0;           //Generate waveform continuously
double waveform[ ] ={1.1,1.2,1.4,2.1,2.2,2.4,4.1,4.2,4.4};
unsigned long taskID;  //Used in subsequent function calls
        .
        .
        .
        .
AOGenerateWaveforms( deviceNumber, "0,2,4", updatesSecond,
          updatesChannel, iterations, waveform, &taskID);
        .
        .
        .
        .
```

The values in the waveform array are applied to channels `0`, `2`, and `4` in the order specified in `waveform` array and are as follows:

```
waveform[0] = 1.1 is applied to channel 0
waveform[1] = 1.2 is applied to channel 2
waveform[2] = 1.4 is applied to channel 4
waveform[3] = 2.1 is applied to channel 0
waveform[4] = 2.2 is applied to channel 2
waveform[5] = 2.4 is applied to channel 4
```

```
waveform[6] = 4.1 is applied to channel 0
waveform[7] = 4.2 is applied to channel 2
waveform[8] = 4.4 is applied to channel 4
```

Here each waveform contains nine elements from the waveform array.

- **AOClearWaveforms.** The *AOClearWaveforms* function clears the waveform generated by *AOGenerateWaveforms* when you pass a `0` for the `iterations` argument in *AOGenerateWaveforms* to generate the waveform continuously.

Digital Input/Output

The *Digital I/O* of a DAQ board contains several digital lines for sensing and controlling multiple digital signals. The digital I/O circuitry on the DAQ board is used for reading and writing data to peripheral equipment, controlling processes, and sending test patterns. The number of digital lines available, the rate at which these lines can send and receive data, and the drive capability of the lines all determine how you can use this feature of the DAQ card. Several digital I/O lines are grouped into a port (four lines comprise a port on the MIO board). The digital I/O port can be configured to read data from or write data to the entire port or from single digital lines within a port. You can set the configuration of the port through *CVI* as input or an output port using the DAQ library function *DIG_Prt_Config*. Some boards, such as MIO E-series, 6533 family, 6602 family, and TIO-10, allow you to configure each digital line individually as an input or output using the DAQ library function *DIG_Line_Config*. In other boards all digital lines within the same port must be configured as either input or output.

When the port is configured as an input port, reading the port returns the values of the digital lines as logical high or low. When configured as an output port, writing to the port sets each digital line of the port to the corresponding digital logical high or low value, enabling you to send data to an external device through these lines. The digital circuitry of the DAQ board can input and output digital signals using transistor-transistor (TTL) logic. The TTL signal is considered logical low if the voltage is between 0 and 0.8 volt and logical high if the voltage lies between 2.0 and 5.5 volts.

There are two categories of digital applications: immediate and timed. *Immediate digital I/O* instantaneously updates or reads a line or port when a call to the digital library function is made. In *timed digital I/O*, external signals are used to control or handshake the data transfers. This ensures that the DAQ board and the instrument are ready to transfer data before starting the operation.

To access digital input/output functions from the easy I/O DAQ library functions, select **Library>>Easy I/O for DAQ>>Digital Input/Output....** You will see four digital output library functions in the displayed function tree panel: *ReadFromDigitalLine, ReadFromDigitalPort, WriteToDigitalLine,* and *WriteToDigitalPort*. The purpose of these library functions is shown by means of code fragments.

- **ReadFromDigitalLine.** The *ReadFromDigitalLine* function reads the logical state of a digital line on the port number specified. The code fragment below reads the state of line 3 on port 1 of the DAQ board and returns the state of the line in the variable `LineState`. The port in this example is assigned a width of 4 bits and set to configure before every read.

```
short device=1;
char PortNumber = "1";   //Port Number on the DAQ
short line =3;           //Line number in the port
short PortWidth = 4;     //Number of lines in a port
long Configure=1;  // 1=Always configure; 0 = Only if necessary
unsigned long LineState;  //State of the read digital line

ReadFromDigitalLine(device, PortNumber, line, PortWidth,
       Configure, &LineState);
```

  Depending on your application, it may not always be necessary to configure a port before a read or write since it adds overhead. You can also combine multiple 4-bit ports into a port of wider width on an MIO (non-E-series) board. If you want to use a port 12 bits wide, you can combine three 4-bit ports by setting the `PortWidth` to 12 in this function. The `PortWidth` must be an even multiple of the physical port width. The port numbers in the combined port must begin by the `PortNumber` specified in this function and must increase consecutively.

- **ReadFromDigitalPort**. The*ReadFromDigitalPort* function reads the data on the digital port that you configure for input. The code fragment below reads the data in the variable `ReadPattern` from the port specified in `PortNumber`. The `ReadPattern` is a decimal number with the state of the digital lines read on the port.

```
short device=1;
char PortNumber = "1"; //Port Number on the DAQ
short PortWidth = 4;   //Number of lines in a port
long Configure=1;  // 1=Always configure; 0 = Only if necessary
unsigned long ReadPattern; //Pattern read from the digital port

ReadFromDigitalPort(device, PortNumber, PortWidth, Configure,
       &ReadPattern);
```

For example, if `ReadPattern` is `9`, it would indicate that the binary pattern `1001` was read, indicating that lines 0 and 3 are in a logical high state.

- **WriteToDigitalLine**. The *WriteToDigitalLine* function outputs the logical state of the digital line on a digital port. In the code fragment below, line 3 is set to a logical high state and the remaining three lines to logical low states on port 1.

```
short device=1;
char PortNumber = "1"; //Port Number on the DAQ
short line =3;          //Line number in the port
short PortWidth = 4;    //Number of lines in a port
long Configure=1;  // 1=Always configure; 0 = Only if necessary
unsigned long LineState=1;  //State of digital line

WriteToDigitalLine(device, PortNumber, line, PortWidth,
       Configure, LineState);
```

- **WriteToDigitalPort**. The *WriteToDigitalPort* function outputs a decimal pattern to a digital port. In the code fragment below, lines 0 and 2 are set to logical high states and the remaining two lines to logical low states on port 2.

```
short device=1;
char PortNumber = "2";  //Port Number on the DAQ
short PortWidth = 4;     //Number of lines in a port
long Configure=1;  // 1=Always configure; 0 = Only if necessary
unsigned long WritePattern =5;  //Pattern to write "0101"

WriteToDigitalPort(device, PortNumber, PortWidth, Configure,
       WritePattern);
```

Counter Fundamentals

Counters on the DAQ board provide the means to manipulate and coordinate various timing parameters and events. A counter is a device that counts digital signal transitions or generates various types of timing signals. A counter event is any TTL rising or falling edge that occurs at the SOURCE input of the counter. The inputs and outputs of the counter are shown in Figure 6–38.

A counter consists of a SOURCE input, a GATE input, an OUT output, and an internal Count Register (Figure 6–38). The signal connected to the SOURCE input is counted for low-to-high or high-to-low transitions. The GATE input controls when the counting is to occur. The OUT signal of the counter is used to generate a single pulse or pulse train with specified phase durations and frequency. You can use these output signals to connect to other counters to increase the counting resolution. The OUT signal of the counter can be configured to generate either a pulse or a toggle state when the internal Count Register reaches it maximum value. This maximum value is referred to as the *terminal count* (TC).

As mentioned earlier, counters can be configured to count low-to-high or high-to-low transitions of the signal connected to the SOURCE input. The internal Count Register of the counter is incremented or decremented with each transition and can be read to retrieve the current count of the signal transitions.

You can use the counter in conjunction with a clock to create timing and counting applications, such as monitoring and analyzing digital waveforms and generating square waveforms. A counter can only measure TTL high and TTL low signals that have pulse widths, and rise and fall times that meet the counter requirements. Figure 6–39 shows a typical TTL signal showing these characteristics.

National Instruments DAQ boards use various types of counters on different DAQ boards:

- **Am9513A System Timing Controller (STC)**. This integrated circuit is found on some legacy National Instruments multifunction DAQ boards. It consists of five 16-bit counters for the timing of DAQ operations

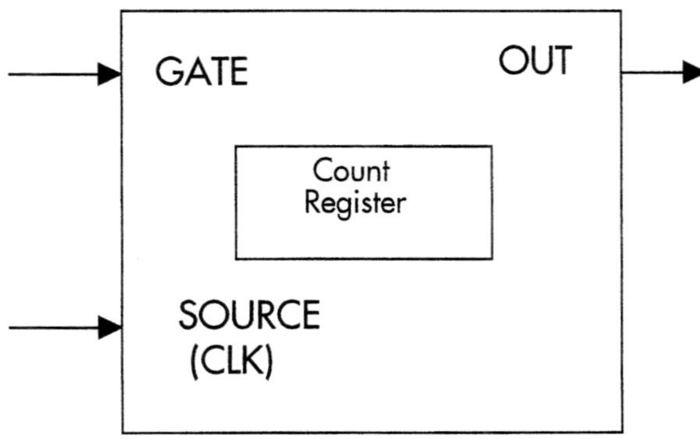

Figure 6–38
Typical Counter

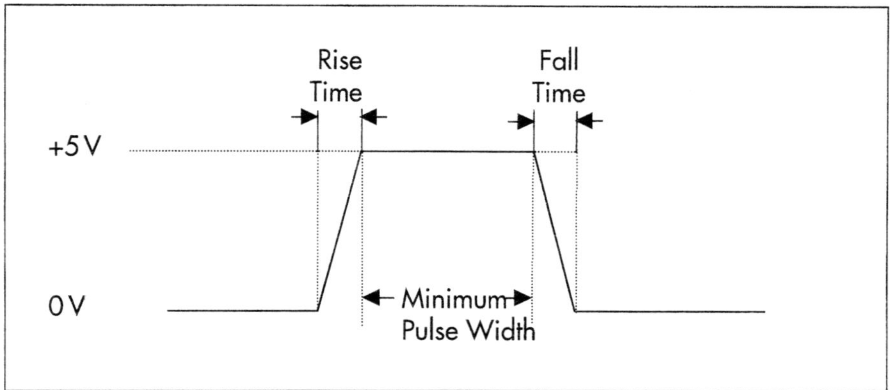

Figure 6–39
Counter Signal Characteristics

and is used in general-purpose timing and counting applications. This integrated circuit is limited by not having enough counters to perform simultaneous analog input, analog output, and frequency measurement.

- **Data Acquisition System Timing Controller (DAQ-STC)**. The DAQ-STC was developed by National Instruments to control timing on E-series MIO boards. It consists of a total of 10 counters; eight are 16-bit counters, four used for analog input timing and four for analog output timing; two are 24-bit 20-MHz counter/timers that are user-available, general-purpose up/down counters.
- **NI-TIO ASIC**. The TIO ASCI developed by National Instruments is used for high-performance counting and timing applications. It is used on 6602 and 445x family boards. The 6602 have two NI-TIO chips per board. Each chip has four 32-bit counters with available timebases of 80 MHz, 20 MHz, and 100 kHz.

Counter Applications

In this section we look at some common counter applications. The counter applications discussed are event counting and timing, pulse generation, pulse width or period measurement, and frequency measurement. To access the Counter Easy I/O DAQ library functions, select **Library>>Easy I/O for DAQ>>Counter/Timer...** .

Event Counting and Timing

To count the rising or falling edges of a TTL signal on the counter's SOURCE pin or the number of cycles of a specified internal timebase signal, the easy I/O DAQ library function *CounterEventOrTimeConfig* is used. You start the count by calling the library function *CounterStart*, stop the count by invoking *CounterStop*, and read the value in the internal Count Register of the counter using *CounterRead* library functions. The code fragment in Figure 6–40 shows how this is accomplished.

At lines 10–14 the *CounterEventOrTimeConfig* is used for configuring the counter(s), as mentioned above. The first argument of this function, `device`, is the number of the DAQ board assigned during setup. The `counterString`

```
1   short device =1;  //Device number of DAQ board
2   char counterString="1"; //Counter used
3   unsigned short CounterSize = 0; //Size of counter to perform the operation;
4   double sourceTimebase = 10000.0; //Timebase value
5   static unsigned long taskID = 0;
6   static int error;
7   long CounterValue;
8   short Overflow;
9
10  error = CounterEventOrTimeConfig (device, counterString, CounterSize,
11                                     sourceTimebase,
12                                     COUNT_CONTINUOUSLY,
13                                     COUNT_ON_RISING_EDGE,
14                                     UNGATED_SOFTWARE_START, &taskID);
15
16  if (error !=0) printf("DAQ Error String : %s\n", GetDAQErrorString(error));
17
18  //Start the counter
19  error = CounterStart(taskID);
20  .
21  .
22  .
23  .
24  //Read counter in an infinite loop until stopped by CounterStop ( )
25  error = CounterRead(taskID, &CounterValue, &Overflow);
26  if (error !=0) printf("DAQ Error String : %s\n", GetDAQErrorString(error));
27  .
28  .
29  .
30
31  //Insert this part in the code when you want to stop counting
32  error = CounterStop(taskID);
33  if (error !=0) printf("DAQ Error String : %s\n", GetDAQErrorString(error))
```

Figure 6–40
CounterEventOrTimeConfig Code Fragment

argument is assigned to the counter number used for counting operation. These counter numbers depend on the device type being used on the DAQ board. For DAQ-STC devices the valid counters are 0 and 1; for Am9513 MIO boards the valid counters are 1, 2, and 5; and for PC-TIO-10 you can use a counter from 1 through 10. The `CounterSize` argument refers to the size of the counter being used to perform the operation. For a device with DAQ-STC counters, the size of the counter is 24 bits. For a device with Am9513 counters, you can use either one 16-bit counter or two counters, for an aggregate of 32 bits. The `sourceTimebase` argument specifies whether the counter should count TTL edges at its SOURCE pin, in which case the value of this argument is `0.0`, or count the TTL edges of an internal clock. In the latter case the value for this argument is entered as a floating-point number greater than `0.0`. The fifth argument of this function (shown here as `COUNT_CONTINUOUSLY`) is the count limit action, for which the choices are:

- **COUNT_UNTIL_TC.** In this mode the counter counts to its maximum (TC) value and sets the overflow status when reached. This option is not available on DAQ-STC.
- **COUNT_CONTINUOUSLY.** In this mode the counter counts continuously and the Am9513 does not set the overflow status at a terminal count. However, the DAQ-STC does set the overflow status.

The sixth argument of this function (shown here as `COUNT_ON_RISING_EDGE`) specifies the edge of the counter source signal for which the counter is incremented. You can specify either `COUNT_ON_RISING_EDGE` or `COUNT_ON_FALLING_EDGE` for this argument.

The seventh argument (shown here as `UNGATED_SOFTWARE_START`) is the `gate mode` and specifies how the signal on the counter's GATE pin is used. You can choose from among the following options:

- **UNGATED_SOFTWARE_START.** This option is used to start counting when the *CounterStart* function is called and ignores the gate signal.
- **COUNT_WHILE_GATE_HIGH.** This option is used to count when the gate signal is TTL high after the *CounterStart* function is called.
- **COUNT_WHILE_GATE_LOW.** This option is used to count when the gate signal is TTL low and after the *CounterStart* function is called.

- **START_COUNTING_ON_RISING_EDGE.** This option is used to count on the rising edge of the TTL gate signal after the *CounterStart* function is called.
- **START_COUNTING_ON_FALLING_EDGE.** This option is used to count on the falling edge of the TTL gate signal after the *CounterStart* function is called.

The `taskID` is the last argument for this function and is the identifier number assigned to this counter operation. It is used in subsequent function calls when referring to the operation.

After you have configured the counter, you start the counting operation by calling the library function *CounterStart* with the `taskID` as its argument, as shown on line 19. The *taskID* is created by the *CounterEventOrTimeConfig* function. The *CounterStart* function is valid only for DAQ-STC and Am9513 counters.

The internal Count Register of the counter is read when the *CounterRead* function is called, as shown on line 25. In your application you may want to obtain this count continuously or only after a certain criterion has been met, such as when the counter reaches the TC value. This function takes the `taskID` as the first argument, returns the count value from the internal Count Register, and sets the overflow flag if the terminal count is exceeded. This function is valid only for DAQ-STC and Am9513 counters.

The *CounterStop* function at line 32 stops the counting operation. This function is valid only for DAQ-STC and Am9513 counters.

Pulse Generation

You can use the counter to generate either a single pulse or continuous pulses (called a *pulse train*) on the OUT pin of a counter. Before explaining the library functions used for generating pulses, some terms need to be explained.

The *duty cycle* of a pulse is defined as

$$\text{Duty Cycle} = \text{Phase 2}/(\text{Phase 1} + \text{Phase 2})$$

or equivalently,

$$\text{Duty Cycle} = \text{Width}/(\text{Delay} + \text{Width})$$

These items are shown in Figure 6–41.

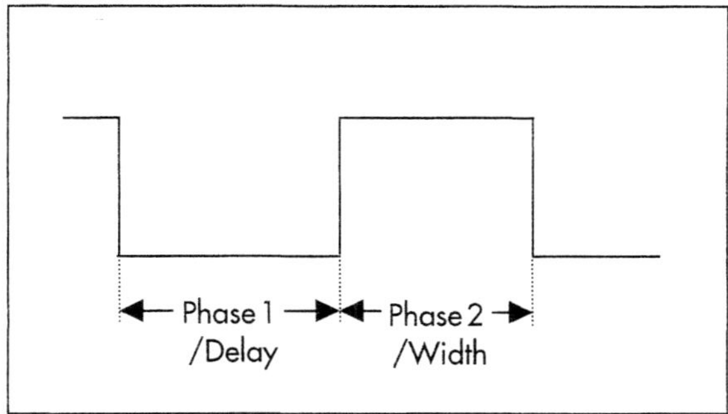

Figure 6–41
Pulse Duty Cycle

The *pulse polarity* determines whether the pulse is high or low. The pulse polarity is high/low if the phase 2 of the pulse is high/low. The high and low pulse polarities are shown in Figure 6–42.

To configure a counter to generate a continuous TTL pulse train on its OUT pin with a specified frequency, duty cycle, gate mode, and pulse polarity, use the easy I/O DAQ library function *ContinuousPulseGenConfig*. The code fragment in Figure 6–43 shows how this function may be used in your application.

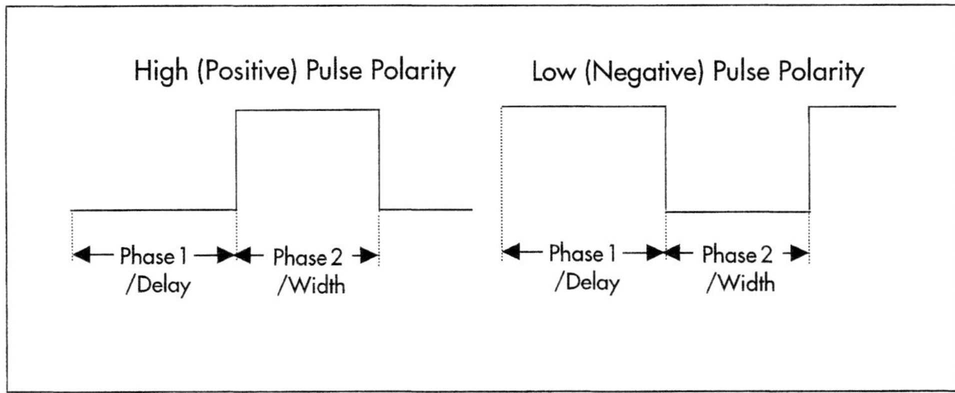

Figure 6–42
Pulse Polarity

```
short device =1;  //Device number of DAQ board
char counterString="1"; //Counter used
double frequency=100.0; //Continuous pulse train frequency
double dutyCycle=0.40;
double actualFrequency; //Achieved frequency
double actualDutyCycle; //Achieved duty cycle
static unsigned long taskID = 0;
static int error;

//Generate continuous TTL pulse train
error = ContinuousPulseGenConfig (device, counterString, frequency, dutyCycle,
                        UNGATED_SOFTWARE_START, POSITIVE_POLARITY,
                        &actualFrequency, &actualDutyCycle, &taskID);

printf("DAQ Error String : %s\n", GetDAQErrorString(error));

error = CounterStart (taskID);
printf("DAQ Error String : %s\n", GetDAQErrorString(error));
```

Figure 6–43
ContinuousPulseGenConfig Code Fragment

The first and second arguments in the *ContinuousPulseGenConfig* function are the device number and the counter string, respectively, used for this operation. These are the same as defined for the library function *CounterEvent-OrTimeConfig*. The third argument is `frequency`, which is the rate at which the pulse train is repeated. The desired duty cycle of the pulse is specified in the `dutyCycle` argument of this function. To generate a uniform square wave, use a value of `0.5` in this argument. A pictorial view of the duty cycle of a pulse was shown in Figure 6–41. The `UNGATED_SOFTWARE_START` argument is used to specify how to use the signal on the counter's GATE pin. The options for this argument are the same as shown for the *CounterEventOrTimeConfig* function in the section *Event Counting and Timing* above. The `POSITIVE_POLARITY` argument specifies the phase 2 polarity of the pulse (Figure 6–42) and can be either `POSITIVE_POLARITY` or `NEGATIVE_POLARITY`, depending on whether phase 2 is high or low. The `actual-Frequency` and `actualDutyCycle` arguments are the frequency and duty cycle values obtained from the hardware and may be different from the values specified due to hardware limitations and range. Again, `taskID` is the reference number assigned to this counter for this configuration and operation and is used in subsequent function calls.

Use the Easy I/O for DAQ library function *DelayedPulseGenConfig* to configure a counter to generate a single TTL pulse or triggered pulse train on the counter's OUT pin with specified pulse delay, pulse width, and pulse polarity. You can indicate the timebase source as internal or external to specify the signal used by the counter to count. The code fragment shown in Figure 6–44 demonstrates how this function can be used in your application.

The first and second arguments in the *DelayedPulseGenConfig* function are the device number and the counter string, respectively, used for this operation, as seen above. The third argument is `pulseDelay`, the desired value of the delay (Phase 1) (see Figure 6–41). The units for this argument are seconds if `timebaseSource` is `USE_INTERNAL_TIMEBASE` and cycles if `timebaseSource` is `USE_COUNTER_SOURCE`. The `pulseWidth` argument is the desired pulse width (Phase 2) (Figure 6–42). Again, the units for this argument are seconds if `timebaseSource` is `USE_INTERNAL_TIMEBASE` and cycles if `timebaseSource` is `USE_COUNTER_SOURCE`. The next argument (`USE_INTERNAL_TIMEBASE`) is the timebase source and defines whether an internal or external signal will be used for the delay and width duration. You have the option of using either `USE_INTERNAL_TIMEBASE` or `USE_COUNTER_SOURCE`. When you select `USE_INTERNAL_TIMEBASE`, the onboard clock is used and the delay and pulse width are timed in seconds. When `USE_COUNTER_SOURCE` is selected, you must connect the timebase signal to the counter SOURCE pin, and the delay and pulse width are timed in cycles of the external timebase signal. `UNGATED_SOFTWARE_START`

```
short device =1;  //Device number of DAQ board
char counterString="1"; //Counter used
double frequency=100.0; //Continuous pulse train frequency
double pulseDelay=1.0; //Phase 1 of pulse
double pulseWidth=1.0; //Phase 2 of pulse
unsigned short pulsePolarity; //Positive or Negative polarity
double actualDelay; //Achieved delay
double actualPulseWidth; //Achieved pulse width
static unsigned long taskID = 0;
static int error;

//Generate a delayed TTL pulse
error = DelayedPulseGenConfig (device, counterString, pulseDelay, pulseWidth,
            USE_INTERNAL_TIMEBASE, UNGATED_SOFTWARE_START,
            pulsePolarity, &actualDelay, &actualPulseWidth, &taskID);

printf("DAQ Error String : %s\n", GetDAQErrorString(error));

error = CounterStart (taskID);
printf("DAQ Error String : %s\n", GetDAQErrorString(error));
```

Figure 6–44
DelayedPulseGenConfig Code Fragment

and pulsePolarity arguments are the same as explained above for the *ContinuousPulseGenConfig* function. This function returns values for the actual delay and actual pulse width that may be different from the values specified, due to hardware limitations. The task identifier (taskID) is obtained from *CounterEventOrTimeConfig* as explained above.

Pulse Measurement

Counters can be used to measure the pulse width or the period of a pulse or pulse train. The Easy I/O for DAQ library function *PulseWidthOrPeriodMeasConfig* configures the specified counter to measure the pulse width or period of the TTL signal connected to the GATE pin of the counter. You can measure the pulse width or period by counting the number of cycles of the timebase specified. The timebase can be either internal or external. The *PulseWidthOrPeriodMeasConfig* function prototype is shown below and its arguments explained.

```
error = PulseWidthOrPeriodMeasConfig (device,
                                counterString, typeOfMeasurement,
                                          sourceTimebase, &taskID);
```

The first two arguments of this function are the same as explained previously. In the typeOfMeasurement argument, you specify the type of pulse width or period measurement you want to make. You can select from the following options:

- **MEASURE_HIGH_PULSE_WIDTH.** The high pulse width is measured from the rising to the falling edge.
- **MEASURE_LOW_PULSE_WIDTH.** The low pulse width is measured from the falling to the rising edge.
- **MEASURE_PERIOD_BTW_RISING_EDGES.** The period between adjacent rising edges is measured.
- **MEASURE_PERIOD_BTW_FALLING_EDGES.** The period between adjacent falling edges is measured.

To use an internal timebase, set the appropriate frequency of the timebase in the function argument. To use an external timebase, connect the signal with known frequency to the SOURCE pin of the counter and select

USE_COUNTER_SOURCE for the sourceTimebase argument in the *PulseWidthOrPeriodMeasConfig* function.

The taskID is the identifier for the configured counter that is used in subsequent function calls. After configuring the counter with the *PulseWidthOrPeriodMeasConfig* function, you start the measurement by calling the *CounterStart* and *CounterRead* library functions to read the counter's internal Count Register. The duration is calculated by dividing the internal Count Register value by the timebase value.

Frequency Measurement

Counters can be used to measure the frequency of an unknown pulse train connected to the SOURCE pin of the counter. This is achieved by counting the rising edges of the signal during a specified period of time. A pulse connected to the GATE pin of the counter controls the time period specified. To measure the frequency, you must connect the counter's GATE pin to the OUT pin of the second counter, referred to as "counter-1" or to an external source of known pulse width. The *CounterMeasureFrequency* function configures one counter, counter-1, to generate a pulse to the GATE of a second counter, the event counter. The event counter counts the transition events on the source signal and calculates the frequency of the source signal by dividing the events counted by the pulse width of the pulse counter-1 sent to the GATE of the event counter.

For example, if you have the pulse train of unknown frequency connected to the SOURCE pin of counter-1, the OUT pin of counter-0 must be connected to the GATE pin of counter-1. To calculate the pulse train frequency, divide the pulse count by the gating period of the known pulse connected to the GATE pin of counter-1.

A list of counters and counter-1 is shown in the *CounterMeasureFrequency* function panel for both Am9513 and DAQ-STC counters. The *CounterMeasureFrequency* function prototype is shown below and its arguments explained.

```
error = CounterMeasureFrequency (device, counterString,
        counterSize, gateWidthSampleTime,
        MaxDelayBeforeGate, Counter1GateMode,
        &actualGateWidth, &Overflow, &Valid, &timeout,
                                             &frequency);
```

The `device` and `counterString` are the board number and counter string, respectively, that receive the pulse train. `counterSize` for the DAQ-STC counter is 24 bits, and for the Am9513 counter could be 16 bits if one counter is used or 32 bits if two counters are used. `gateWidthSampleTime` is the width in seconds of the pulse used to gate the signal. `MaxDelayBeforeGate` is the maximum expected delay between the time the function is called and the start of the gating pulse. A timeout error occurs if the gate signal does not start within this time. `Counter1GateMode` is the gate mode for counter-1. The options for the *gate mode* for this function are the same as those for *CounterEventOrTimeConfig* shown above.

`actualGateWidth` is the length in seconds of the gating pulse that is returned from the hardware. It may differ from the `gateWidthSampleTime` value specified. An `Overflow` value of 1 indicates counter overflow. A 0 value indicates normal operation. A `Valid` argument value of 1 indicates that there was no counter overflow and the operation was completed successfully. If the time limit expires during the function call, the `timeout` value is set to 1. It is possible for a timeout and a valid measurement to occur at the same time. A timeout does not produce a function error. The last argument returns the value of `frequency` of the pulse train connected to the SOURCE pin of the counter.

Summary

This chapter introduced you to some of the basic hardware and software features of data acquisition. You saw the basic MIO DAQ board hardware architecture and how to select a DAQ board appropriate for your application, install it, configure it, and acquire data using the various components of the DAQ board. You were introduced to some of the easy I/O DAQ library functions and their use, demonstrated by means of code fragments. Data acquisition is a vast topic, but here we only touched on the basics to get you started. As you start to write sophisticated data acquisition applications, you can look into the sample programs that are included in the `samples` folder included with *CVI* and experiment with some of the more elaborate DAQ library functions.

Library Function Prototypes and Definitions

This section lists alphabetically the *CVI* library functions that were introduced in this chapter.

AOClearWaveforms Function

The *AOClearWaveforms* function clears the waveforms generated by *AOGenerateWaveforms* when you pass `0` for the `iterations` argument in *AOGenerateWaveforms*. The prototype for *AOClearWaveforms* is shown below and its arguments explained in Table 6–1.

```
short status = AOClearWaveforms (unsigned long taskID);
```

AOUpdateChannel Function

The *AOUpdateChannel* function applies the specified voltage to a single analog output channel. Its prototype is shown below and its arguments explained in Table 6–2.

```
short status = AOUpdateChannel (short device,
                          char singleChannel[], double value);
```

Table 6–1 AOClearWaveforms *Function*

Input/ Output	Name	Type	Description
Input	*taskID*	unsigned long	task identifier returned from *AOGenerateWaveforms* function
Output	*status*	short	zero indicates success; negative number indicates error; positive number means operation executed with possible side effects; refer to NI-DAQ On-line Help topic *Status Codes Summary* for a listing of error codes

Table 6–2 AOUpdateChannel *Function*

Input/ Output	Name	Type	Description
Input	*device*	short	logical device number assigned during configuration
	singleChannel	char[ ]	string representing the single analog output channel to which voltage is applied
	value	double	voltage applied to the analog output channel
Output	*status*	short	zero indicates success; negative number indicates error; positive number means operation executed with possible side effects; refer to NI-DAQ On-line Help topic *Status Codes Summary* for a listing of error codes

AOUpdateChannels Function

The *AOUpdateChannels* function applies the specified voltage to analog output channels specified in the channel string. Its prototype is shown below and its arguments explained in Table 6–3.

```
short status = AOUpdateChannels (short device,
                                 char  channelString[],
                                 double valueArray[]);
```

AOGenerateWaveforms Function

The *AOGenerateWaveforms* function generates a timed waveform of the voltage data on the analog output channels specified. Its prototype is shown below and its arguments explained in Table 6–4.

```
short status = AOGenerateWaveforms (short device,
                  char  channelString[],
                  double updatesSecond,
                  int updatesChannel, int iterations,
                  double waveforms, unsigned long *taskID);
```

Table 6–3 AOUpdateChannels *Function*

Input/ Output	Name	Type	Description
Input	*device*	short	logical device number assigned during configuration
	channelString	char[ ]	string representing the analog output channels to which voltages are applied from the *valueArray* argument
	valueArray	double[ ]	voltages applied to the analog output channels in the order specified in the array
Output	*status*	short	zero indicates success; negative number indicates error; positive number means operation executed with possible side effects; refer to NI-DAQ On-line Help topic *Status Codes Summary* for a listing of error codes

Table 6–4 AOGenerateWaveforms *Function*

Input/ Output	Name	Type	Description
Input	*device*	short	logical device number assigned during configuration
	channelString	char[ ]	string representing the analog output channels receiving the output values from the *waveforms* argument
	updatesSecond	double	number of updates per second
	updatesChannel	integer	number of updates that compose a waveform for a particular channel
	iterations	integer	number of waveform iterations to perform; pass 0 to generate waveform continuously
	waveforms	double[ ]	values to be applied to channels in *channelString;* for an explanation, see discussion in the section *Analog Output*

(continued)

Table 6–4 AOGenerateWaveforms *Function (continued)*

Input/ Output	Name	Type	Description
Output	*taskID*	unsigned long*	identifier for waveform generation
	status	short	zero indicates success; negative number indicates error; positive number means operation executed with possible side effects; refer to NI-DAQ On-line Help topic *Status Codes Summary* for a listing of error codes

ContinuousPulseGenConfig Function

The *ContinuousPulseGenConfig* function configures a counter to generate a continuous TTL pulse train on its OUT pin. Its prototype is shown below and its arguments explained in Table 6–5.

```
short status =  ContinuousPulseGenConfig (short device,
                  char  counter[],double frequency,
                  double dutyCycle, unsigned short gateMode,
                  unsigned short pulsePolarity,
                  double *actualFrequency,
                  double *actualDutyCycle,
                  unsigned long *taskID);
```

CounterEventOrTimeConfig Function

The *CounterEventOrTimeConfig* function configures one or two counters to count edges in the signal on the counter SOURCE pin specified or the number of cycles of an internal timebase signal. Its prototype is shown below and its arguments explained in Table 6–6.

```
short status = CounterEventOrTimeConfig (short device,
                  char  counter[],unsigned short counterSize,
                  double sourceTimebase,
                  unsigned short countLimitAction,
                  short sourceEdge, unsigned short gateMode,
                                      unsigned long *taskID);
```

Table 6–5 ContinuousPulseGenConfig *Function*

Input/ Output	Name	Type	Description
Input	*device*	short	logical device number assigned during configuration
	counter	char[]	counter used for this operation
	frequency	double	frequency of continuous pulse train
	dutyCycle	double	pulse duty cycle; duty cycle of 0.5 generates a square wave
	gateMode	unsigned short	how signal on counter's GATE pin is used; for an explanation, see description in the section *Event Counting and Timing*
	pulsePolarity	unsigned short	polarity of Phase 2 of pulse; for Phase 2 high, use `POSITIVE_POLARITY`; for Phase 2 low, use `NEGATIVE_POLARITY`
Output	*actualFrequency*	double*	actual frequency of pulse train
	actualDutyCycle	double*	actual duty cycle of pulse
	taskID	unsigned long*	identifier for waveform generation
	status	short	zero indicates success; negative number indicates error; positive number means operation executed with possible side effects; refer to NI-DAQ On-line Help topic *Status Codes Summary* for a listing of error codes

Table 6–6 CounterEventOrTimeConfig *Function*

Input/ Output	Name	Type	Description
Input	*device*	short	logical device number assigned during configuration
	counter	char[ ]	counter used for this operation

(continued)

Table 6–6 CounterEventOrTimeConfig *Function (continued)*

Input/ Output	Name	Type	Description
	counterSize	unsigned short	determines the size of counter used to perform operation; for a discussion of counter size, see the section *Event Counting and Timing*
	sourceTimebase	double	to count TTL edges at its SOURCE pin, enter 0.0; to count TTL edges of an internal clock, enter the timebase for the internal clock
	countLimitAction	unsigned short	action to take when counter reaches terminal count; options are: ■ COUNT_UNTIL_TC ■ COUNT_CONTINUOUSLY
	sourceEdge	short	edge of the counter source or timebase signal on which to increment count; options are: ■ COUNT_ON_RISING_EDGE ■ COUNT_ON_FALLING_EDGE
	gateMode	unsigned short	specify how signal on counter's GATE pin is used; for an explanation, see description in the section *Event Counting and Timing*
Output	*taskID*	unsigned long*	identifier for waveform generation
	status	short	zero indicates success; negative number indicates error; positive number means operation executed with possible side effects; refer to NI-DAQ On-line Help topic *Status Codes Summary* for a listing of error codes

CounterMeasureFrequency Function

The *CounterMeasureFrequency* function measures the frequency of a TTL signal on the counter SOURCE pin specified by counting the rising edges of the signal during a certain time. Its prototype is shown below and its arguments explained in Table 6–7.

```
int short = CounterMeasureFrequency(short device,
            unsigned short counterSize,
            double gateWidth_SampleTime_inSec,
            double maxDelayBeforeGate_Sec,
            unsigned short counter1GateMode,
            double *actualGateWidth_Sec,
            short *overflow, short *valid,
            short *timeout,
            double *frequency);
```

CounterRead Function

The *CounterRead* function reads the counter identified by the task identifier. This function is used for DAQ-STC and Am9513 counters only. Its prototype is shown below and its arguments explained in Table 6–8.

```
short status = CounterRead (unsigned long taskID);
```

CounterStart Function

The *CounterStart* function starts the counter identified by the task identifier. This function is used for DAQ-STC and Am9513 counters only. Its prototype is shown below and its arguments explained in Table 6–9.

```
short status = CounterStart (unsigned long taskID);
```

CounterStop Function

The *CounterStop* function stops the counter identified by the task identifier. This function is used for DAQ-STC and Am9513 counters only. Its prototype is shown below and its arguments explained in Table 6–10.

```
short status = CounterStop (unsigned long taskID);
```

Table 6–7 CounterMeasureFrequency *Function*

Input/ Output	Name	Type	Description
Input	*device*	short	logical device number assigned during configuration
	counterSize	unsigned short	determines the size of counter used to perform operations; for discussion on counter size, see the section *Event Counting and Timing*
	gateWidth_Sample Time_inSec	double	length of the pulse in seconds to gate the signal
	maxDelayBeforeGate_ Sec	double	maximum expected delay between the time the function is called and the start of the gating pulse; timeout occurs if the gate signal does not start in this time
	counter1GateMode	unsigned short	gate mode for counter-1; for a description of the options for this argument, see the section *Frequency Measurement*
Output	*actualGateWidth_Sec*	double*	length of pulse in seconds of the gating pulse returned from hardware
	overflow	short*	if counter overflows during counting, a `1` is returned, otherwise `0` is returned
	valid	short*	returns a value of `1` if the measurement completes without overflow
	timeout	short*	returns a value of `1` if the time limit expires during function call; this value does not produce an error
	frequency	double*	frequency value returned
	status	short	zero indicates success; negative number indicates error; positive number means operation executed with possible side effects; refer to NI-DAQ On-line Help topic *Status Codes Summary* for a listing of error codes

Table 6–8 CounterRead *Function*

Input/ Output	Name	Type	Description
Input	*taskID*	unsigned long	counter task identifier
Output	*status*	short	zero indicates success; negative number indicates error; positive number means operation executed with possible side effects; refer to NI-DAQ On-line Help topic *Status Codes Summary* for a listing of error codes

Table 6–9 CounterStart *Function*

Input/ Output	Name	Type	Description
Input	*taskID*	unsigned long	counter task identifier
Output	*status*	short	zero indicates success; negative number indicates error; positive number means operation executed with possible side effects; refer to NI-DAQ On-line Help topic *Status Codes Summary* for a listing of error codes

Table 6–10 CounterStop *Function*

Input/ Output	Name	Type	Description
Input	*taskID*	unsigned long	counter task identifier
Output	*status*	short	zero indicates success; negative number indicates error; positive number means operation executed with possible side effects; refer to NI-DAQ On-line Help topic *Status Codes Summary* for a listing of error codes

DelayedPulseGenConfig Function

The *DelayedPulseGenConfig* function configures a counter to generate a delayed TTL pulse train or triggered pulse train on its OUT pin. Its prototype is shown below and its arguments explained in Table 6–11.

```
short status = DelayedPulseGenConfig (short device,
                    char  counter[],
                    double pulseDelay_s_or_cycles,
                    double pulseWidth_s_or_cycles,
                    unsigned short timebaseSource,
                    unsigned short gateMode,
                    unsigned short pulsePolarity,
                    double *actualDelay,
                    double *actualPulseWidth,
                    unsigned long *taskID);
```

DIG_Line_Config Function

The *DIG_Line_Config* function configures the specified line for input or output. This function is supported by DIO-32HS, PC-TIO-10, and MIO E-series boards only. Its prototype is shown below and its arguments explained in Table 6–12.

```
short status =  DIG_Prt_Config (short Board, short Line,
                                             short Direction);
```

DIG_Prt_Config Function

The *DIG_Prt_Config* function configures the specified port for input, output, or bidirectional. This function is used on all boards that support digital I/O. Its prototype is shown below and its arguments explained in Table 6–13.

```
short status =  DIG_Prt_Config (short Board, short Port,
                            short Handshaking, short Direction);
```

Table 6–11 DelayedPulseGenConfig *Function*

Input/ Output	Name	Type	Description
Input	*device*	short	logical device number assigned during configuration
	counter	char[ ]	counter used for this operation
	pulseDelay_s_or_cycles	double	desired duration of delay (Phase 1) before start of pulse; if *timebaseSource* is `USE_INTERNAL_TIMEBASE`, the unit is in seconds; if *timebaseSource* is `USE_COUNTER_SOURCE`, the unit is in cycles
	pulseWidth_s_or_cycles	*double*	desired duration of pulse (Phase 2) after the delay; if *timebaseSource* is `USE_INTERNAL_TIMEBASE`, the unit is in seconds; if *timebaseSource* is `USE_COUNTER_SOURCE`, the unit is in cycles
	timebaseSource	unsigned short	whether an internal or external signal will be used for the delay and width duration; options are using either `USE_INTERNAL_TIMEBASE` or `USE_COUNTER_SOURCE`; see the section *Pulse Generation*
	gateMode	unsigned short	how signal on counter GATE pin is used; for an explanation, see the section *Event Counting and Timing*
	pulsePolarity	unsigned short	polarity of Phase 2 of pulse; for Phase 2 high, use `POSITIVE_POLARITY`; for Phase 2 low, use `NEGATIVE_POLARITY`
Output	*actualDelay*	double*	actual delay of pulse train
	actualPulseWidth	double*	actual pulse width

(continued)

Table 6–11 DelayedPulseGenConfig *Function (continued)*

Input/ Output	Name	Type	Description
	taskID	unsigned long*	identifier for waveform generation
	status	short	zero indicates success; negative number indicates error; positive number means operation executed with possible side effects; refer to NI-DAQ On-line Help topic *Status Codes Summary* for a listing of error codes

Table 6–12 DIG_Line_Config *Function*

Input/ Output	Name	Type	Description
Input	*Board*	short	logical device number assigned during configuration
	Line	short	specified *Line* to configure
	Direction	short	I/O direction of the port; options are: 0 input 1 output
Output	*status*	short	zero indicates success; negative number indicates error; positive number means operation executed with possible side effects; refer to NI-DAQ On-line Help topic *Status Codes Summary* for a listing of error codes

GroupByChannel Function

The *GroupByChannel* function reorders an array of data from group by scan mode into grouped by channel mode. Its prototype is shown below and its arguments explained in Table 6–14.

```
long status = GroupByChannel (double Array [],
                    long number_ofScans,
                    unsigned long Number_of_Channels);
```

Table 6–13 DIG_Prt_Config *Function*

Input/ Output	Name	Type	Description
Input	*Board*	short	logical device number assigned during configuration
	Port	short	port number to configure:
	Handshaking	short	0 disable handshaking; 1 enable handshaking handshaking is used only for the following boards: ■ DIO-24 ■ DIO-96 ■ Lab and 1200 series boards ■ AT-MIO-16DE-10 ■ AT-MIO-16D use 0 for all other boards
	Direction	short	I/O direction of the port; options are: 0 input 1 output 2 bidirectional see the function panel for a description of boards and ports that can be bidirectional
Output	*status*	short	zero indicates success; negative number indicates error; positive number means operation executed with possible side effects; refer to NI-DAQ On-line Help topic *Status Codes Summary* for a listing of error codes

Table 6–14 GroupByChannel *Function*

Input/ Output	Name	Type	Description
Input	*Array*	double[ ]	group by scan array is converted to group by channel in place
	number_ofScans	long	number of scans to acquire
	Number_of_ Channels	unsigned long	number of channels that were scanned
Output	*status*	long	zero indicates success; negative number indicates error; positive number means operation executed with possible side effects; refer to NI-DAQ On-line Help topic *Status Codes Summary* for a listing of error codes

nidaqAICreateTask Function

The *nidaqAICreateTask* function creates an analog input task. Its prototype is shown below and its arguments explained in Table 6–15.

```
short status = nidaqAICreateTask (char channelPath[],
                        long taskType, long *number_ofChannels,
                                                   long taskID);
```

nidaqAIDestroyTask Function

The *nidaqAIDestroyTask* function destroys the task, making the task identifier number invalid for use in other function calls. Its prototype is shown below and its arguments explained in Table 6–16.

```
long status = nidaqAIDestroyTask (long taskID);
```

Table 6–15 nidaqAICreateTask *Function*

Input/ Output	Name	Type	Description
Input	*channelPath*	char[ ]	channels from which you want to acquire data

	taskType	long	options are:
			`kNidaqPointByPoint`
			`kNidaqWaveformCapture`
			for an explanation, see the section *Analog Input* in *DAQ Library Functions*

Output	*number_of Channels*	long*	number of channels in `channelPath` argument
	status	short	zero indicates success; negative number indicates error; positive number means operation executed with possible side effects; refer to NI-DAQ On-line Help topic *Status Codes Summary* for a listing of error codes

Table 6–16 nidaqAIDestroyTask *Function*

Input/ Output	Name	Type	Description
Input	*taskID*	long	task identifier for task to be destroyed
Output	*status*	long	zero indicates success; negative number indicates error; positive number means operation executed with possible side effects; refer to NI-DAQ On-line Help topic *Status Codes Summary* for a listing of error codes

nidaqAIRead Function

The *nidaqAIRead* function retrieves the analog input data from the analog input storage buffer. Its prototype is shown below and its arguments explained in Table 6–17.

```
long status = nidaqAIRead (long taskIDIn,
                           char channelPath[],
                           long number_ofScans,
                           double timeLimit_sec,
                                 double data []);
```

nidaqAIScanOp Function

The *nidaqAIScanOp* function acquires a finite number of scans from analog input channels. Its prototype is shown below and its arguments explained in Table 6–18.

```
long status = nidaqAIScanOp (char channelPath[],
                     long number_ofScans,
                     double scanRate, double highLimit,
                     double lowLimit, long fillMode,
                     double *actualScanRate, double data[]);
```

Table 6–17 nidaqAIRead *Function*

Input/ Output	Name	Type	Description
Input	*taskIDIn*	long	task identifier created in *nidaqAICreateTask*
	channelPath	char[ ]	channels from which to acquire data
	number_ofScans	long	number of scans to acquire
	timeLimit_sec	double	maximum time to wait for the data to read
Output	*data*	double	waveform data from each channel in *channelPath*
	status	long	zero indicates success; negative number indicates error; positive number means operation executed with possible side effects; refer to NI-DAQ On-line Help topic *Status Codes Summary* for a listing of error codes

Table 6–18 nidaqAIScanOp *Function*

Input/ Output	Name	Type	Description
Input	*channelPath*	char[ ]	channels from which you want to acquire data
	number_ofScans	long	number of scans to acquire
	scanRate	double	number of scans per second to acquire data
	highLimit	double	upper limit of input signal range
	lowLimit	double	lower limit of input signal range
	fillMode	long	specify how to group data array; options are: ■ `kNidaqGroupByChannel`—group by channels ■ `kNidaqGroupByScan`—group by scans for an explanation, see the section *Analog Input*
Output	*actualScanRate*	double*	actual scan rate used in acquisition
	data	double[ ]	retrieves waveform data from each channel specified in *channelPath*
	status	long	zero indicates success; negative number indicates error; positive number means operation executed with possible side effects; refer to NI-DAQ On-line Help topic *Status Codes Summary* for a listing of error codes

nidaqAISinglePointOp Function

The *nidaqAISinglePointOp* function acquires a single value from an analog input channel. Its prototype is shown below and its arguments explained in Table 6–19.

```
long status = nidaqAISinglePointOp (char channelPath[],
                        double highLimit,
                           double lowLimit, double *data);
```

Table 6–19 nidaqAISinglePointOp *Function*

Input/ Output	Name	Type	Description
Input	*channelPath*	char[ ]	channels from which you want to acquire data
	highLimit	double	sets the upper limit of input signal range
	lowLimit	double	sets the lower limit of input signal range
Output	*data*	double*	retrieves a single data value from a single channel specified in *channelPath*
	status	long	zero indicates success; negative number indicates error; positive number means operation executed with possible side effects; refer to NI-DAQ On-line Help topic *Status Codes Summary* for a listing of error codes

nidaqAISingleScanOp Function

The *nidaqAISingleScanOp* function retrieves a single scan of values from analog input channels. Its prototype is shown below and its arguments explained in Table 6–20.

```
long status = nidaqAISingleScanOp (char channelPath[],
                        double highLimit,
                        double lowLimit, double data[]);
```

nidaqAIStart Function

The *nidaqAIStart* function starts data acquisition on the analog input task identifier specified. Its prototype is shown below and its arguments explained in Table 6–21.

```
long status = nidaqAIStart (long taskIDIn);
```

Table 6–20 nidaqAISingleScanOp *Function*

Input/ Output	Name	Type	Description
Input	*channelPath*	char[ ]	channels from which you want to acquire data
	highLimit	double	sets the upper limit of input signal range
	lowLimit	double	sets the lower limit of input signal range
Output	*data*	double[ ]	retrieves a single data value from each channel specified in *channelPath*
	status	long	zero indicates success; negative number indicates error; positive number means operation executed with possible side effects; refer to NI-DAQ On-line Help topic *Status Codes Summary* for a listing of error codes

Table 6–21 nidaqAIStart *Function*

Input/ Output	Name	Type	Description
Input	*taskIDIn*	long	task identifier created in *nidaqAICreateTask*
Output	*status*	long	zero indicates success; negative number indicates error; positive number means operation executed with possible side effects; refer to NI-DAQ On-line Help topic *Status Codes Summary* for a listing of error codes

nidaqAIStop Function

The *nidaqAIStop* function stops data acquisition on the analog input task identifier specified. Its prototype is shown below and its arguments explained in Table 6–22.

```
long status = nidaqAIStop (long taskIDIn);
```

Table 6–22 nidaqAIStop *Function*

Input/ Output	Name	Type	Description
Input	*taskIDIn*	long	task identifier created in *nidaqAICreateTask*
Output	*status*	long	zero indicates success; negative number indicates error; positive number means operation executed with possible side effects; refer to NI-DAQ On-line Help topic *Status Codes Summary* for a listing of error codes

nidaqGetErrorString Function

The *nidaqGetErrorString* function returns the string with a description of the error code. Its prototype is shown below and its arguments explained in Table 6–23.

```
char *errorString = nidaqAIStop (long taskIDIn,
                                  long errorNumber);
```

PulseWidthOrPeriodMeasConfig Function

The *PulseWidthOrPeriodMeasConfig* function configures the counter specified to measure the pulse width or period of a TTL signal connected to its GATE pin. This function is used with DAQ-STC and Am9513 counters only. Its prototype is shown below and its arguments explained in Table 6–24.

```
short status = PulseWidthOrPeriodMeasConfig (short device,
                char counter[],
                unsigned short typeOfMeasurement,
                double sourceTimebase,
                unsigned long *taskID);
```

Table 6–23 nidaqGetErrorString *Function*

Input/ Output	Name	Type	Description
Input	*taskIDIn*	long	task identifier created in *nidaqAICreateTask*
	errorNumber	long	error code to obtain description
Output	*errorString*	char	string containing the description of error code

Table 6–24 PulseWidthOrPeriodMeasConfig *Function*

Input/ Output	Name	Type	Description
Input	*device*	short	logical device number assigned during configuration
	counter	char[ ]	counter used for this operation
	typeOfMeasurement	unsigned short	type of pulse width or period measurement to make; for various options, for a description, see the section *Pulse Measurement*
	sourceTimebase	double	to count TTL edges at its SOURCE pin, enter `USE_COUNTER_SOURCE`; to count TTL edges of an internal clock, enter the timebase for the internal clock
Output	*taskID*	unsigned long*	identifier for the configured counter to use in subsequent function calls
	status	short	zero indicates success; negative number indicates error; positive number means operation executed with possible side effects; refer to NI-DAQ On-line Help topic *Status Codes Summary* for a listing of error codes

ReadFromDigitalLine Function

The *ReadFromDigitalLine* function reads the logical state of digital line on a port that is configured as input. Its prototype is shown below and its arguments explained in Table 6–25.

```
short status = ReadFromDigitalLine (short device,
                char portNumber[], short line,
                short portWidth, long configure,
                unsigned long *lineState);
```

Table 6–25 ReadFromDigitalLine *Function*

Input/ Output	Name	Type	Description
Input	*device*	short	logical device number assigned during configuration
	portNumber	char[ ]	digital port number
	line	short	individual bit or line to be used within the port for I/O
	portWidth	short	total width in bits of the port; for an explanation, see the section *Digital Input/Output*
	configure	long	specifies when to configure port before reading; options are: 0 only when necessary, if device, port number, port width has changed since last call 1 always configure before every read
Output	*lineState*	unsigned long*	state of digital line; returned values are: 1 logical high 0 logical low
	status	short	zero indicates success; negative number indicates error; positive number means operation executed with possible side effects; refer to NI-DAQ On-line Help topic *Status Codes Summary* for a listing of error codes

ReadFromDigitalPort Function

The *ReadFromDigitalPort* function reads the digital port that is configured as input. Its prototype is shown below and its arguments explained in Table 6–26.

```
short status = ReadFromDigitalPort (short device,
               char portNumber[], short portWidth,
               long configure,
               unsigned long *pattern);
```

Table 6–26 ReadFromDigitalPort *Function*

Input/ Output	Name	Type	Description
Input	*device*	short	logical device number assigned during configuration
	portNumber	char[ ]	digital port number
	portWidth	short	total width in bits of the port; for an explanation, see the section *Digital Input/Output*
	configure	long	when to configure port; options are: 0 only when necessary, if device, port number, port width has changed since last call 1 always configure before every read
Output	*pattern*	unsigned long*	data read from the specified digital port
	status	short	zero indicates success; negative number indicates error; positive number means operation executed with possible side effects; refer to NI-DAQ On-line Help topic *Status Codes Summary* for a listing of error codes

WriteToDigitalLine Function

The *WriteToDigitalLine* function writes the logical state of a digital line on a port that is configured as output. Its prototype is shown below and its arguments explained in Table 6–27.

```
short status = WriteToDigitalLine (short device,
                   char portNumber[], short line,
                   short portWidth, long configure,
                          unsigned long lineState);
```

Table 6–27 WriteToDigitalLine *Function*

Input/ Output	Name	Type	Description
Input	*device*	short	logical device number assigned during configuration
	portNumber	char[ ]	digital port number
	line	short	individual bit or line to be used within the port for I/O
	portWidth	short	total width in bits of the port; for an explanation, see the section *Digital Input/Output*
	configure	long	when to configure port; options are: 0 only when necessary, if device, port number, port width has changed since last call 1 always configure before every write
Output	*lineState*	unsigned long	state of digital line; options are: 1 logical high 0 logical low
	status	short	zero indicates success; negative number indicates error; positive number means operation executed with possible side effects; refer to NI-DAQ On-line Help topic *Status Codes Summary* for a listing of error codes

WriteToDigitalPort Function

The *WriteToDigitalPort* function writes a digital pattern to the digital port that is configured as output. Its prototype is shown below and its arguments explained in Table 6–28.

```
short status = WriteToDigitalPort (short device,
                char portNumber[], short portWidth,
                long configure, unsigned long pattern);
```

Table 6–28 WriteToDigitalPort *Function*

Input/ Output	Name	Type	Description
Input	*device*	short	logical device number assigned during configuration
	portNumber	char[ ]	digital port number
	portWidth	short	total width in bits of the port; for an explanation, see the section *Digital Input/Output*
	configure	long	when to configure; options are: 0 only when necessary, if device, port number, port width has changed since last call 1 always configure before every read
Output	*pattern*	unsigned long	data to write to digital port
	status	short	zero indicates success; negative number indicates error; positive number means operation executed with possible side effects; refer to NI-DAQ On-line Help topic *Status Codes Summary* for a listing of error codes

7

CREATING AND USING FUNCTION PANELS

Chapter Highlights

- Purpose of a Function Panel
- Creating a Function Tree
- Creating a Function Panel
- Testing the Function Panel Functions
- Function Panel Controls
- Summary

This chapter will show you how to create a function tree and include classes and functions to the function tree. You will see how to create a function panel using various types of function panel controls and adding help text to the function panel and the controls. In this chapter you are stepped through creating a function tree with two function panels using algebraic equations. These functions are tested using the Interactive Execution Window (IW). You are introduced to the various types of controls that you can use on function panels and shown how to interchange control types with ease.

Purpose of a Function Panel

A function panel is a graphical interface to the functions of the *CVI* libraries, instrument driver functions, or user-defined functions that allow you conveniently to select the function arguments on the function panel and include the function into your source code. Using the function panel, you can view and test these functions before inserting them in your code. The function panel template consists of function arguments with a description of their functionality and their use. The template consists of the correct number and order of the arguments that are required for the function. As you enter the values in the function panel controls, the bottom section of the function panel (called the *code box generated*) is automatically filled in by *CVI* with the values and selections that you enter. If you need help with the definitions and properties of the function or a certain argument of the function, just click with the right mouse button on the function panel or in the argument control box to see their purpose and use. A sample function panel is shown in Figure 7–1.

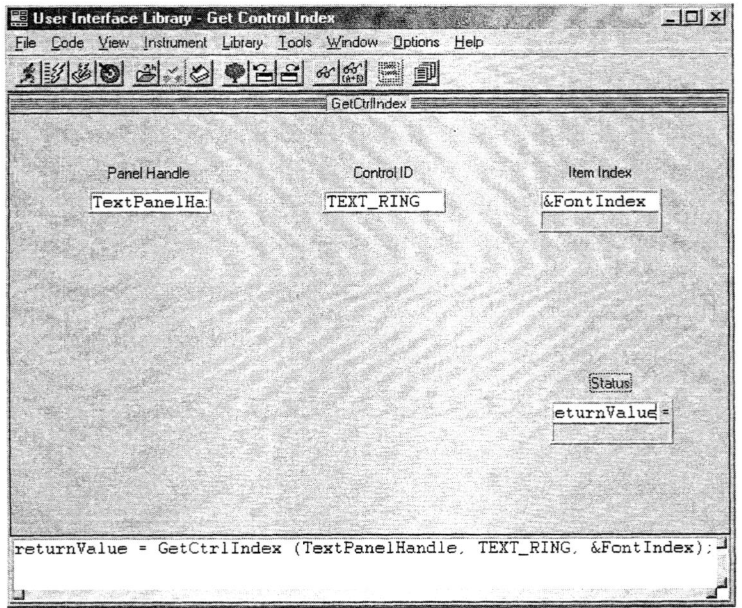

Figure 7–1
Sample Function Panel

After you have entered the data in the function panel template, you can insert the function with the specified arguments into your source code at the location of the cursor by selecting **Code>>Insert Function Call** (shortcut is <Ctrl –I>) from the function panel. The same can be accomplished by clicking on the **Insert Function Call** icon on the function panel toolbar.

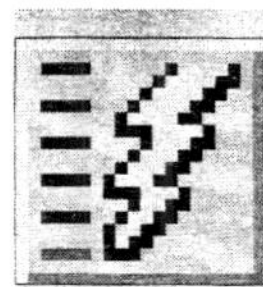

*The **Insert Function Call** icon is similar to selecting **Code>> Insert Function Call** from the function panel menu.*

If you want to test the function before inserting it in your code, you can run and test it by executing the **Interactive Execution Window** by selecting the **Code>>Run Function Panel** from the function panel. Click on the **Run Function Panel** icon on the function panel toolbar.

*The **Run Function Panel** icon is similar to selecting **Code>>Run Function Panel** from the function panel menu.*

You can access the library function panel or the instrument driver by selecting the **Library** menu or the **Instrument** menu, respectively. You can choose a function panel by selecting from the **Select Function Panel**, a sample of which is shown in Figure 7–2.

The functions are grouped as a hierarchy in a multilevel structure called a *function tree*. In a function tree the functions are arranged into various *classes,* by grouping the functions that perform similar kinds of operations together. For example, all the functions that perform the operations related to panels are in one class, pop-up panel functions in another, menu control operations in another, and so on. When you select the class from the **Select Function Panel**, you can traverse through the list of classes until you find the function panel you need. The **Select Function Panel** gives you a choice to view the functions in a different format. You can check the **Flatten** box on the bottom of the **Select Function Panel** (Figure 7–2) to see a list of all function panels in the library selected. You can also view the function panels in alphabetical order by checking the **Alphabetize** box in Figure 7–2.

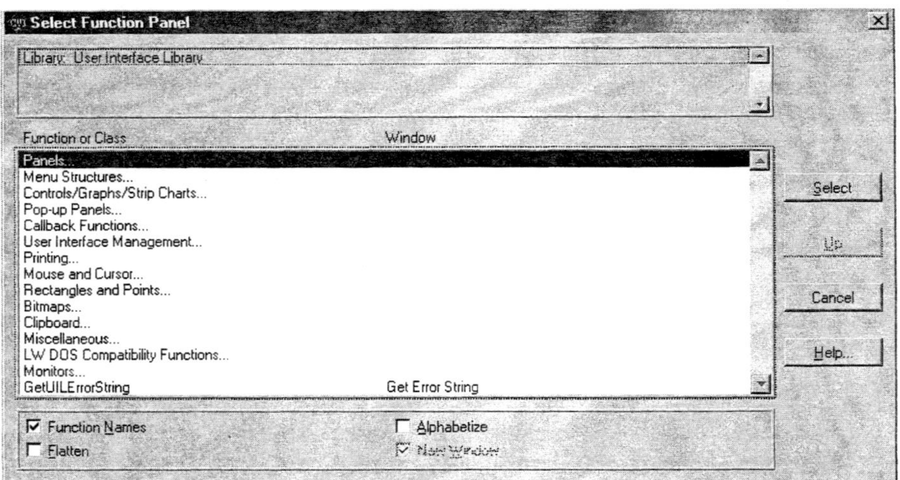

Figure 7–2
Sample **Select Function Panel**

To view the function panel from your source code, move the cursor over the function and select **View>>Recall Function Panel** or (<Ctrl-P>). The function panel containing the arguments specified will be displayed. You can change the argument in the function panel and reinsert it into the source code if you desire. You can always enter the function by typing it directly into the source code.

Creating a Function Tree

In this section you will learn to create a function tree and add your own functions to this tree. From the function tree you will be able to access the function panels for these functions, add the function arguments, and insert the function in your source code. These functions will form part of a static library that you will be able to load into your project(s) and use in your code.

Let us begin by creating a new function tree. Open a new *CVI* project and name it `MyFunctions.prj`. From the **File** menu in the **Project** window, select **New>>Function Tree (*.fp)**. A blank **Function Tree Editor** window is displayed, as shown in Figure 7–3.

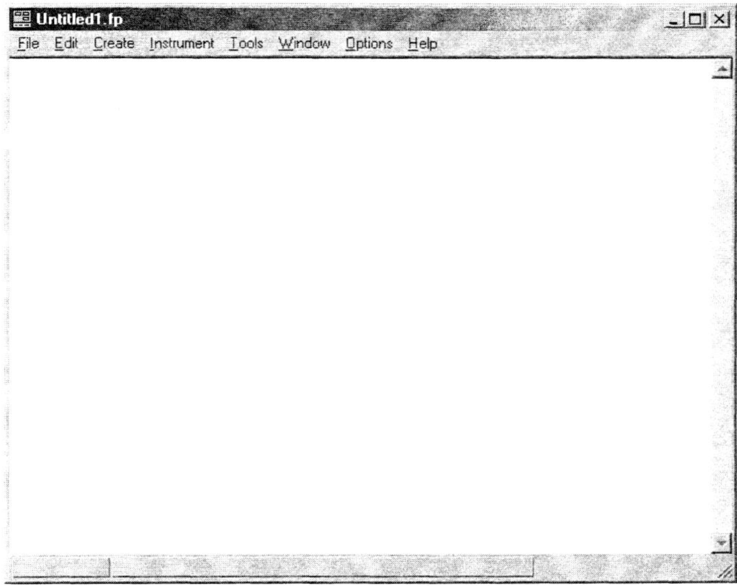

Figure 7–3
Blank **Function Tree Editor** Window

Select **Create>> Instrument** from the **Function Tree Editor** to display Figure 7–4. In the **Name:** box, enter the name you would use for the function tree (up to 40 characters long). In this example, type the name as `MyFunctions`. *CVI* automatically adds a prefix (up to eight characters) to all the function names that you include in the **Prefix:** box. This prefix helps you identify all functions related to a certain instrument or function group. *CVI* adds an underscore character as a separator between the prefix and the function name. Enter my as the prefix in the **Prefix:** box without the underscore character. Click on the **OK** command button.

A **Function Tree Editor** window with the function tree name is created as shown in Figure 7–5. To create a function class from the **Function Tree Editor** menu, select **Create>>Class** and a **Create Class Node** dialog box appears. Enter the name in the **Name:** box as `Algebraic Functions`. This is the class name you will use for the function panels that are created here. When you click **OK** on the **Create Class Node** box, this name is added to the **Function Tree Editor** window below `MyFunctions`.

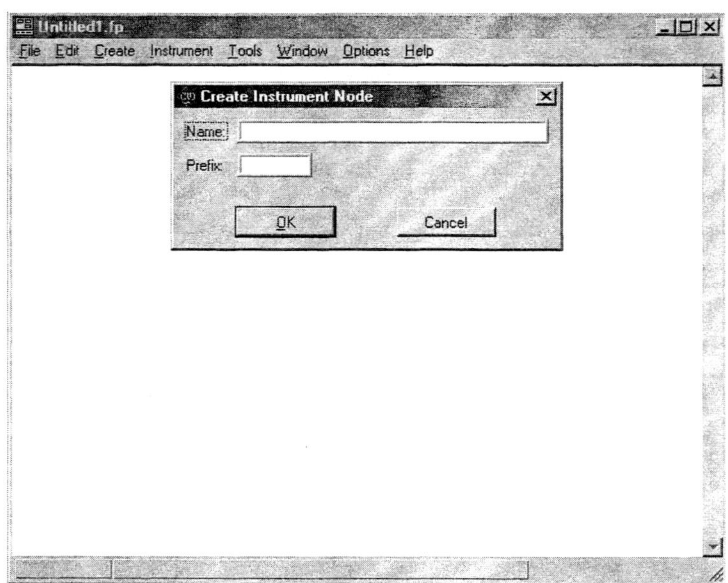

Figure 7–4
Create Instrument Node Window

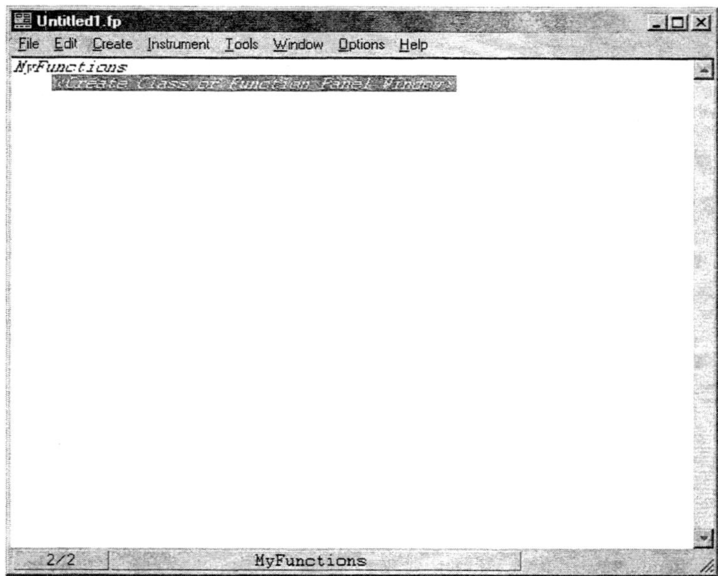

Figure 7–5
`MyFunctions` Function Tree Window

Highlight the line **<Create Class or Function Panel Window>** located just below `Algebraic Functions` and select **Create>>Function Panel Window....** A **Create Function Panel Window Node** dialog box will popup as shown in Figure 7–6. Enter the name of the function in the **Name:** dialog box as `Quadratic Equation Roots`, and in the **Function Name:** dialog box enter `SolveQuadratic`. When you select **OK**, the function name is added to the **Function Panel Window** just below `Algebraic Functions`. You will use this function to calculate the roots of a quadratic equation.

Similarly create another function by entering `Factorial` in the **Name:** box and `Factor` in the **Function Name:** box. The `Factor` function will calculate the factorial value of the number specified. The completed function tree with these two functions will now look similar to Figure 7–7.

To save the function panel, select **Save .FP File As...** from the **File** menu on the function panel and save the file as `MyFunctions.fp`. You have included the functions for the MyFunctions function tree. In the next section you will create the function panel for each of these functions and add code for these functions.

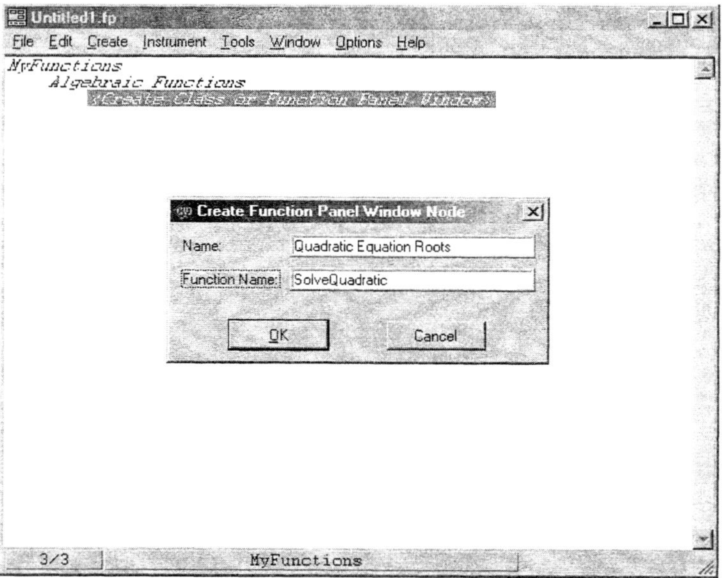

Figure 7–6
Create Function Panel Window Node Window

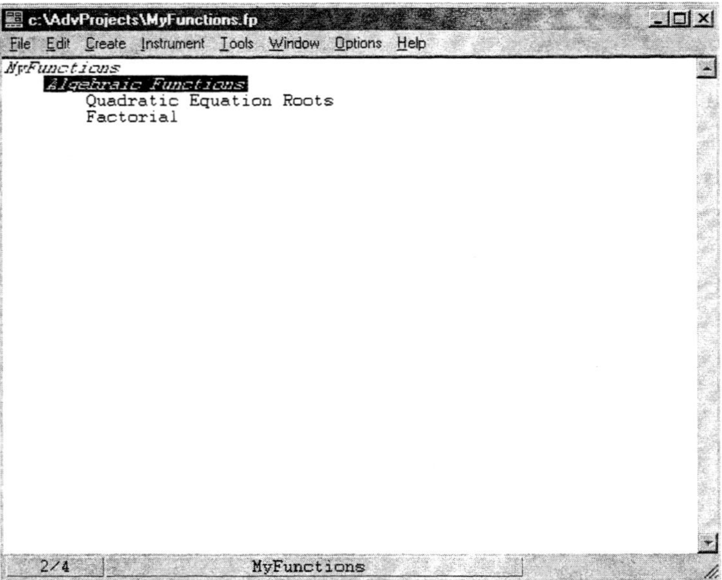

Figure 7–7
Complete Function Tree Window

Creating a Function Panel

In this section you will learn to create the function panels for the functions that you added to the function tree. If not already created, create the project and name it `MyFunctions.prj`. Load `MyFunctions.fp` file by selecting **Instrument>>Load**. You can add the `MyFunctions.fp` file to the project window from **Edit>>Add Files to Project>>Instrument[*.fp]**. From the project window select **File>>Open>> Function Tree [*.fp]**, and double-click on `MyFunctions.fp` file. A function tree window will be displayed with the class and function names you created above. To create the function panel for `SolveQuadratic`, double-click on this function to bring up a blank function panel as shown in Figure 7–8. Notice that the blank function panel title bar contains the name of the function appended with the prefix `my_` that you specified when you created the **Instrument Node** above. Also notice that the code box generated at the bottom of this panel has the name of the function with no arguments at this time. These arguments will appear when we add controls to this function panel.

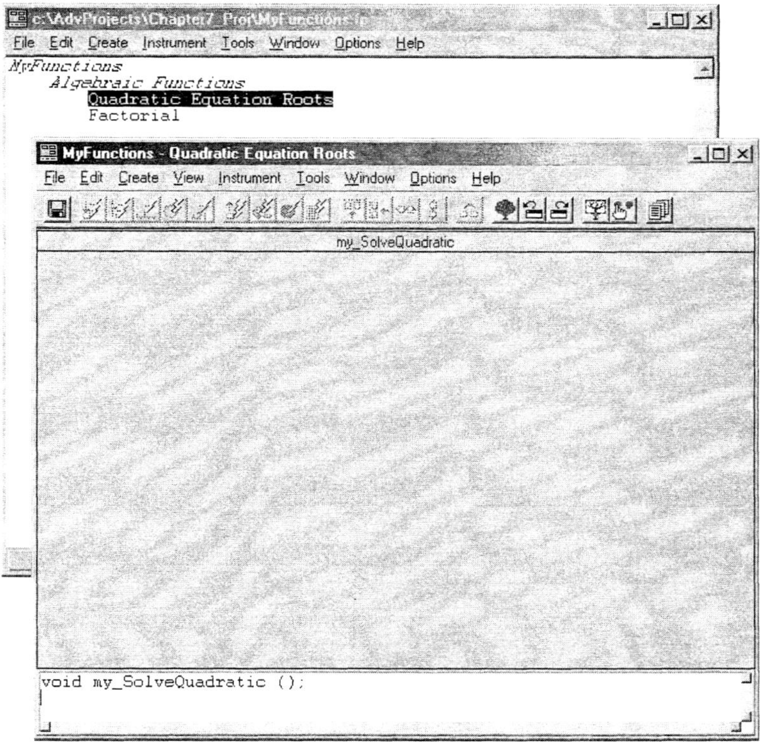

Figure 7–8
Blank Function Panel

The *my_SolveQuadratic* function will require three *input control* boxes to receive the coefficients of the quadratic equation, and the two roots calculated for the quadratic equation will be placed in the *output control* box. The function operation status is sent to the *return value control* box.

Before creating these controls on the function panel, let us understand the purpose of these controls. The *input control* receives the values that you input to the function. You can assign a default value to this control which appears when the function initially loads. You will see how this is done when you create this control on the function panel. The *output control* receives the value calculated by the function. The *output control* is included as a formal parameter in the function argument list. The *return value control* is usually the result of the function operation that appears on the left side of the function call.

We are now ready to add the controls for the function arguments on the function panel for *my_SolveQuadratic* function. To create an input control on the function panel, select **Create>>Input...** and a **Create Input Control** dialog box will appear (see Figure 7–9). Alternatively, you can also click on the **Create Input...** icon on the function panel toolbar.

*The **Create Input...** icon is similar to selecting **Create>> Input...** from the function panel menu.*

In the **Create Input Control** dialog box, **Control Label:** is the label that you want to appear for this control. Enter `First Coefficient` in this box. The **Parameter Position:** is the location of the control value in the function argument list. Leave this value as `1` since you will be using this value as the first argument in the function. When you click on the **Data Type:** control box you get a list of all the data types available. Set the **Data Type:** to **double**. The **Default**

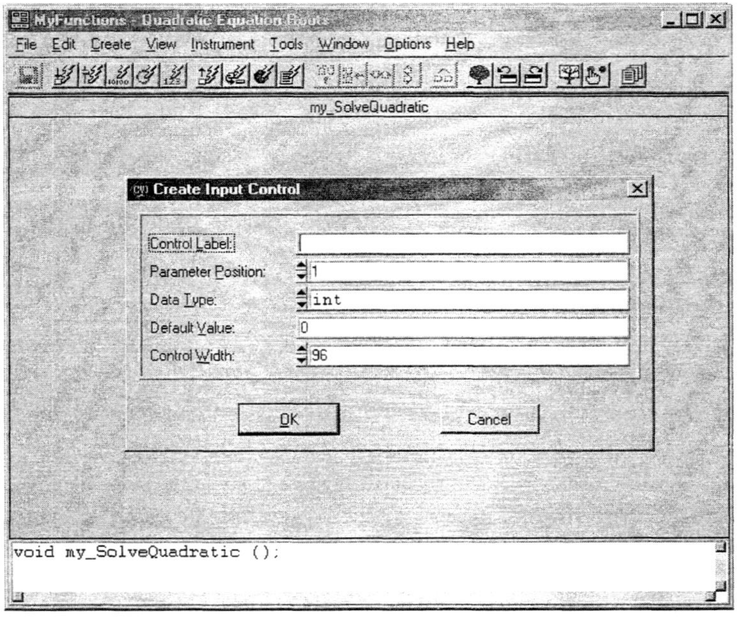

Figure 7–9
Create Input Control Dialog Box

Value: control box is the value that is displayed when the function panel first loads. Enter the value `1.0` in this box. The **Control Width:** is the width of this control in pixels. The range for this control is between 24 and 2048, with a default value of 96. Leave this control set to its default value. Once you have made the selections, click **OK** and the input control box will appear on the top left side of the function panel. You can use the mouse to drag and drop it to a different location on the function panel if you wish.

Similarly add two more input controls with **Control Labels** `Second Coefficient` and `Third Coefficient`. The **Parameter Position** will increment to the next-higher number automatically when you select this control. Set the remaining parameters the same as for `First Coefficient`. The function panel will now look like Figure 7–10.

Let us now add the two output controls to the function panel. These controls will receive the roots of the quadratic equation. Select **Create>>Output...** to display the **Create Output Control** dialog box as shown in Figure 7–11. Alternatively, you can click on the **Create Output...** icon on the function panel toolbar.

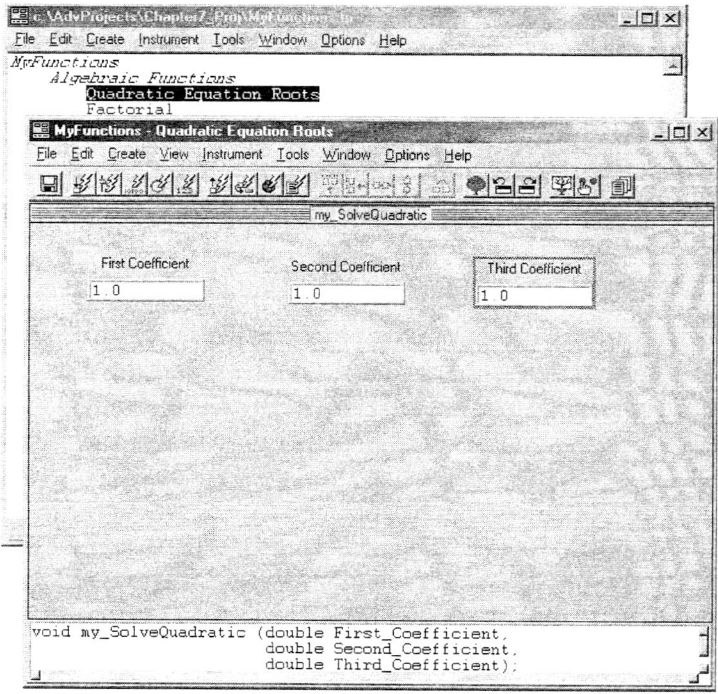

Figure 7–10
Function Panel with Input Control Boxes

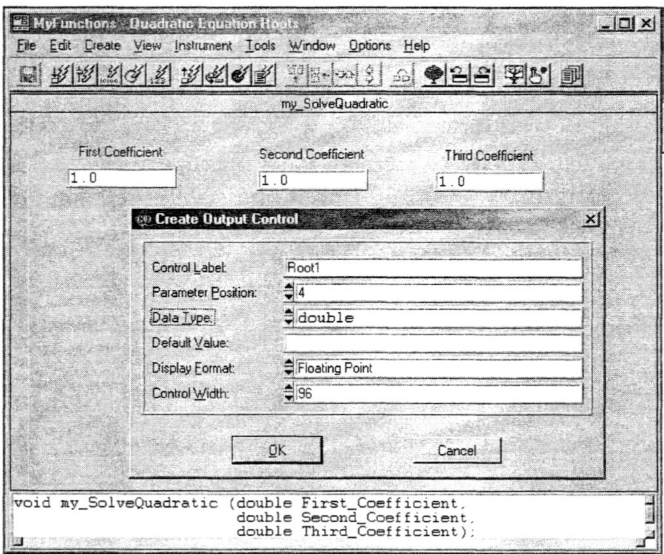

Figure 7–11
Create Output Control Dialog Box

*The **Create Output...** icon is similar to selecting **Create>> Output...** from the function panel menu.*

Enter `Root1` in the **Control Label:** dialog box, and select **double** from the **Data Type:** dialog box. Leave the **Default Value:** dialog box blank. **Display Format:** is the format in which the output control displays the output values. You have a choice to display integers, longs, shorts, and characters in decimal, hexadecimal, octal, or ASCII. The doubles and floats can be displayed in either scientific or floating-point notation. Select **Floating Point** for this dialog box and then select **OK**. The output control will be added to the function panel. Drag it to a position below the **First Coefficient** input control on the panel.

Create a second output control and enter `Root2` in the **Control Label:** dialog box. Leave the rest of the parameters the same as the `Root1` control.

*The **Create Return Value...** icon is similar to selecting **Create>> Return Value...** from the function panel menu.*

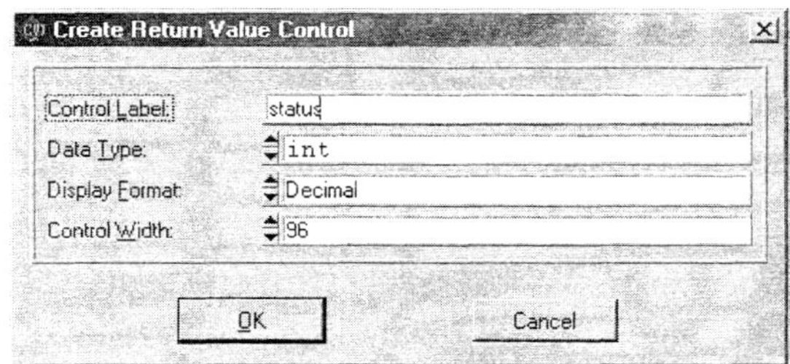

Figure 7–12
Create Return Value Control Dialog Box

To create a return value control for this function, select **Create>>Return Value...** or the **Create Return Value** icon from the function panel toolbar. A **Create Return Value Control** dialog box is displayed as shown in Figure 7–12.

In the **Control Label:** dialog box, enter `status`, and select `int` for the **Data Type:** and `Decimal` for the **Display Format:**. When you select **OK**, the control is added to the function panel. The completed function panel for *my_SolveQuadratic* function looks similar to Figure 7–13, with the default values shown in the controls and the function arguments added in the code generation box at the bottom of the panel. You can scroll on the code box to see the completed code for this function panel. From the **File** menu, select **Save .FP File** to save the function panel.

Similarly, you can create the function panel for *my_Factor,* which has one input control, one output control, and a return value control. Create an input control and enter `Number` in the **Control Label:** box. Select **unsigned int** as **Data Type:** for this control. Create an output control and enter `Factorial` in the **Control Label:** box. Select double as **Data Type:** for this control. Leave the remaining controls at their default values. The completed *my_Factor* function panel will now look like Figure 7–14.

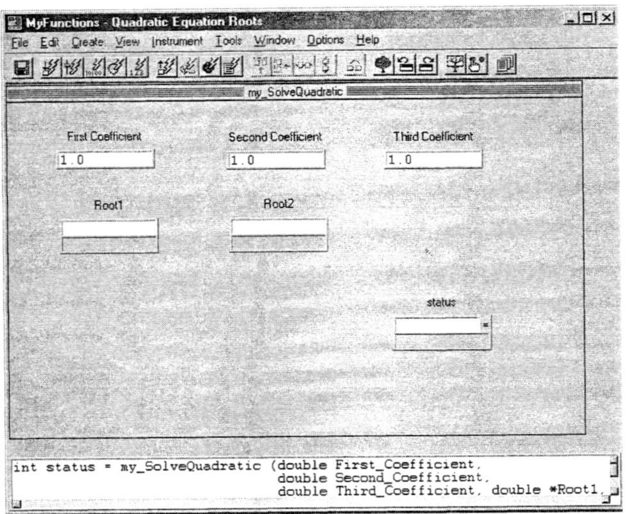

Figure 7–13
my_SolveQuadratic Function Panel

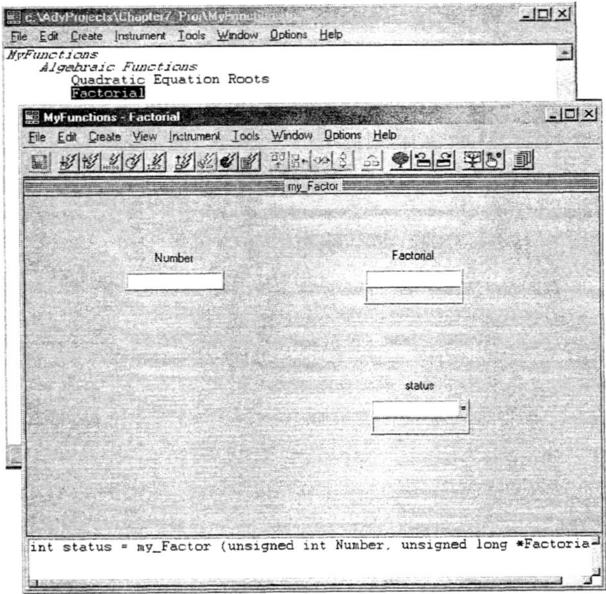

Figure 7–14
my_Factor Function Panel

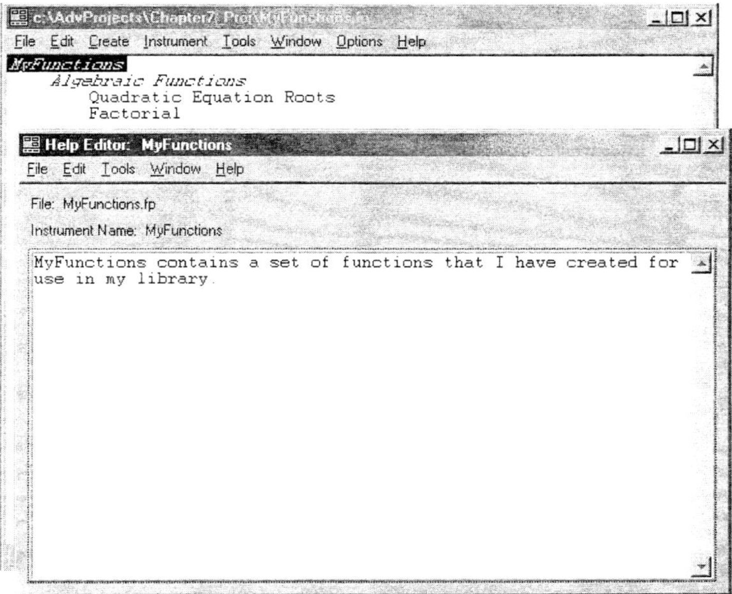

Figure 7–15
MyFunctions Help Window

Before we leave the topic of creating the function panel, let us add help text to the function panel and its controls. Help text serves to define the function, its prototype, and the function arguments adequately. To add help to `MyFunctions`, open the function panel using the **File>>Open>>Function Panel [*.fp]** from the **Project** window and select `MyFunctions.fp`. The function tree window will be displayed. Highlight `MyFunctions` and select **Edit>>Edit Help** from the menu. A blank window will be displayed in which you can enter the help text to explain the purpose of the functions in this function tree. Type in the help text in this window for `MyFunctions`. This may look similar to Figure 7–15.

To add help text to the class node, highlight *Algebraic Functions* from the function tree and select **Edit>>Edit Help** from the menu. Enter the help text that will describe the functions in this class. An example of this is shown in Figure 7–16. Save by selecting **Save .FP File** from the **File** menu.

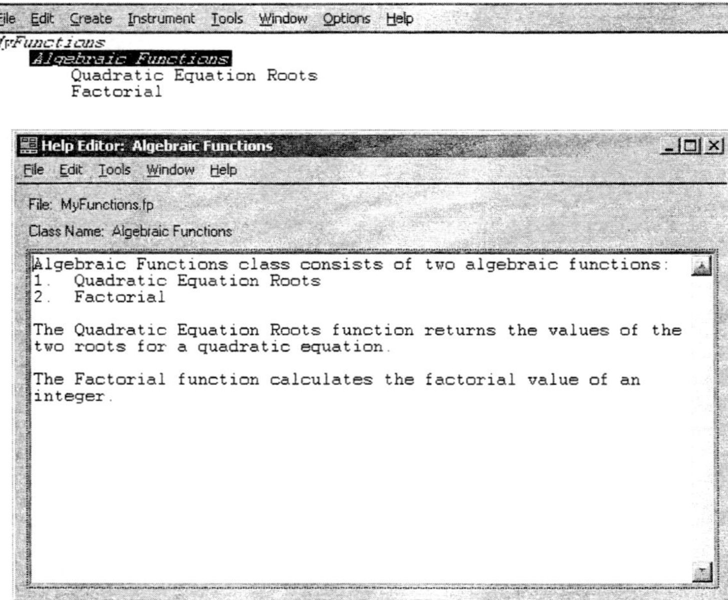

Figure 7–16
MyFunctions Class Help Window

It is advisable to add help text to the function panels and controls that you have created to enable the user to understand the purpose of the function and the use of the controls placed on the panel. To add help text to the function panel, double-click the function name in the function tree to bring up its function panel. For *CVI* version 5.5 and earlier, right-click on the panel window or the control to open a **Help Editor** window for the function or the control and type in the help information. For *CVI* versions following 5.5, when you right-click on the panel window, you get a context menu, from where you select the **Function Help** menu item to open the **Help Editor** window. Enter help text for the function. To add help information for the control, right-click on the control to get a context menu, from where you select the **Control Help** menu item to open the **Help Editor** window.

After you have completed the function panels, you need to create the header file and the source code file for these functions. Open a blank header file by selecting **File>>New>>Include (*.h)...** from the project window and type the prototypes for the two functions created above. The header file for these two functions is shown in Figure 7–17. Save the file as `MyFunctions.h`.

```
#include <cvidef.h>

int CVIFUNC my_SolveQuadratic (double firstCoefficient, double secondCoefficient,
        double thirdCoefficient, double *root1,  double *root2);

int CVIFUNC my_Factor (unsigned int number, double *factorial);
```

Figure 7–17
MyFunctions Header File

The type `CVIFUNC` is defined in `cvidef.h` as

```
#define CVIFUNC __stdcall
```

That is *CVI*'s calling convention for library and DLL functions. You can also generate the header file for the functions from the **Function Tree Editor** by opening `MyFunctions.fp` from the **File** menu of the project window and selecting **Options>>Generate Function Prototypes** from the **Function Tree Editor**.

You can now write the source code for the functions by creating a new file, `MyFunctions.c`. The listing for `MyFunctions.c` is shown in Figure 7–18.

Note that you can also generate the "skeleton" source code automatically for a single panel function or for the entire driver by selecting **Tools>>Generate New Source For Function Tree** from the **Function Tree Editor**.

The formula used to find the roots of the quadratic equation

$$ax^2 + bx + c = 0$$

is

$$x = \frac{-b \pm \sqrt{b^2 - 4ac}}{2a}$$

This formula is not accurate and you may get a round-off error when `ac` is very small compared to b^2. To use a more accurate formula, refer to the book *Numerical Recipes in C: The Art of Scientific Computing*. If you desire, you can use the more accurate formula from the book in your source code here. In the function *my_SolveQuadratic*, the formula shown above is used to calculate the roots of the quadratic equation, outputting the roots in the variables `root1` and `root2` of your program. These roots are displayed in the function panel output control boxes when you run the IW, as you will see in the next section. Notice that in the source code there is error checking to determine if the first coefficient is a nonzero value. If a zero value is entered for this coefficient, the

```
/*****************************************************

 File Name: MyFunctions.c

*****************************************************/

#include <ansi_c.h>
#include "MyFunctions.h"

//Function to solve quadratic equation
int CVIFUNC my_SolveQuadratic (double firstCoefficient, double secondCoefficient,
                               double thirdCoefficient, double *root1,
                               double *root2)
{
int status=0;
double sqrtVal;
if ( firstCoefficient==0) //first coefficient cannot be zero in quadratic equation
        return status =-1;
//Check for negative square root
else if ( (secondCoefficient * secondCoefficient -
     4 *firstCoefficient* thirdCoefficient ) < 0)
          return status =-2;
else  //Calculate the roots
   {
   sqrtVal= sqrt(secondCoefficient * secondCoefficient -
                    4 *firstCoefficient* thirdCoefficient);
   *root1 = (-secondCoefficient + sqrtVal) /(2 *firstCoefficient);
   *root2 = (-secondCoefficient - sqrtVal) /(2 *firstCoefficient);
    return status =0;
    }

} //my_SolveQuadratic


//Factorial calculations
int CVIFUNC my_Factor (unsigned int number, double *factorial)
{
   int status=0, i;
   double result;


      //Factorial of 0 is 1
      if (number == 0)
      {
        *factorial =1;
        return status=0;
      }
      //Number is too large
      else if (number > 170 )
```

Figure 7–18
MyFunctions Source Code File *(continued)*

```
        return status=-1;
      else
      {  //Number is in valid range
                result =1;
                for ( i=1; i<=number; i++)
                result = result *i;

                *factorial=result;
                return status =0;
      }
}//my_Factor
```

Figure 7–18
MyFunctions Source Code File *(continued)*

equation is not a quadratic equation and the function returns a value `-1`. The function returns a value `-2` if the expression `b`^{2} – `4ac` is negative in the quadratic equation formula, since the square root of a negative value is not a real number. If the operation is successful, the function returns a value zero.

The function *my_Factor* calculates the factorial value for an unsigned integer entered in the first argument of the function. To prevent overflow, specify a number less than `170`. Entering a number greater than `170` would cause the function to return a `-1`.

The status codes for these functions are entered in the help description for the status control boxes on the function panel. You can view the description by right-clicking on the **status** control box on the function panels.

Testing the Function Panel Functions

You have created a function tree with function panels for the two algebraic functions above. You need to test these functions to gain confidence in their results. The IW is one of the means by which you can test the functions. Load `MyFunctions.prj` and double-click on `MyFunctions.fp` in the **Project** window. Select *SolveQuadratic* to test this function. The function panel for calculating the roots of the quadratic equation will be displayed. Let us find the roots for the quadratic equation

$$x^2-5x+6=0$$

Enter a `1.0` in the **First Coefficient** control box, `-5.0` in the **Second Coefficient** control box, and `6.0` in the **Third Coefficient** control box. Click on the **Run Function Panel** icon on the function panel toolbar. *CVI* creates temporary variables for **Root1**, **Root2,** and the **status** control boxes and displays the values calculated below these controls, as you can see in Figure 7–19. The **status** value is `0` if the operation is successful. You can experiment with this function panel by inserting different valid and erroneous values in the input control boxes and verifying the results calculated and the return status.

Similarly, bring up the *Factor* function panel from the function panel. To test this function from the IW, enter `6` in the **Number** control box and click on the **Run** icon. Again, *CVI* will create temporary variables for **Factorial** and **status** controls and display the results below these control boxes, as shown in Figure 7–20.

Once you know that the functions you created are working correctly, you can create an object file of the source code for future use. To create an object file, open `MyFunctions.c` and from the **Source Editor** select **Options >> Create Object File....** A dialog box will pop up giving you the choice of selecting either **Generate object file for current compatibility mode** or **Generate object file for all compilers**. Select the option you want for your application. When you click on **OK** in the dialog box, the object file will be created as `MyFunctions.obj`. When you want to use these functions, be

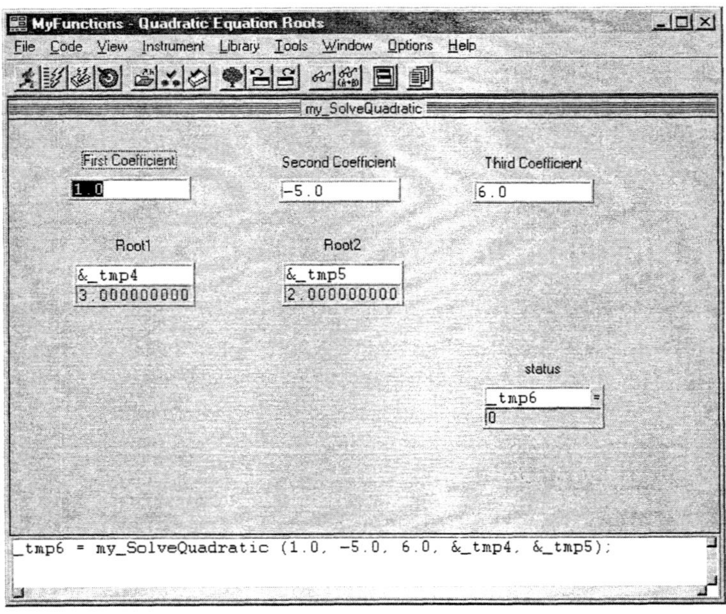

Figure 7–19
Testing *my_SolveQuadratic* Function Panel Using IW

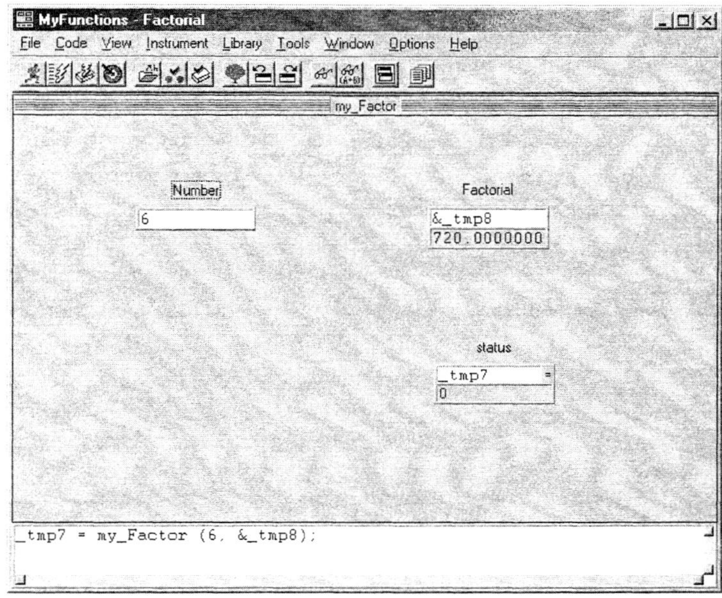

Figure 7–20
Testing *my_Factor* Function Panel Using IW

sure to include the following files in your project window: `MyFunctions.h`, `MyFunctions.fp`, and `MyFunctions.obj`.

Function Panel Controls

In the section *Creating a Function Panel* you learned how to create the input control, output control, and return value control on the function panel. There are six other controls that you can create on the function panel: numeric control, slide control, binary control, ring control, global variable control, and message control. Each of these controls is explained in the following sections.

Numeric Control

A *numeric control* is an input control in which you can change the values by clicking the up and down arrows located on the left side of the numeric control or use the <up> and <down> arrows on the keyboard. The numeric control is shown in Figure 7–21. To create the numeric control, select

Create>>Numeric... from the function panel menu, or click on the **Create Numeric...** icon on the function panel toolbar. The **Create Numeric Control** dialog box is displayed in Figure 7–22, in which you enter your selections and select **OK** to create the numeric control on the function panel.

*The **Create Numeric...** icon is similar to selecting **Create>> Numeric...** from the function panel menu.*

Control Label: allows you to specify a label you assign to this control that appears above the numeric control. The **Parameter Position:** is the location of this argument in the function argument list. When you click on the **Data Type:** control, you will see a pull-down menu from which you can choose one of the following data types:

int
short
char
unsigned int
unsigned short
unsigned char
double
float

Display Format: is the same as that for **Create Output Control** in Figure 7–11. When either **double** or **float** is selected as the **Data Type:**, the **Precision:** dialog box is enabled. Here you set the number of digits to be displayed to the right of the decimal point.

Selecting the **Value Set...** command button brings up the **Edit Value Set** dialog box shown in Figure 7–23. In this dialog box you enter the **Minimum:** and **Maximum:** values for the numeric control. In the **Inc Value:** dialog box, enter

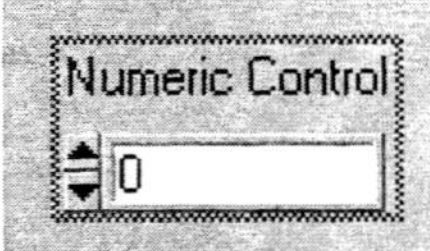

Figure 7–21
Numeric Control

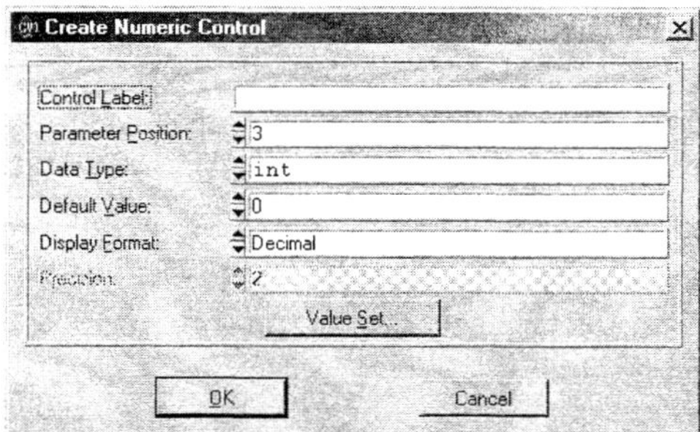

Figure 7–22
Create Numeric Control Dialog Box

the number by which you want the value to increment/decrement on the numeric control when you click on the up/down arrows of the numeric control.

Slide Control

Slide control, an input control, is shown in Figure 7–24. This control specifies a parameter value depending on the position of the marker on the slide control. To create this control, select **Create>>Slide...** or click on the **Create Slide...** icon on the function panel toolbar and enter your selections in the **Edit Slide Control** dialog box shown in Figure 7–25.

*The **Create Slide...** icon is similar to selecting **Create>> Slide...** from the function panel menu.*

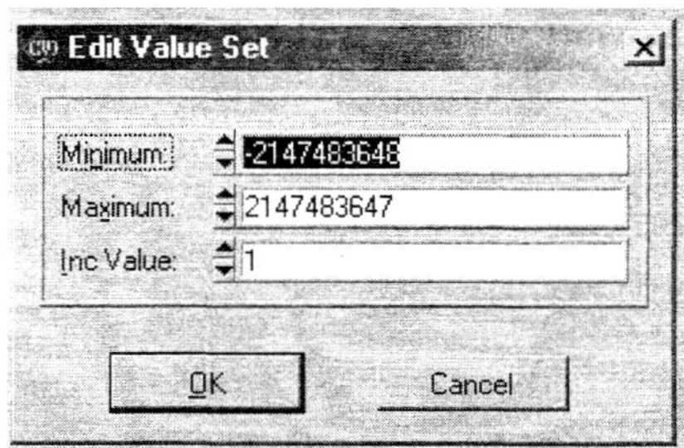

Figure 7–23
Edit Value Set Dialog Box

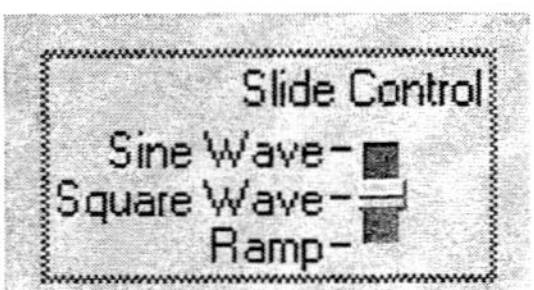

Figure 7–24
Slide Control

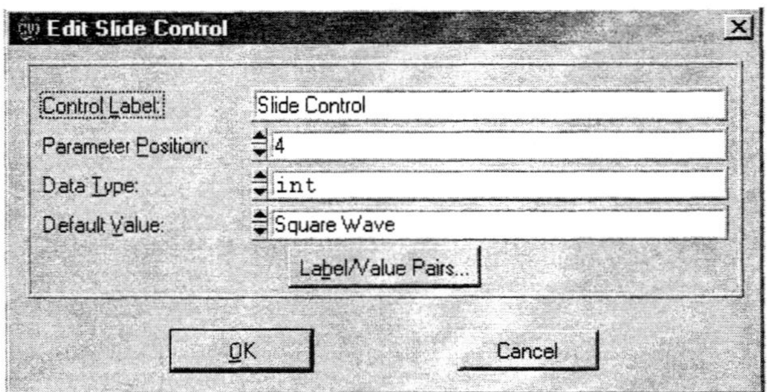

Figure 7–25
Edit Slide Control Dialog Box

The **Control Label:**, **Parameter Position:**, and **Data Type:** dialog boxes are the same as those you have seen for the other controls above. **Default Value:** is the position of the slide marker when the function panel is first loaded. You can click on the ring up/down arrows to find the default value after you have entered the label in the **Edit Label/Value Pairs...** list box. When you select the **Label/Value Pairs...** command button, Figure 7–26 is displayed where you enter the **Label:** and **Value:** pair in the list box. Selecting **OK** will return you to the **Edit Slide Control** dialog box. Selecting **OK** again will create the control on the function panel.

Binary Control

Binary control is an input control that can take one of two values, depending on the switch being in the up or the down position (see Figure 7–27). This control is created when you select **Create>>Binary...** or when you click on the **Create Binary...** icon on the function panel toolbar and enter your selections in the **Create Binary Control** dialog box shown in Figure 7–28.

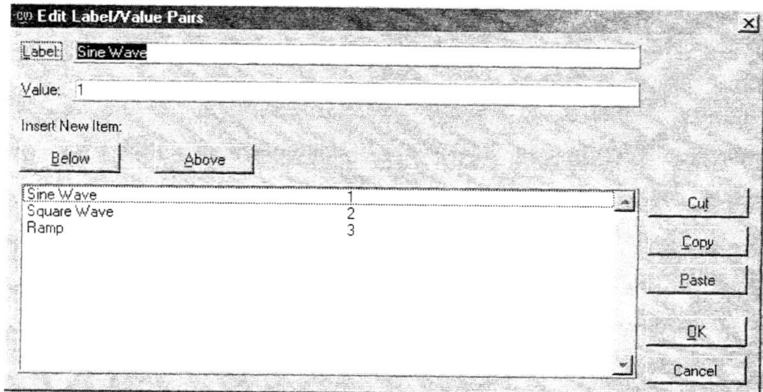

Figure 7–26
Slide Control **Label/Value Pairs...** Box

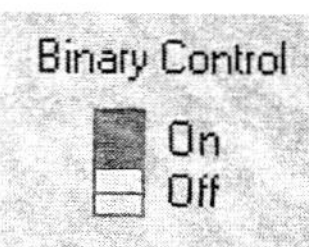

Figure 7–27
Binary Control

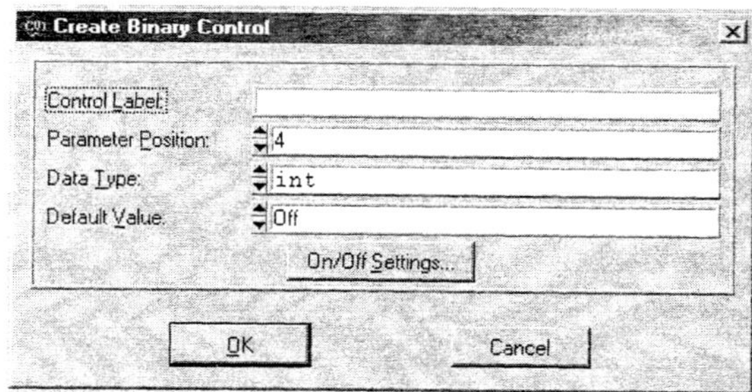

Figure 7–28
Create Binary Control Dialog Box

*The **Create Binary...** icon is similar to selecting **Create>> Binary...** from the function panel menu.*

The **Control Label:**, **Parameter Position:**, and **Data Type:** dialog boxes are the same as those you have seen for the other controls above. In the **Default Value:** dialog box you have a choice of setting it to either **On** or **Off**. When you select the **On/Off Settings...** command button from the **Create Binary Control** dialog box, Figure 7–29 is displayed, in which you can set the label for the two control states and their associated values. The **ON Text:** refers to the upper position of the binary switch, and the **OFF Text:** to the lower position. In the **ON Value:** and **OFF Value:** boxes you can enter a value, a constant name, or a valid C expression to associate with these switch positions.

Ring Control

The *ring control* allows you to select one item at a time from among the options in the pull-down list. This ring control is shown in Figure 7–30. *Ring control* is created when you select **Create>>Ring...** or when you click on the **Create Ring** icon on the function panel toolbar and enter the data in the **Create Ring Control** dialog box shown in Figure 7–31.

*The **Create Ring...** icon is similar to selecting **Create>> Ring...** from the function panel menu.*

The **Control Label:**, **Parameter Position:**, and **Data Type:** dialog boxes are the same as those you have seen for the controls above. The **Default Value:** is the position of the ring control when the function panel is first loaded.

Edit On/Off Settings

ON Text: On

OFF Text: Off

ON Value: 1

OFF Value: 0

OK Cancel

Figure 7–29
Edit On/Off Settings Dialog Box

Figure 7–30
Ring Control

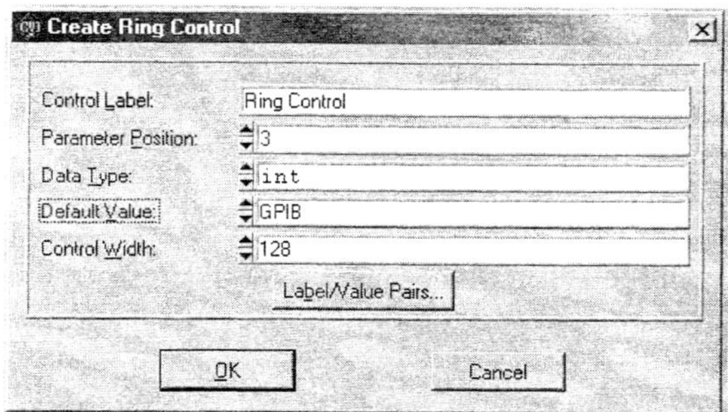

Figure 7–31
Create Ring Control Dialog Box

Select the **Label/Value Pairs...** command button from the **Create Ring Control** to create a list and assign a value to the labels in the list. A figure similar to the **Edit Label/Value Pairs** for the slide control dialog box (Figure 7–26) will be displayed. You can enter the appropriate label/value pairs in the list box.

Global Variable

The *global variable* control displays the value of the global variable defined in your *CVI* source code when the function panel is operated. The global variable control is shown in Figure 7–32. The global variable control is created when you select **Create>>Global Variable...** from the function panel menu or click on the **Create Global Variable...** icon on the toolbar and enter data in the **Edit Create Global Variable Control** dialog box shown in Figure 7–33.

In the **Global Variable Name:** dialog box, enter the name of the global variable whose contents you would like to monitor on the panel.

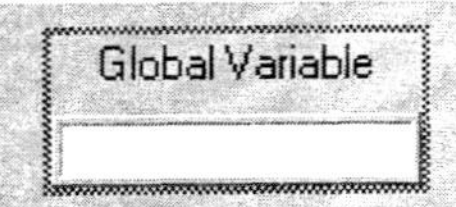

Figure 7–32
Global Variable Control

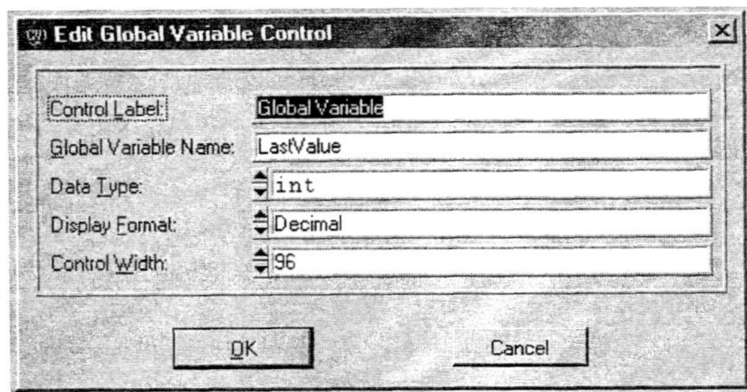

Figure 7–33
Edit Global Variable Control Dialog Box

The ***Create Global Variable...*** *icon is similar to selecting* ***Create>> Global Variable...*** *from the function panel menu.*

Message Control

The *message control* is used to display a text message on the function panel. The purpose of this control is to provide information to the user while the function panel is opened. It may be useful in instances when you want to alert the user to the limitations of this function. The message control looks similar to Figure 7–34, containing the text that you entered in the message control. You can create a message control by selecting **Create>>Message...** from the function panel menu or **Create Message...** icon on the toolbar. Enter your message in the **Message Text:** box in the **Create Message Control** dialog box shown in Figure 7–35.

Figure 7–34
Message Control

Figure 7–35
Create Message Control Dialog Box

*The **Create Message...** icon is similar to selecting **Create>> Message...** from the function panel menu.*

Before we leave the topic of the function panel, you should know that you have an option to change the control type on the function easily to a different control type. For example, if you have a slide control on the panel and later wanted to change it to a ring control with the same label/pair values, you do not have to delete the slide control and create a new ring control and reenter the data. An easier way to do this is to highlight the slide control on the function panel (for example) and select **Edit>>Change Control Type**. A **Change Input Control Type** selection box (Figure 7–36) will appear. Highlight the control that you want to change to (**Ring** in this example) and select **OK**. A new ring control will be placed on the function panel with the same label/pair values as those of the slide control replaced.

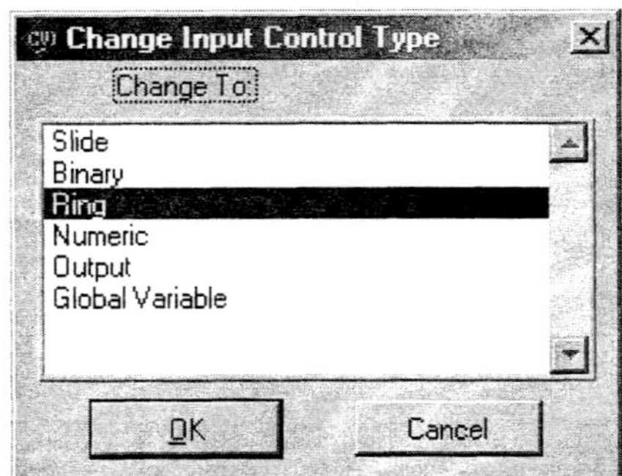

Figure 7–36
Change Input Control Type Selection Box

Summary

The creation and purpose of the function tree and the function panel were explained. You were introduced to the creation of nine different types of controls that can be used on the function panel. You were shown how to add help information to the function panel and the controls to make your function panels more useful to the user. You learned how to create the header, the source code file, and the object file for the functions and test them using the IW. You saw how you could switch from one control to another by the use of a function panel command.

8

CREATING INSTRUMENT DRIVERS

Chapter Highlights

- Introduction
- Creating an Instrument Driver
- Generating Driver Files Review
- Using the Attribute Editor
- Editing High-Level Instrument Driver Functions
- Deleting High-Level Instrument Driver Functions
- Adding High-Level Instrument Driver Functions
- Creating Instrument Driver Documentation
- Testing the Instrument Driver
- Summary

The basic features of instrument driver architecture are introduced in this chapter. You will see how to create an instrument driver using Interchangeable Virtual Instruments (IVI) and discuss the features of the IVI and *VXIplug&play* architectures. You are stepped through creating the instrument driver for the power supply using the Instrument Driver Development Wizard. The various types of files generated by the wizard are explained. You are shown the use of the Attribute Editor to add, delete, or modify the instrument driver attributes. You are also shown how to add, edit, and delete the high-level instrument driver functions and modify the source file appropriately. Finally, you are shown how to create documentation for the instrument driver both as a text file and as Windows On-line Help, and to test the operations of the instrument driver created.

Introduction

Instrument drivers are collections of high-level reusable routines that perform the useful functionality of communicating with a physical instrument. Instrument drivers may not be limited to an instrument device only. A noninstrument driver such as a software utility that performs a certain specific task can also be packaged as an instrument driver. You saw an example of the software utility instrument driver `legend.fp` used in `project2-2.prj` in *Chapter 2, Plotting on Graph Controls.* For communicating with a physical instrument the high-level routines are grouped so that they can be used for the purpose of communicating with hardware, to perform low-level functions to send and retrieve useful information to/from the hardware. The instrument drivers usually communicate to a physical instrument via the GPIB, VXI, or serial interface. When using high-level instrument driver routines the user does not have to be aware of the communication protocol of the instrument since it is implemented by low-level communication routines in the instrument driver. The instrument driver saves the user time by using the already created generic instrument driver functions in his/her applications. Instrument drivers consist of graphical function panels that facilitate using the instrument driver functions from where the user can select and enter the appropriate information in the function panel to activate a particular instrument driver task. Function panels were introduced in *Chapter 7, Creating and Using Function Panels.*

There are two types of instrument driver architectures: *VXIplug&play* and Interchangeable Virtual Instruments (IVI). The *VXIplug&play* architecture uses Virtual Instrumentation Software Architecture (VISA), which uses standard data types to define the arguments of the instrument driver functions. Using this feature makes the instrument driver portable between different operating systems and programming languages. All communication with the instruments is performed using VISA. You saw the features of VISA in *Chapter 5, VXI Communication Using VISA.* Instrument drivers using the IVI architecture are an enhancement of the *VXIplug&play* model, which also uses VISA but allows for higher performance and instrument simulation. *VXIplug&play* uses direct instrument I/O to query and modify instrument settings, whereas IVI drivers do the same through attributes.

The IVI Foundation, created by National Instruments together with several other companies in 1998, established standards for the IVI instrument driver. The IVI architecture consists of an IVI engine that works in conjunction with the instrument driver to control the reading and writing of instrument settings to and from the instrument as *attributes*. An attribute represents an engine setting or a driver option. The IVI engine checks these settings for valid range values in the *range tables* for each instrument attribute when communicating with the instrument (if the range checking attribute is enabled). If the value is within the valid range in the range table, the instrument driver generates the *callback function* to update the instrument attribute by sending a command string to the instrument. If the value is outside the range table, the instrument driver returns an error code. If the instrument driver query status is enabled, the check status callback function reads the status register of the instrument for error conditions. Similarly, if state caching and simulation are enabled, appropriate callback functions are invoked to implement these conditions. Once you have debugged the instrument driver functions you can disable range and status checking to speed your application.

For more features of the IVI instrument drivers, see *Chapter 2, IVI Architecture Overview* of *LabWindows/CVI Instrument Driver Developers Guide*. Some of the main features of IVI instrument drivers are:

- **Instrument Interchangeability**. This allows for changing instruments without modifying the software. This is based on the IVI Foundation standard APIs for common types of instruments, such as DMM, oscilloscope, function/arbitrary waveform generator, DC power supply, switch, and more, that are divided into class drivers. The usefulness of the class drivers lies in users being able to configure their test systems by exchanging instruments of the same class irrespective of the manufacturer or of the hardware interface used.
- **Instrument Simulation**. IVI instrument drivers can simulate instrument operations when a physical instrument is not available during instrument driver development. When the simulation mode is set in your application program, simulated data is returned to your application. During simulation all parameters are checked for valid instrument ranges to determine if the instrument is configured properly.
- **Instrument State-Caching**. IVI instrument drivers automatically cache the current state of the instrument and eliminate redundant I/Os to the instrument. This leads to an improvement in the performance of the instrument driver, since instrument I/O is performed only when the instrument settings differ from those requested by the function.

- **Multithread Safety**. IVI instrument drivers can be used in multi-threaded applications by locking an IVI session using the IVI library function *Ivi_*`LockSession` upon entry and unlocking it by calling *Ivi_*`UnlockSession` when exiting. These functions are called from the instrument driver functions *Prefix_*`LockSession` and *Prefix_*`Unlock-Session`, where *Prefix* refers to the alphanumeric characters that you assigned to the instrument when creating the instrument driver. This is discussed in the section *Creating an Instrument Driver* below. Using the lock and unlock mechanism ensures that no other execution thread can interfere with the session while it is locked and modify the instrument's setting(s).

You will now step through creating an instrument driver. Following are the steps recommended for creating IVI instrument drivers.

1. Create the instrument driver files using the **Instrument Driver Development Wizard**.
2. Edit the instrument driver attributes and modify the write and read callback functions (if necessary).
3. Edit the high-level instrument driver functions (if necessary).
4. Delete the high-level instrument driver functions not used by the instrument (if necessary).
5. Add new attributes and functions (if necessary).
6. Create the instrument driver documentation and Windows Help files.
7. Test the instrument driver.

You will see how these steps are implemented in the following sections.

Creating an Instrument Driver

The simplest way to create an instrument driver is to use the *CVI* **Instrument Driver Development Wizard**. The **Instrument Driver Development Wizard** automates the procedure to create all the files necessary for the instrument driver. The wizard will generate the following files:

- An instrument function panel (`.fp`) file, which defines the function tree and the function panels, with corresponding help text

- An instrument driver source (`.c`) file, containing the source code for the driver functions
- An instrument driver include (`.h`) file, containing function declarations, constant definitions, and external declarations of global variables
- A (`.sub`) file that contains the instrument driver attributes and their possible values, whose contents you can view when you use certain instrument driver function panels and modify using the **Attribute Editor** (created only for instrument drivers using the attribute model, e.g., IVI drivers). The **Attribute Editor** is explained below in the section *Using the Attribute Editor*.

The files created by the wizard use the *driver name* as the base name, with the appropriate file extensions shown above. The driver name is the *instrument prefix* that you assign when you create the instrument driver, as you will see later in this section.

Let us go through the procedure to create the instrument driver for a Triple Output DC Power Supply, E3661A. (Hewlett-Packard manufactured this power supply, but it is now sold by Agilent Technologies.) Start *CVI* and from the **Project** window, select the **Tools>>Create IVI Instrument Driver** to launch the **Instrument Driver Development Wizard**. A **Welcome** window is displayed, as shown in Figure 8–1.

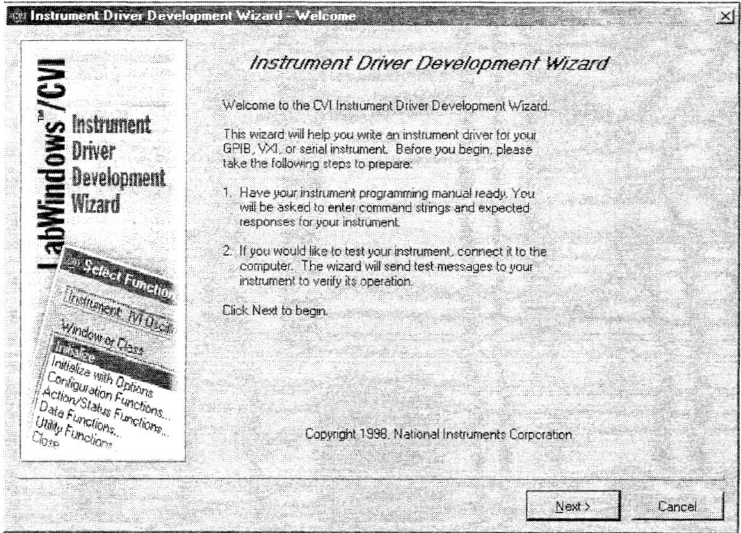

Figure 8–1
Instrument Driver Development Wizard: **Welcome Window**

Click on the **Next>** command button to bring up the **Select an Instrument Driver** dialog window, as shown in Figure 8–2. Note that in all wizard windows the **Help** command button is available to provide context-sensitive help for that window, explaining what to enter in the dialog boxes.

To create a driver based on an existing driver, select the radio button next to **Create Driver Based on Existing Driver**. The **Existing .fp File** dialog box at the bottom of this window is enabled, in which you must enter the pathname of the function panel (`.fp`) file for the existing driver. The wizard copies the existing driver `.fp`, `.c`, `.h`, and `.sub` files to the new **Target Directory** that you will specify later in the dialog window shown in Figure 8–3. These files are copied with the new instrument driver name that you entered in the **Instrument Prefix** dialog box in Figure 8–3. Note that the location of the files given here is for *CVI* 5.5 and earlier versions. Later versions of *CVI* have different locations where these files are generated and where they are read from.

In this example you will be creating a new instrument driver. Check the **Create New Driver** radio button in Figure 8–2. From the **I/O Interface** list box select the type of I/O interface you will use to communicate with your instrument. Select **GPIB** for this exercise. The wizard consists of a list of predefined class templates that define complete driver architecture with functions

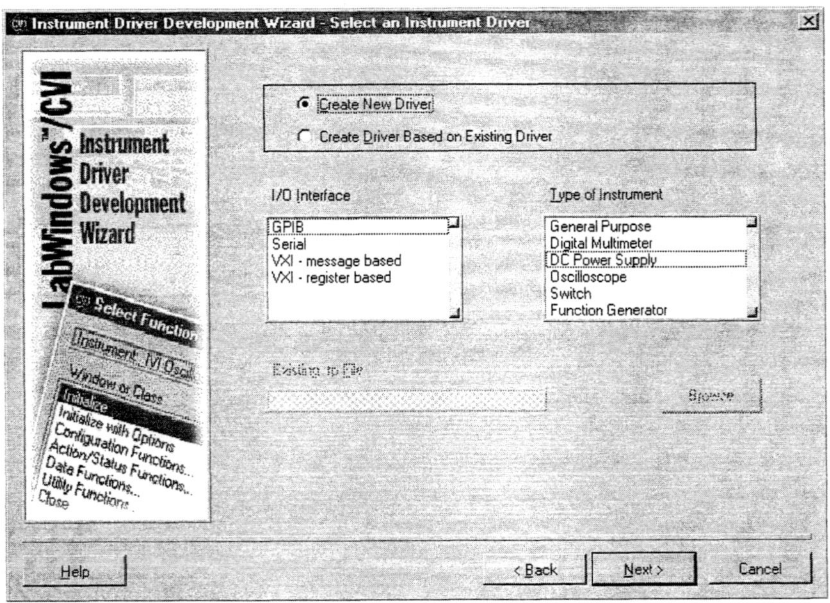

Figure 8–2
Instrument Driver Development Wizard: **Select an Instrument Driver**

and attributes for the instruments shown in the **Type of Instrument** list box depending on the **I/O Interface** selected. When you select any of the class templates, the **Instrument Driver Development Wizard** generates the skeleton driver files with functions and attributes that *VXIplug&play* and IVI require. You should use the **General Purpose** class template (shown at the top of this list box) only when there are no predefined instrument class templates for your instrument. The other class templates are to be used with the appropriate instruments shown in the list. Here we will be creating an instrument driver for a DC Power Supply.

Select **DC Power Supply** in the **Type of Instrument** list box and click on **Next>** command button to display the **General Information** dialog window (Figure 8–3). In the **Instrument Name** dialog box, enter the name of the instrument for which you are developing the driver. Enter `Triple Output DC Power Supply`. This name will be used to identify your instrument driver and will appear in the function panel tree and on the **Instrument** menu when you load the instrument driver from the **Project** window. Enter `E3631A` in the **Instrument Prefix** dialog box. The **Instrument Prefix** must have eight characters or less. This prefix will be used for the entire instrument driver files and appended to all driver functions to identify them uniquely. In this example the files that comprise the instrument driver (when generated)

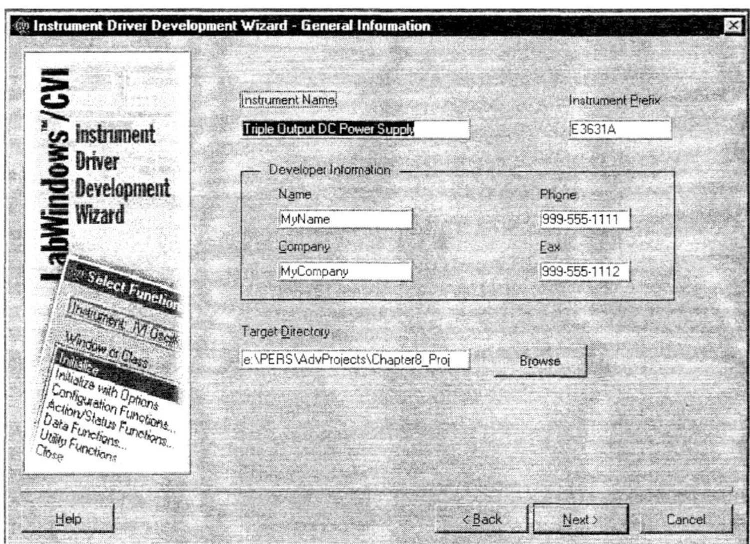

Figure 8–3
Instrument Driver Development Wizard: **General Information** Window

would consist of `E3631A.c`, `E3631A.h`, `E3631A.fp`, and `E3631A.sub`. The driver function names would be, for example, `E3631A_init`. Do not type the underscore "_" separator in the **Instrument Prefix** dialog box since *CVI* adds it to the prefix before appending it to the function name.

In the **Developer Information** section on the **General Information** dialog window (Figure 8–3), you can add your personal information. This information will appear in the description section at the top of the driver source file. Click the **Browse** command button to select the directory for the new instrument driver files. When you select the directory, it is entered in the **Target Directory** dialog box.

Click on the **Next>** command button; the **General Command Strings** dialog window appears as shown in Figure 8–4. In the **Default Setup Command** dialog box enter the command string to set the instrument to a default state at power-on. Enter the command string as `APPL P6V,3.5,1.5`. This command string uses the *Standard Commands for Programmable Instrumentation* (SCPI) format. This command string will initialize the power supply's +6V output to 3.5 volts and set the current to 1.5 amperes. For SCPI commands and their syntax, refer to your instrument's user's manual. Enter the channel string in the **Channel List String** dialog box. The channel string identifies the channels used on your instrument. If you have multiple channels on your instrument,

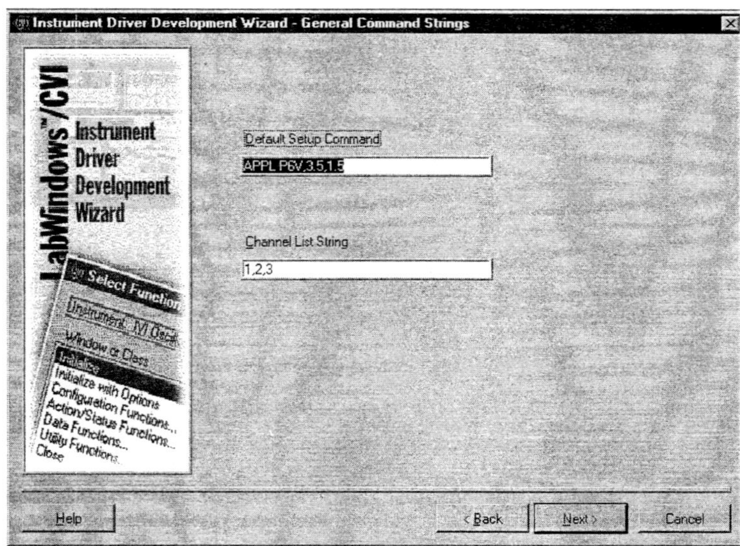

Figure 8–4
Instrument Driver Development Wizard: **General Command Strings** Window

you can enter them in the dialog box separated by commas. If you wish, you can also use the channel names from the front panel of your instrument for better comprehension when referencing the channels in the driver functions. Enter `1,2,3` in the **Channel List String** dialog box since this DC power supply consists of three outputs (+6V, +25V, and –25V) that are referenced using these channel numbers. "1" selects the +6V output, "2" selects the +25V output, and "3" selects the –25V output for this power supply.

Click on the **Next>** command button and Figure 8–5, the **Standard Operations** dialog window, is displayed.

The **Standard Operations** dialog window allows you to select the standard operations of your instrument. E3631A DC power supply supports all the standard operations shown in this window; therefore, you need to put check mark in all the boxes. The command string your instrument requires to trigger these operations will be entered in the next few dialog windows.

Click on the **Next>** command button to bring up Figure 8–6, the **ID Query** dialog window. Enter `*IDN?` (default value) in the **ID Query Command** dialog box to obtain the instrument's identification string. In the **Expected ID Query Response** dialog box, enter the following instrument identification string:

```
HEWLETT-PACKARD,E3631A,0,1.4-5.0-1.0
```

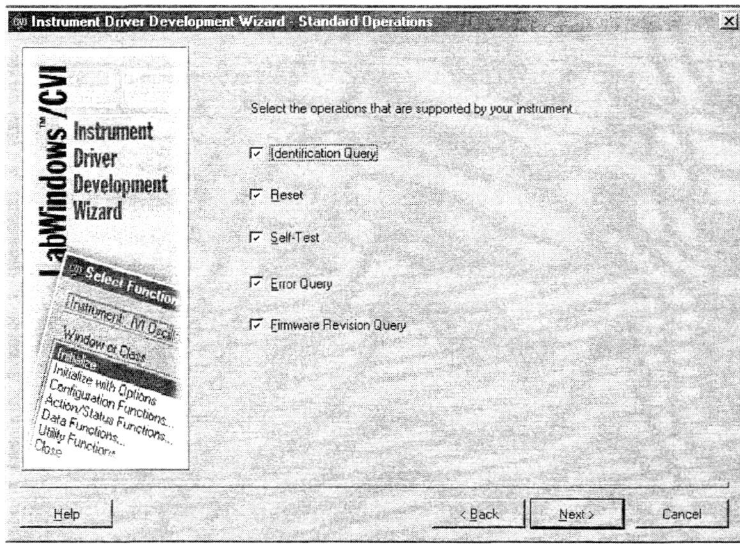

Figure 8–5
Instrument Driver Development Wizard: **Standard Operations** Window

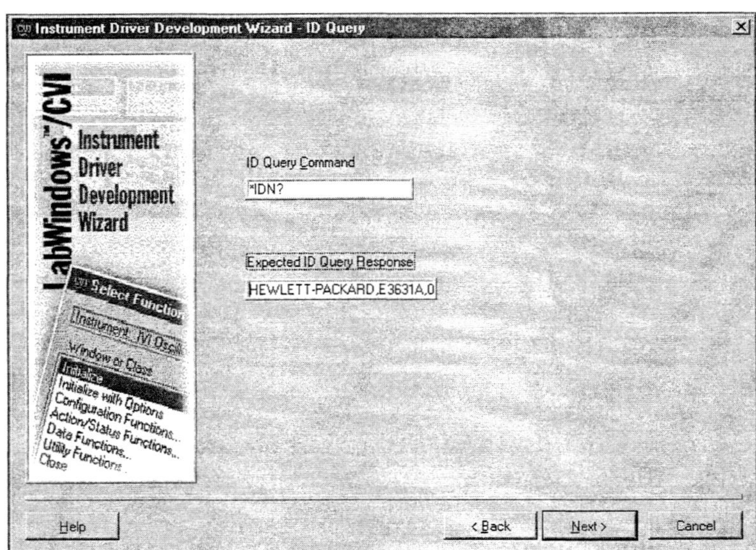

Figure 8–6
Instrument Driver Development Wizard: **ID Query** Window

You obtain the instrument identification string from the instrument user's manual. The driver uses this string to compare the output received from the instrument to determine if it is communicating with the correct instrument.

Click the **Next>** command button to display Figure 8–7. Enter `*RST` (default string) in the **Reset Command** dialog box. This command resets the instrument to its power-on state.

Click on **Next>** to display the **Self Test** dialog window, shown in Figure 8–8. Enter `TST?` (default string) in the **Self-Test Command** dialog box to perform its internal self-test and return the pass/fail result. In the **Self-Test Response Contents** control box you can select the instrument's response contents to return the results of the self-test. Some instruments return just a pass or fail code as "0" or "1", while others return a message, and others return a code and a message. If your instrument uses none of these, you have a choice of selecting **Custom** from the pull-down menu to create the appropriate return output. This power supply returns a code only; therefore, select **Self-Test Code** from this control box. Depending on your selection, the **Format String** indicator box automatically displays the appropriate string format that VISA uses to interpret the instrument's response in the **Format Choices** and **Format String** dialog boxes.

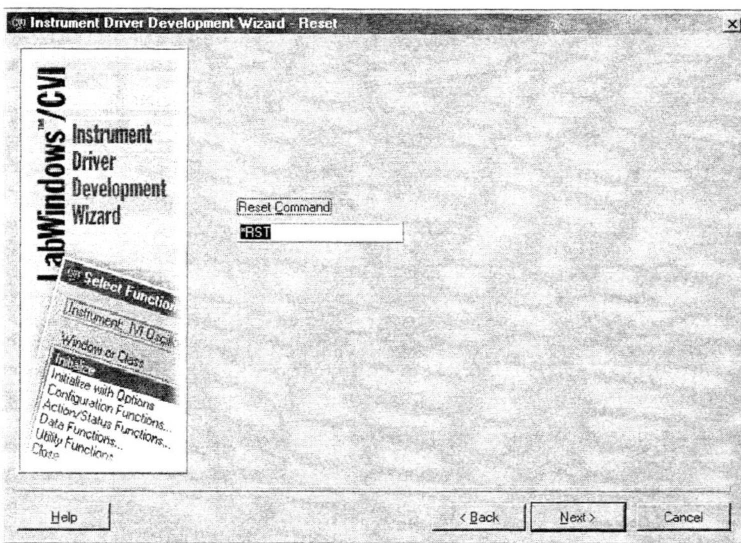

Figure 8–7
Instrument Driver Development Wizard: **Reset Command** Window

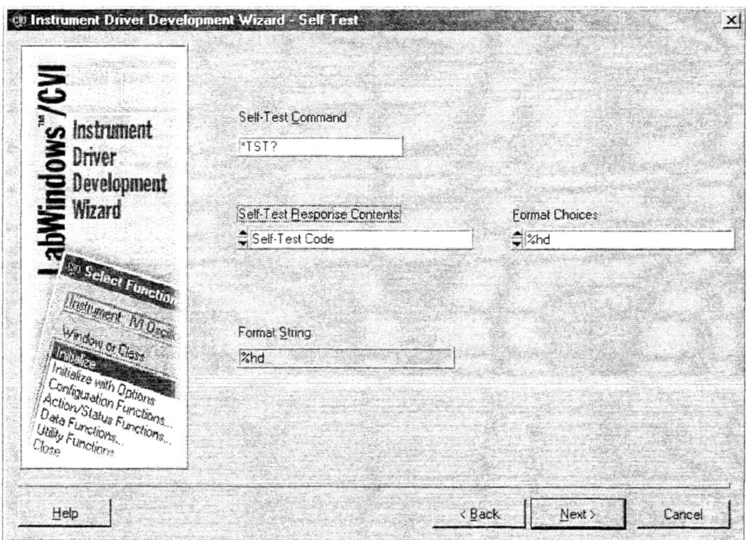

Figure 8–8
Instrument Driver Development Wizard: **Self-Test** Window

When you click on the **Next>** command button, the **Error Query** dialog window is displayed (Figure 8–9). The **Error Query Command** string for this instrument is `:SYST:ERR?`. This command string reads an error from the instrument's error queue. The **Error Query Response Contents** pull-down box allows you to select the format of your instrument response to the error messages. When you check on this box, you have a choice of **Error Code**, **Error Message**, **Error Code and Message,** or **Custom**. Select the appropriate format as indicated in your instrument's user's manual. For E3631A the error code and a message are returned in the response buffer using the VISA string format indicated in the **Format String** box. Select **Error Code and Message** from the **Error Query Response Contents** box. The **Format Choices** are set automatically to the correct string format based on the selection you make in the **Error Query Response Contents** box.

Click on the **Next>** command button to display the **Revision** dialog window shown in Figure 8–10. To obtain the firmware revision of the instrument, enter `*IDN?` (default string) in the **Revision Command** dialog box. Select the **Format Choices** as shown in Figure 8–10. The **Format String** indicator box gives the format string for VISA to interpret the instrument's response. This format string ignores everything up to the third comma. The remainder of the response is read up to the linefeed to indicate the firmware revision returned by the instrument.

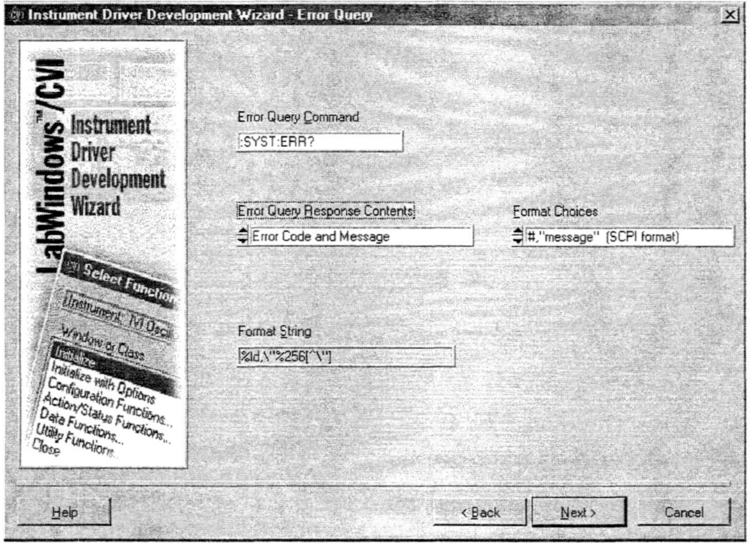

Figure 8–9
Instrument Driver Development Wizard: **Error Query** Window

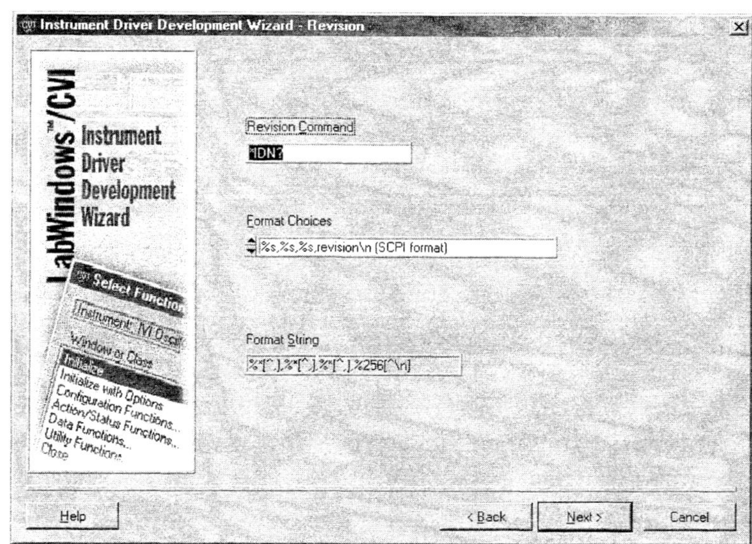

Figure 8–10
Instrument Driver Development Wizard: **Revision** Window

Click on the **Next>** command button. The **Test** dialog window is displayed as in Figure 8–11. The dialog window shown here is for *CVI* version 5.5. For later *CVI* versions a **Resource Descriptor** combo box where the user can enter any input replaces the **GPIB Address** box. Here we create the instrument driver using the **GPIB Address** control box. In the **GPIB Address** control box, enter the GPIB address as displayed on your instrument's front panel at power-on or obtain this address from the user's manual. In the **Reset Delay(s)** control box enter the number of seconds to delay after performing the instrument reset. In the **Self Test Delay(s)** control box enter the time taken by this power supply to perform the self-test. Before you run the test you need to have your instrument powered-on and connected to the interface you selected in Figure 8–2 (GPIB, in this case). Click on the **Run Tests** command button to test your instrument.

After the instrument has completed running the tests, the **Test Results** window is displayed (Figure 8–12), indicating the test operations and the results of these operations. Be sure that there are no errors and that all the responses are valid. After viewing the test results, select the **Done** command button to return to the previous window. If you found errors in the test run, click on the **<Back** command button to return to the appropriate window to

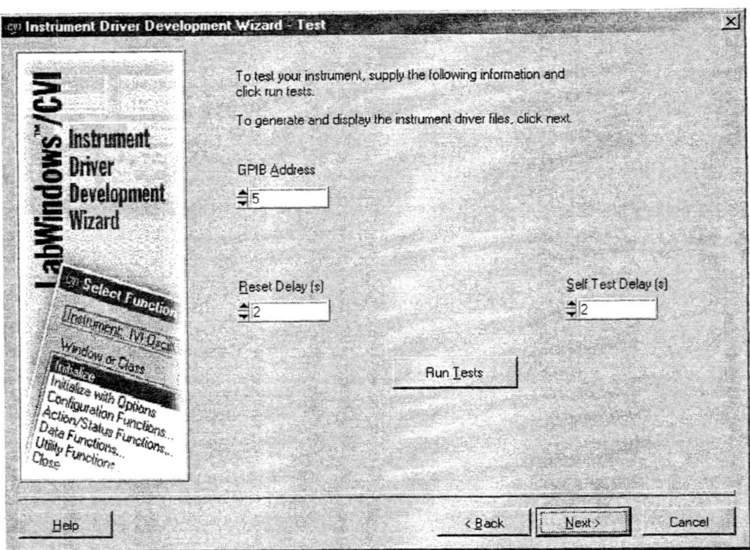

Figure 8–11
Instrument Driver Development Wizard: **Test** Window

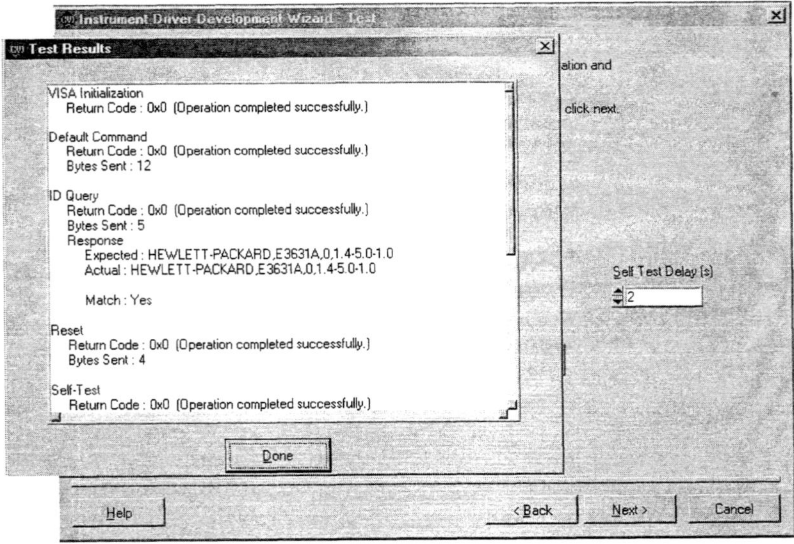

Figure 8–12
Instrument Driver Development Wizard: **Test Results** Window

enter the correct information. Return to the **Test** window to rerun the tests and verify the test results.

When you are satisfied with the results, click on the **Next>** command button in the **Test** window. Using the information you provided, the wizard generates the instrument driver files. Figure 8–13 is displayed, indicating that the driver files have been generated successfully. As mentioned earlier, the following instrument driver files are created in the **Target Directory** that you specified in Figure 8–3: `E3631A.c`, `E3631A.h`, `E3631A.fp`, and `E3631A.sub`. These instrument driver files will implement all the required IVI and *VXIplug&play* functions shown below.

The *VXIplug&play* standard requires the following seven functions to be implemented by each instrument driver.

*Prefix*_init

*Prefix*_close

*Prefix*_reset

*Prefix*_self_test

*Prefix*_revision_query

*Prefix*_error_query

*Prefix*_error_message

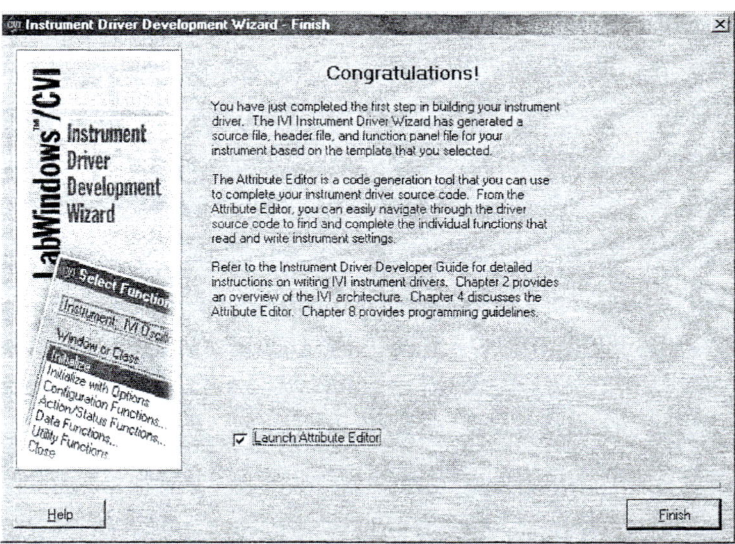

Figure 8–13
Instrument Driver Development Wizard: **Finish** Window

Prefix appended to the function name is the instrument prefix for the instrument driver that you specified in the wizard (Figure 8–3); for this instrument driver, they are E3661A_init, E3661A_close, ...

The IVI standard calls for the following additional functions:

*Prefix*_InitWithOptions
*Prefix*_IviInit
*Prefix*_IviClose
*Prefix*_LockSession
*Prefix*_UnlockSession
*Prefix*_GetErrorInfo
*Prefix*_ClearErrorInfo
*Prefix*_SetAttribute`<type>`
*Prefix*_GetAttribute`<type>`
*Prefix*_InvalidateAllAttributes
*Prefix*_GetNextCoercionRecord
*Prefix*_ReadInstrData
*Prefix*_WriteInstrData

where `<type>` in the *Prefix*_SetAttribute and *Prefix*_GetAttribute functions refers to the following six VISA data types that the instrument driver must assign to each attribute: `ViInt32`, `ViReal64`, `ViString`, `ViBoolean`, `ViSession`, and `ViAddr`. This implies that there are six *Prefix*`_SetAttribute` and six *Prefix*`_GetAttribute` functions, each corresponding to a different data type. The `ViAddr` data type is used only for attributes that are hidden from the instrument driver user.

The instrument driver created by the wizard has all the functions and attributes that are common to the DC Power Supply class template. You will need to modify the attributes and source code in the driver files created to suit your specific power supply.

One of the ways you can modify the attributes is through the **Attribute Editor**. The **Attribute Editor** is invoked when you leave the box next to **Launch Attribute Editor** checked on the **Finish** dialog window (Figure 8–13) and click on the **Finish** command button. The operations of the **Attribute Editor** are discussed in the section *Using the Attribute Editor*.

Generating Driver Files Review

In this section we look at the files generated by the instrument driver wizard. As mentioned above, the instrument driver generates the instrument function panel (`.fp`) file, instrument driver source (`.c`) file, instrument driver include (`.h`) file, and the (`.sub`) file. Let us look at the contents of these files.

Function Panel File

The function panel file generated for the power supply is shown in Figure 8–14. This file, when opened, shows all the functions in a hierarchical function tree that are created by the instrument driver wizard for the power supply. The purpose of these function categories is discussed below.

Initialize Functions

Initialize functions allow you to initialize the software before you communicate with the instrument. There are two functions, *Prefix_init* and *Prefix_InitWithOptions,* that can initialize an IVI instrument driver *session*. When you initialize an IVI session, the IVI engine creates a data structure to store all the information for this session. Using the initialization functions, you have a choice of performing an ID query and sending a reset to the instrument identified in the *resource name* of the function argument. *Resource name* is a string used to identify the instrument with the hardware interface and the address. This is the same as the *Instrument Descriptor* discussed in the section *Basics of Programming with VISA* in *Chapter 5, VXI Communication Using VISA*. The resource name for this instrument driver is `GPIB::5::INSTR`. Both these initialization functions return the handle of the initialized session that identifies the session uniquely and is used in subsequent function calls. The *Prefix_InitWithOptions* has an additional string argument allowing you to set the following options: simulation mode, range checking, state-caching mechanism, and status checking.

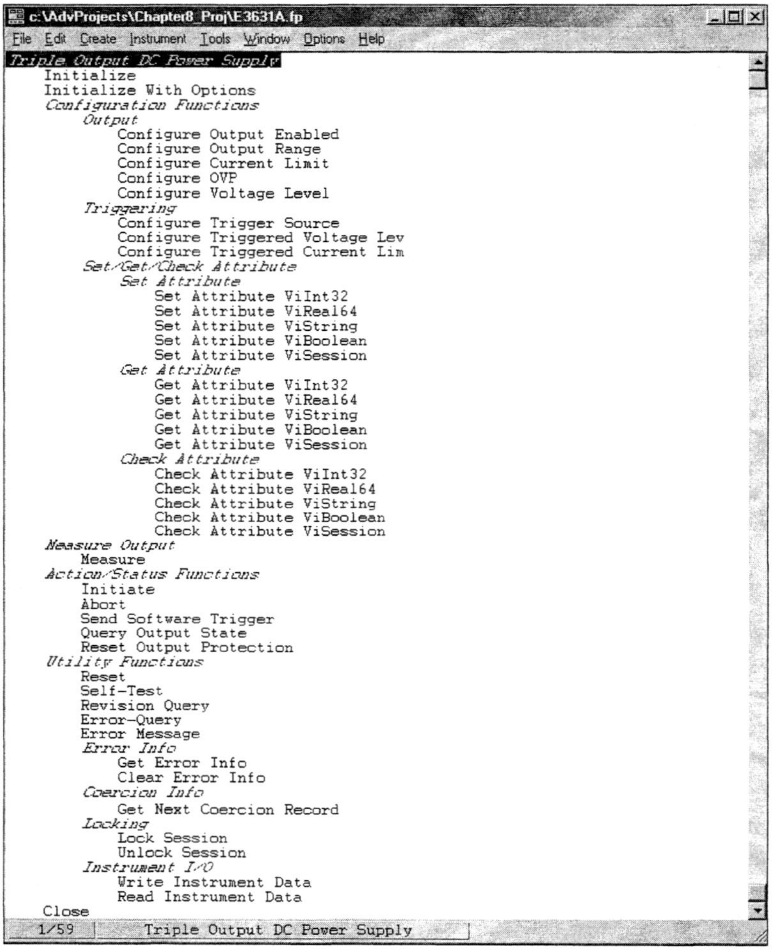

```
c:\AdvProjects\Chapter8_Proj\E3631A.fp
File Edit Create Instrument Tools Window Options Help
Triple Output DC Power Supply
    Initialize
    Initialize With Options
    Configuration Functions
        Output
            Configure Output Enabled
            Configure Output Range
            Configure Current Limit
            Configure OVP
            Configure Voltage Level
        Triggering
            Configure Trigger Source
            Configure Triggered Voltage Lev
            Configure Triggered Current Lim
        Set/Get/Check Attribute
            Set Attribute
                Set Attribute ViInt32
                Set Attribute ViReal64
                Set Attribute ViString
                Set Attribute ViBoolean
                Set Attribute ViSession
            Get Attribute
                Get Attribute ViInt32
                Get Attribute ViReal64
                Get Attribute ViString
                Get Attribute ViBoolean
                Get Attribute ViSession
            Check Attribute
                Check Attribute ViInt32
                Check Attribute ViReal64
                Check Attribute ViString
                Check Attribute ViBoolean
                Check Attribute ViSession
    Measure Output
        Measure
    Action/Status Functions
        Initiate
        Abort
        Send Software Trigger
        Query Output State
        Reset Output Protection
    Utility Functions
        Reset
        Self-Test
        Revision Query
        Error-Query
        Error Message
        Error Info
            Get Error Info
            Clear Error Info
        Coercion Info
            Get Next Coercion Record
        Locking
            Lock Session
            Unlock Session
        Instrument I/O
            Write Instrument Data
            Read Instrument Data
    Close
1/59 | Triple Output DC Power Supply
```

Figure 8–14
Power Supply Function Panel Tree

Configuration Functions

Configuration Functions are used to perform certain instrument operations. The number and type of configuration functions and subclasses depend on the instrument driver class template used to create the instrument driver. This class contains functions to configure the various instrument settings. This class may include setting the output of the instrument, configuring the triggering system of the instrument, or to query, modify, and validate the individual attributes of the instrument using the *Prefix_GetAttribute, Prefix_SetAttribute,* and *Prefix_CheckAttribute* functions for each of the instrument driver attributes data types. These functions, in turn, call the IVI library functions *Ivi_GetAttribute, Ivi_SetAttribute,* and *Ivi_CheckAttribute.* For a description of the IVI library functions, see the On-line Help.

Measure Output Functions

Measure Output Functions obtain the value of the measured output signal generated by the instrument.

Action/Status Functions

Action/Status Functions are used for initiating and aborting the instrument operations and to obtain the current or pending status of the instrument.

Utility Functions

Utility Functions consist of a variety of functions that are used to perform on the instrument. These functions are: sending a reset, obtaining the revision query from the instrument's firmware, obtaining instrument-specific error information, and translating the error code received from the instrument to a user-readable form. These functions also include getting and clearing the error information, getting the next coercion record from the IVI engine, invalidating the state of all attributes, locking/unlocking the instrument from interference from other execution threads, and reading and writing data to the instrument I/O directly.

Close Function

The *close function* is used to close the instrument driver session and deallocate the system resources.

Source File

The instrument driver *Source File* consists of code for the functions generated by the wizard and is grouped in logical categories for easy navigation in the source code. The section at the top of the instrument driver source file has suggestions on how to modify the code inside the function body. Read these suggestions, as you will need them to modify the various instrument driver functions.

Include File

The *Include File* contains the defined constant and constant names for the attributes and attribute values used by the instrument driver. This file also includes the function prototypes used in the instrument driver source file. You need to change this file only when you add, delete, or modify functions in your driver or add new attribute values. In the sections *Deleting High-Level Instrument Driver Functions* and *Adding High-Level Instrument Driver Functions* we explain how to make these changes.

.sub File

The *.sub File* has information about instrument driver attributes and their values. You can view this information using the *SetAttribute* or *GetAttribute* function panels. To do so, bring up the function panel for one of these functions in the Configuration Functions class in your driver and click on the **Attribute ID** control on the function panel. The **Select Attribute Constant** dialog window is displayed in Figure 8–15. To view the values of the attributes, select the attribute in the **Attributes:** list box, and the possible values of the attributes are displayed in **Attributes help:** list box in the lower half of the dialog window. National Instruments does not recommend editing this file directly. It should be edited using the **Attribute Editor** only.

You can modify the `.sub` file by using the **Attribute Editor** that is explained in the next section.

Using the Attribute Editor

The **Attribute Editor** is used to modify and navigate through instrument driver files that were created by the **Instrument Driver Development Wizard**. The **Attribute Editor** analyzes the contents of the *Prefix_InitAttributes* function in the driver source file and range tables in the `.c` file, and the contents of the `.sub` and `.h` files for the driver. The purpose of the *Prefix_InitAttributes*

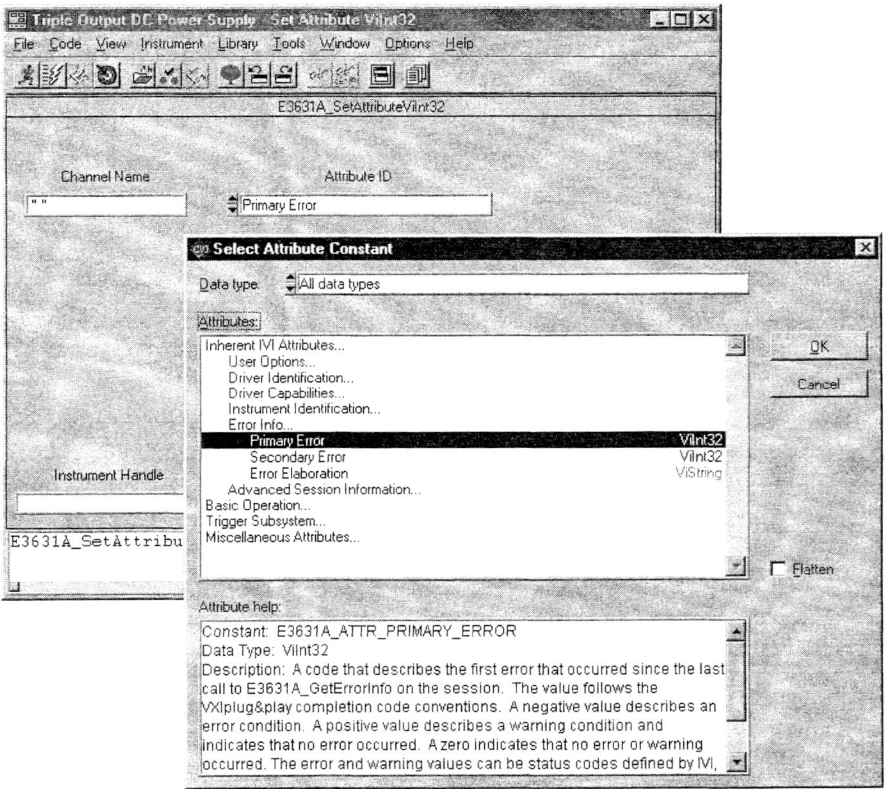

Figure 8–15
Select Attribute Constant Dialog Window

function is to add attributes to the IVI session, to initialize instrument attributes, and to set attribute invalidation dependencies.

You can use the **Attribute Editor** to add, edit, and delete the instrument attributes. You can also add and edit the range table(s) to define valid values for the attributes using the **Attribute Editor**. The IVI engine uses the range table to validate and coerce values for the attribute. Refer to the section *Range Tables* in *Chapter 2, IVI Architecture Overview* of *LabWindows/CVI Instrument Driver Developers Guide,* for a complete discussion of range tables.

As you saw above, the **Attribute Editor** can be launched from the **Finish** window of the wizard. You can also invoke the **Attribute Editor** by selecting

Tools>>Edit Instrument Attributes from a **Source Editor, Function Tree Editor**, or **Function Panel Editor** window. For the power supply instrument driver created by the wizard, the **Edit Driver Attributes** window is displayed in Figure 8–16. This figure is shown for *CVI* version 5.5. The later versions of *CVI* may not contain all the command buttons shown in this figure. There are significant changes to subdialog boxes, and the attributes no longer expand to show the callback functions. You should experiment with the various features of this window of the later versions of *CVI*. The features of the **Attribute Editor** for *CVI* 5.5 are explained below.

The **Edit Driver Attributes** window lists all the driver attributes created by the wizard in the **Instrument Attributes:** list box on the left side of this window in hierarchical order, with the group labels shown for the first and second levels. This information is read from the instrument driver .sub file created by the wizard.

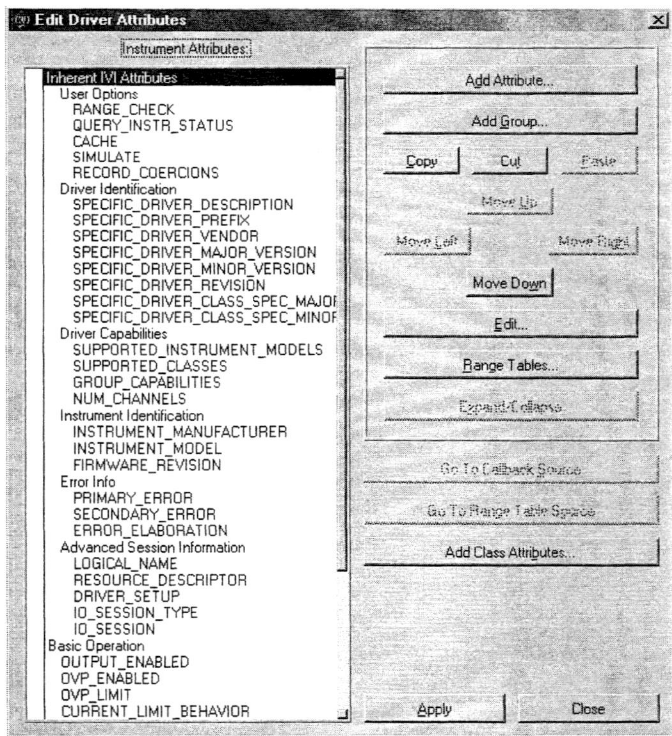

Figure 8–16
Instrument Driver Development Wizard: **Edit Driver Attributes** Window

Attribute Editor Controls

In this section you will see how to use the command buttons and other features of the **Attribute Editor** window. You will see how to add an attribute, add a group, move the attribute in the **Instrument Attributes:** list box, expand/collapse an attribute, view and modify the range tables, and add a new class attribute.

To add a new attribute to the list box, click on the **Add Attribute...** command button to display a blank **Edit Attribute** dialog window similar to Figure 8–17. You can enter the appropriate information for the new attribute in the dialog boxes in this window. To understand the use of these dialog boxes, refer to the section *Adding and Editing Instrument Attributes* in *Chapter 4, Attribute Editor* of *LabWindows/CVI Instrument Driver Developers Guide*.

To create a new group label in the **Instrument Attributes:** list box (Figure 8–16), click on the **Add Group...** command button. An empty **Edit Group** dialog window will appear similar to Figure 8–18, where you can enter the label and the text describing the group's purpose.

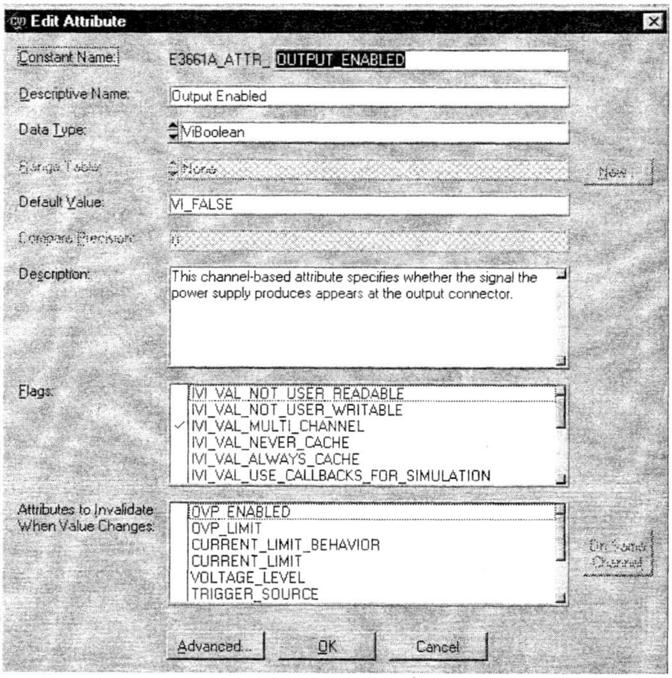

Figure 8–17
Attribute Editor: **Edit Attribute** Window

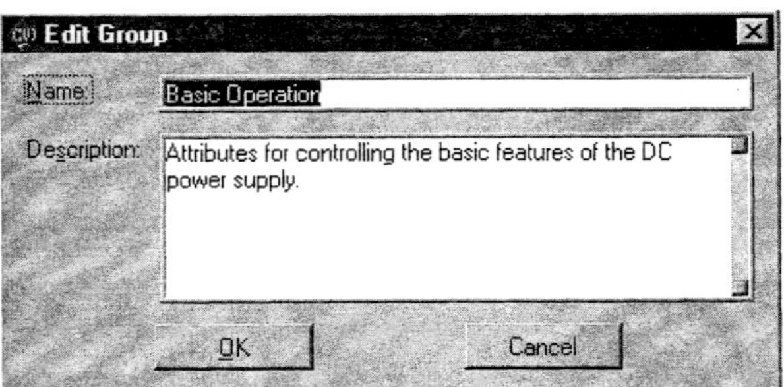

Figure 8–18
Attribute Editor: **Edit Group** Window

The attributes can be moved up and down in the list box by the **Move Up** and **Move Down** command buttons.

To indent the attributes left or right in the list box to enhance grouping and readability, use the **Move Left** or **Move Right** command buttons.

To edit an attribute, select the attribute and click on the **Edit...** command button. You can also edit an attribute by double clicking on the attribute or pressing <Enter> on the attribute. The **Edit Attribute** dialog window is displayed (Figure 8–17), from where you can edit the attribute information.

You can similarly edit a group label by highlighting the group label and clicking on the **Edit...** command button (Figure 8–16), pressing <Enter>, or double-clicking on the group label. The **Edit Group** dialog window is displayed as shown in Figure 8–18. Here you can change the displayed information.

The **Range Tables...** command button (Figure 8–16) is used to display a list of range tables in the driver source file in the **Range Tables** dialog window. You can add a new range table, edit an existing range table, and cut, copy, and paste the range table names in the list box. Refer to the section *Adding and Editing Range Tables* in *Chapter 4, IVI Architecture Overview* of *LabWindows/CVI Instrument Driver Developers Guide,* for a description of how to edit the range tables.

Use the **Expand/Collapse** command button (Figure 8–16) to expand or contract the attribute selected. You can also expand or contract the attribute by selecting the attribute and pressing <Space bar>. Expanding attribute results in a list of different types of attribute callback functions appearing under the attribute name. Some of the attribute callback types have a check mark next to them, indicating that the **Attribute Editor** found the name of a callback function in the *Ivi_AddAttribute* for the attribute or in a call to a function that references the attribute. The *Ivi_AddAttribute* function creates a new attribute of the type specified for an instrument session. You can disassociate a callback function from the attribute by removing the check mark. To do so, click on the check mark or press the <Space bar>. If you add the check mark to associate a callback type with the attribute that did not have a check mark initially, the **Attribute Editor** will associate a default callback function name with the attribute. When you select the **Apply** command button in the **Attribute Editor** dialog window (Figure 8–16), the **Attribute Editor** inserts the skeleton code for the new callback function in the driver's source file.

You can jump to the callback function body in the instrument driver's source code by selecting the callback function in the **Instrument Attributes:** list box and clicking **Go To Callback Source** command button. This operation will close the **Attribute Editor** dialog window; therefore, the **Attribute Editor** prompts you to save the changes before proceeding. The **Go To Callback Source** command button is enabled when you select the callback function for a certain attribute in the **Instrument Attributes:** list box; otherwise, it remains disabled.

Highlight the attribute and select **Go To Range Table Source** command button to go to a range table in the driver source file. This command button will be enabled only if the selected attribute has an associated range table. Again, this operation will close the **Attribute Editor** dialog window; therefore, the **Attribute Editor** prompts you to save the changes before proceeding.

If you want to add class attributes that you had previously deleted or a new version of class definition contains additional attributes, click on the **Add Class Attributes** command button. The **Attribute Editor** searches the header file for the `#include` statement for a class driver header file. If the appropriate `#include` statement is not found, the **Add Class Attribute** command button appears dimmed; otherwise, the **Attribute Editor** creates a list of class attributes currently not in the list box using the predefined instrument template for the class that you selected in Figure 8–2. If there is no new class attribute to be included, a pop-up panel displays a message to that effect.

The **Apply** command button is used to update the instrument driver files with the changes made in the **Attribute Editor**.

Note that the **Inherent IVI Attributes** group listed in the **Instrument Attributes:** list box (Figure 8–16) consists of the attributes common to all IVI instrument drivers that are called *inherent* attributes. You cannot edit, expand, cut, or copy an inherent attribute.

Editing High-Level Instrument Driver Functions

The **Instrument Driver Development Wizard** creates a function panel (`.fp`) file that includes the high-level instrument driver functions. The functions for the Triple Output DC Power Supply were shown in the form of a function tree in Figure 8–14.

You can edit the high-level functions in the function tree to modify the function source code to suit your particular instrument requirements. When the instrument driver creates the source code, it contains instructions and sample source code on how to modify the instrument-specific segment of code. These instructions appear in the source code that starts with "=CHANGE:" and ends with "END=CHANGE=". To edit a high-level function using the function tree, open the function panel file `E3631A.fp` created by the wizard from **File>> Open>>Function Tree (.fp)** to display the function tree as shown in Figure 8–19.

In the function tree, select the function to edit. For this example, select the *ErrorMessage* function and right-click on it to display the context menu as shown in Figure 8–19.

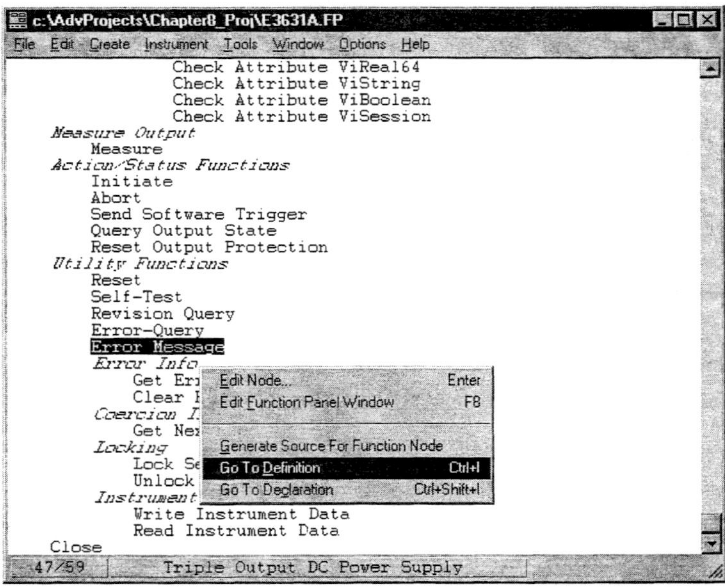

Figure 8–19
Function Tree Editor Context Menu

To modify the source code for this function, go to the function body in the source code by selecting the **Go To Definition** menu item on the context menu. The source code for *E3631A_error_message* function appears as shown in Figure 8–20.

Notice that the section marked between "=CHANGE:" and "END=CHANGE=" within comments between lines 17 and 22 has recommendations with sample source code. You need to modify the source code between

```
1   /*****************************************************************************
2    * Function: E3631A_error_message
3    * Purpose:  This function translates the error codes returned by this
4    *           instrument driver into user-readable strings.
5    *
6    *           Note:  The caller can pass VI_NULL for the vi parameter.  This
7    *           is useful if one of the init functions fail.
8    *****************************************************************************/
9   ViStatus _VI_FUNC E3631A_error_message (ViSession vi, ViStatus errorCode,
10                                          ViChar errorMessage[256])
11  {
12      ViStatus    error = VI_SUCCESS;
13
14      static      IviStringValueTable errorTable =
15          {
16
17          /*=CHANGE:=============================================================*
18              Insert instrument driver specific error codes here.  Example:
19
20              {E3631A_ERROR_TOO_MANY_SAMPLES,    "Sample Count cannot exceed 512."},
21
22           *=========================================================END=CHANGE=*/
23    IVIDCPWR_ERROR_CODES_AND_MSGS,
24              {VI_NULL,                              VI_NULL}
25          };
26
27      if (vi)
28          Ivi_LockSession(vi, VI_NULL);
29
30          /* all VISA and IVI error codes are handled as well as codes in the table */
31      if (errorMessage == VI_NULL)
32          viCheckParm( IVI_ERROR_INVALID_PARAMETER, 3, "Null address for Error Message");
33
34      checkErr( Ivi_GetSpecificDriverStatusDesc(vi, errorCode, errorMessage, errorTable));
35
36  Error:
37      if (vi)
38          Ivi_UnlockSession(vi, VI_NULL);
39      return error;
40  }
```

Figure 8–20
E3631A_error_message Driver Created Source Code

the comments per modification instructions. After you have made the changes to the source code, remove the "=CHANGE:" and "END=CHANGE=" comment lines and the modification instructions. Your source code should now look like as in Figure 8–21.

You can similarly modify the remaining functions in the source file by doing a search on "=CHANGE:" and editing the code according to the instructions given.

```
/*****************************************************************************
 * Function: E3631A_error_message
 * Purpose:  This function translates the error codes returned by this
 *           instrument driver into user-readable strings.
 *
 *           Note:  The caller can pass VI_NULL for the vi parameter.  This
 *           is useful if one of the init functions fail.
 *****************************************************************************/
ViStatus _VI_FUNC E3631A _error_message (ViSession vi, ViStatus errorCode,
                                         ViChar errorMessage[256])
{
    ViStatus    error = VI_SUCCESS;

    static      IviStringValueTable errorTable =
        {
            {E3631A_ERROR_RESET_PROT_NOT_SUPPORTED,     "The Reset Output  \
                                                         Protection is not supported."},
            {E3631A_ERROR_TRIGGER_ABORT_NOT_SUPPORTED, "The Abort trigger is not \
                                                                   supported."},
            IVIDCPWR_ERROR_CODES_AND_MSGS,
            {VI_NULL,  VI_NULL}
        };

    if (vi)
        Ivi_LockSession(vi, VI_NULL);

        /* all VISA and IVI error codes are handled as well as codes in the table */
    if (errorMessage == VI_NULL)
        viCheckParm( IVI_ERROR_INVALID_PARAMETER, 3, "Null address for Error Message");

    checkErr( Ivi_GetSpecificDriverStatusDesc(vi, errorCode, errorMessage, errorTable));

Error:
    if (vi)
        Ivi_UnlockSession(vi, VI_NULL);
    return error;
}
```

Figure 8–21
E3631A_error_message Modified Source Code

Deleting High-Level Instrument Driver Functions

The instrument driver may create high-level functions that are not used by your instrument. You may want to delete these functions. To do so you will have to delete the function definitions from the source file, delete the function prototypes in the header file, and delete the function panel from the function panel file.

The following steps need to be implemented.

1. In the function tree, right-click on the function that you want to delete. This will display the context menu (Figure 8–19). Select **Go To Declaration** menu item. This will take you to the function declaration in the instrument's header (`.h`) file.
2. Delete the declaration.
3. Right-click anywhere in the header file to bring up the context menu. Select the **Edit Function Tree** menu item from the context menu to go back to the function tree.
4. Right-click on the same function that you selected in step 1. The context menu is displayed. Select the **Go To Definition** menu item from the context menu to jump to the function definition in the instrument driver's source file.
5. Delete the entire function code.
6. Right-click anywhere in the source file to bring up the context menu. Select the **Edit Function Tree** menu item from the context menu to return to the function tree.
7. To delete the function name from the function panel, select **Edit>>Cut**.
8. Repeat steps 1 through 7 to delete all the functions one at a time. If you wish, you can also make all the edits in each file at the same time.

Adding High-Level Instrument Driver Functions

You may want to add more functions to the instrument driver to include some special instrument functions that were not created by the wizard. Here we go through the steps that will enable you to do so.

1. Open the instrument driver function panel file.
2. On the function tree displayed, move the mouse cursor to the location below where you want to add the new function.
3. To create a new class for the function, select **Create>>Class...** and add the class name in the **Name:** text box of the **Create Class Node** dialog window that is displayed.
4. To add a function to the already existing class (or the class created above in step 3), select **Create>>Function Panel Window**. The **Function Panel Selection** dialog window is displayed. In the **Name:** text box enter the name to display in the function tree. In the **Function Name:** text box add the actual function name that will be added to the instrument driver source file.
5. Create the function panel for this function as explained in the section *Creating a Function Panel* in *Chapter 7, Creating and Using Function Panels*.
6. From the function tree select the newly created function and right-click to display the context menu.
7. From the context menu select the **Generate Source For Function Node** menu item. This will create a function prototype in the instrument driver header file and the function definition in the instrument driver source file.
8. Right-click on the function tree to display the context menu and select the **Go To Definition** menu item. This will take you to the function body in the instrument driver source file.
9. Add the appropriate source code in the function body.

Creating Instrument Driver Documentation

This section will show you how to create the documentation for your instrument driver. *CVI* creates two types of documentation for your instrument driver: a text file with `.doc` extension and Windows Help with `.hlp` extension.

Creating the Instrument Driver Text File

To create an instrument driver text file, open the function panel file and select **Options>>Generate Documentation...**. A **Generate Documentation** dialog window is displayed (Figure 8–22). Highlight the programming language for which you like to generate documentation and click on **OK**. A *Prefix.doc* file is generated which can be used as a programmer's reference manual where *Prefix* is the function panel file base name (E3631A in this example). Even though the file extension is `.doc`, it is not a Microsoft Word document but a plain American Standard Code for Information Interchange (ASCII) text file. To view the document generated, click on `E3631A.doc`. This document contains a brief description of the instrument, assumptions about using the instrument drivers, the structure of the document, function tree layout with the function panel, and function names. Explanation of the functions appears alphabetically, with a description of the function and its prototype, a description of each argument, and a list of possible error codes. This document is not fully complete since it wants you to add the information relevant to your instrument by modifying the text marked between the delimiters "=CHANGE:" and "END=CHANGE=". Save the file after you have made the changes, keeping the same base file name.

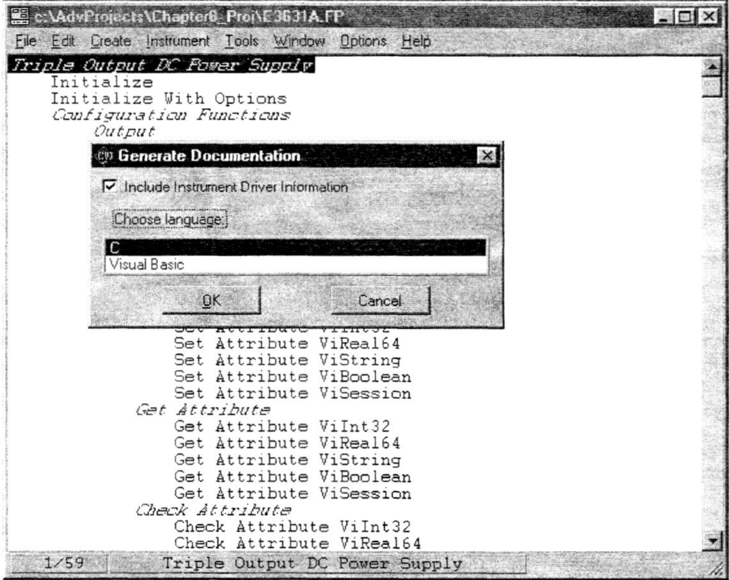

Figure 8–22
Generate Documentation Dialog Window

Creating the Instrument Driver Windows Help

To create an instrument driver Windows Help, open the function panel file and select **Options>>Generate Windows Help....** A **Generate Windows Help** dialog window appears as shown in Figure 8–23. Check-mark the boxes and select the language for which you want to generate the Windows Help. When you click **OK**, a successful completion message appears indicating that the help file *Prefix*.hlp (`E3631A.hlp` in this case) is generated in the same directory as the function panel file. As before, *Prefix* is the function panel file base name.

To view the contents of the `E3631A.hlp` file, double-click on the file name to display the On-line Help window.

Testing the Instrument Driver

As a final step you need to test all the instrument driver functions after you have modified the source code and the driver attributes. Create a *CVI* project to test all the settings of the instrument driver functions and run the application to

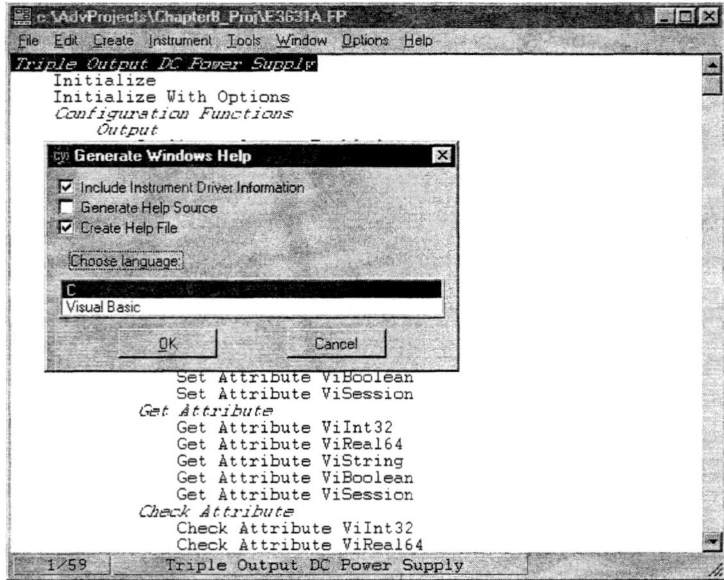

Figure 8–23
Generate Windows Help Window

verify that there are no errors. In the *CVI* project, add an `E3631A.fp`, `E3631A.c`, or `E3631A.lib` file. Recall that you can create the `.lib` file from the **Project** window by setting the **Target Type** to **Static Library** in the **Build** menu. Give the project the same base name of the `.lib` file you are creating (`E3631A`). In the source file of your application, add the include file `E3631A.h`. You should also create a standalone application implementing all the settings of the instrument driver functions and verify that all the functions are executed without any errors.

An intuitive exercise would be to create the front panel (soft panel) of the instrument with all the controls on the physical instrument (if the instrument has a front panel). When you operate the controls from the "soft panel," the instrument connected should respond using the driver software. This would be similar to operating the instrument manually by using the controls on the instrument's front panel.

Summary

In this chapter you obtained a taste of creating instrument drivers. You were stepped through the mechanics of creating an instrument driver for the power supply and shown how to modify the source code and the driver attributes. Creating instrument drivers is a vast topic and is covered in the *LabWindows/CVI Instrument Driver Developers Guide* manual. To fully understand this topic you need to familiarize yourself by reading this manual and practice creating instrument drivers for your instrument(s). As you create the instrument drivers, you will need to understand the IVI library functions given in On-line Help.

9

OPENGL

Chapter Highlights

- Introduction
- OpenGL Project
- Source Code Analysis
- OpenGL Properties Panel
- Summary
- Library Function Prototypes and Definitions

This chapter introduces you to creating OpenGL applications using *CVI*'s OpenGL instrument driver. The rudimentary concepts and terms of OpenGL are explained as necessary. This chapter does not explain the details of using OpenGL. For that you need to obtain information on OpenGL from other sources. The *CVI*'s OpenGL library functions are explained as used in the project created in this chapter. Using the project you will see how to control the plotted data in three dimensions on OpenGL control. Using the mouse and keyboard, you will be able to rotate, zoom, and pan the data plotted. You are shown how to save data to a print file or send it to a printer. You are introduced to the OpenGL properties pop-up panel, which allows you to change the attributes of the OpenGL control.

Introduction

OpenGL ("GL" stands for Graphics Libraries) is a collection of software libraries that interact with the graphics hardware of your computer to render color images of moving two-and three-dimensional objects at high speeds. It is a cross-platform standard that works on Windows, MacOS, Linux, and UNIX operating systems. OpenGL (OGL) was developed in 1992 by Silicon Graphics (SGI) and its standard maintained and updated by the OpenGL Architecture Review Board (ARB). The OpenGL ARB comprises of a consortium of leading computer graphics companies, such as 3Dlabs, Compaq, Evans & Sutherland, Hewlett-Packard, IBM, Intel, Intergraph, NVIDIA, and SGI.

Some of the basics of OpenGL and definition of the terms used in OpenGL are noted here. OpenGL interface consists of about 250 library functions (approximately 200 functions are part of the core OpenGL library and another 50 belong to the OpenGL Utility Library [GLU]). These functions allow you to specify the objects (*models)* and operations to *render* interactive graphical images. *Rendering* is the process by which a computer creates images from models. To build an OpenGL *model,* you use a small set of *geometric primitives*—points, line segments, and polygons to draw into a *framebuffer* with a specified mode that can be set and changed independently. The purpose of a *framebuffer* is to hold all the information that a graphic display needs to control the color and intensity of the pixels. A *pixel* (picture element) is defined as the smallest element that can be displayed on the screen and is a set of a number of bits defining the attributes of the pixel.

Geometric primitives are drawn with one or more vertices. A *vertex* can be a point or an intersection of two line segments. Starting with the vertices, you can build a line segment, a triangle, a polygon, or a combination of other complex geometric objects. Each vertex datum consists of positional coordinates, colors, *normal vector* (normal), and texture coordinates that are transmitted using the OpenGL library functions. A *normal vector* is a vector pointing in a direction perpendicular to the surface. Texture coordinates are used for assigning *texel* (texture element) in the texture map to a particular vertex of the object. Applying a texture map to a primitive gives that object a more realistic look.

OpenGL is designed to work on different windowing and operating systems, and therefore there are no commands implicit to OpenGL to control the opening of a window or to receive commands from the keyboard or mouse. The OpenGL Utility Kit (GLUT) is as a set of library functions written to bridge this gap. GLUT is used for opening windows and receiving inputs from the input devices. In addition, GLUT enhances operation of the OpenGL libraries by adding commands to create more complicated three-dimensional objects in addition to the simple geometric primitives such as points, lines, and polygons.

This introduction highlighted some of the basic features of OpenGL. To study OpenGL in detail, you need to read the references in the Bibliography. Other OpenGL terms and their use will be explained as they occur in this chapter. The purpose of the chapter is to introduce you to *CVI's* OpenGL instrument driver functions and their use in creating and controlling OpenGL applications created in *CVI*. This is shown in the next section by means of a project.

OpenGL Project

`project9.prj` uses the *CVI*'s OpenGL instrument driver and shows you some of the *CVI*'s OpenGL library functions for controlling and plotting data on the OpenGL control. To use the OpenGL library functions, you need to add the `OpenGL 3D Plotting Control` instrument driver (`cviogl.fp`) to your project. This file is located in the `toolslib\custctrl` folder, but for convenience has been copied to the `Chapter9_Proj` folder. When you run `project9.prj`, the data is loaded from the file you select from the ring control on the GUI and plotted on the OpenGL control. By using the keyboard and the mouse, you will be able to rotate, pan, and zoom the data plotted to view various orientations of your plot.

Load and run `project9.prj` to display the GUI showed in Figure 9–1. Let us look at some of the controls on this GUI. When you click on the **Select File to Plot** ring control, names of data files that contain the data are displayed and the data plotted on the **DATA PLOT** control. For example, these data files may contain data that you may have acquired from a data acquisition source, a mathematical model that you are trying to analyze, or from some other data that you are trying to view in three dimensions using OpenGL. The data is plotted using the colors that correspond to the magnitude of data. This color representation is shown by means of a color map on **Color Map** control during execution. The control labeled **DATA PLOT** is a picture control that you create by selecting **Create>>Picture** from the **User Interface Editor**. This picture control is converted to an OpenGL control programmatically, as you will see later.

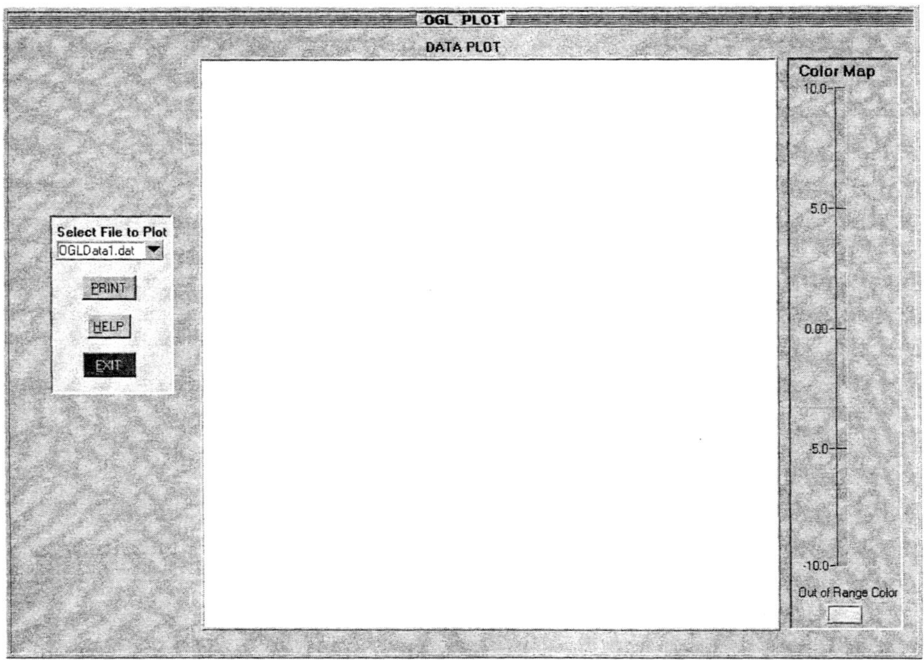

Figure 9–1
`project9` GUI

When you select a data file, a figure similar to Figure 9–2 is displayed with the data plotted on the **DATA PLOT** control.

You can view this plot using various orientations. To rotate the plot, hold down the left mouse button and move the mouse over the data plotted. To zoom in or zoom out, hold the <Ctrl> key down on the keyboard and with the left mouse button pushed down, move the mouse on the **DATA PLOT** control toward you to zoom in and away from you to zoom out. To pan the plot, hold down the <Shift> key, and with the left mouse button pushed down, move the mouse in the direction you want to move the plot. To view the instructions mentioned above for using OpenGL controls, click on the **HELP** command button to display Figure 9–3.

Experiment with the data plotted by rotating, zooming, and panning, and notice how convenient it is to view the plot from various angles. After you have selected a view of your data, you can print it or send it to a file by selecting the **PRINT** command button.

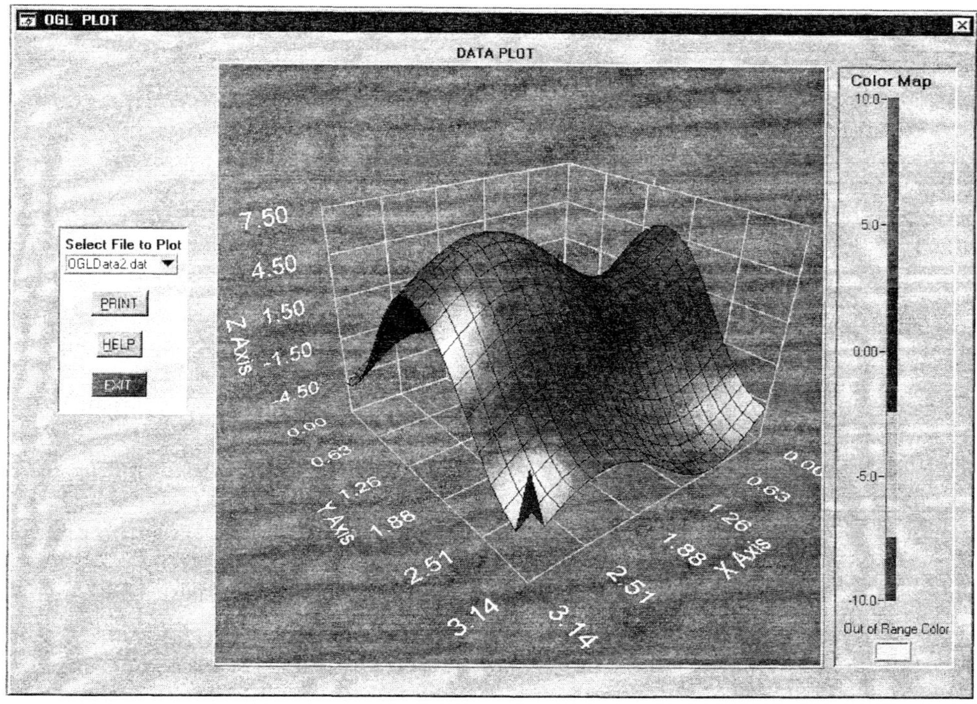

Figure 9–2
`project9` Data Plot

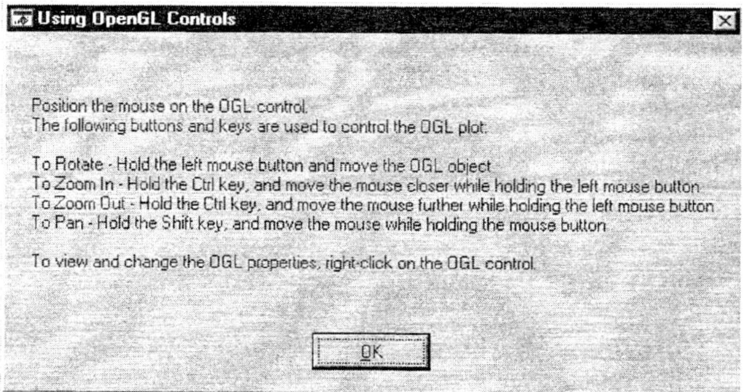

Figure 9–3
`project9` Using **OpenGL Controls** Help Panel

Source Code Analysis

In this section we examine the project's source code and see how *CVI*'s OpenGL library functions are used.

Header and *main* Function

The project's header files and the *main* function are listed in Figure 9–4. The *CVI* OpenGL header file, `cviogl.h`, at line 4 contains the data types, the macros, and *CVI*'s OpenGL function prototypes that are used with the OpenGL instrument driver.

```
1   #include <userint.h>
2   #include <ansi_c.h>
3   #include "project9.h"
4   #include "cviogl.h"
5   #include <formatio.h>
6
7   #define POINTS_X  21
8   #define POINTS_Y  21
9
10  #define PI  3.1415927
11
12  static int OGLPanelHandle;
13  static int OGLControl_ID;
14  static int SetOGLAttributes(void);
15  static int PlotOGLData(void);
16  static char buf[256];
17  double ReadArray[ POINTS_Y][ POINTS_X];
18  char  FileName[MAX_FILENAME_LEN];
19  static ColorMapEntry MapColor[5];
20
21  //Function prototypes
22  void CreateColorMap (int lowColor, int MediumLow, int medColor,int MediumHigh, int highColor);
23  int CreateColorScale (void);
24
25  //Help Message
26  #define HELP_MSG "\
27  \n\
28  \n\
29  Position the mouse on the OGL control.\n\
30  The following buttons and keys are used to control the OGL plot:\n\
31  \n\
32  To Rotate - Hold the left mouse button and move the OGL object \n\
33  To Zoom In - Hold the Ctrl key, and move the mouse closer while holding the left mouse button \n\
```

Figure 9–4
`project9` Header and *main* Function Listing *(continued)*

```
34    To Zoom Cut - Hold the Ctrl key, and move the mouse further while holding the left mouse
35    button\n\
36    To Pan - Hold the Shift key, and move the mouse while holding the mouse button \n\
37    \n\
38    To view and change the OGL properties, right-click on the OGL control.    \n\
39    \n\
40    "
41    //Program entry point
42    int main (int argc, char *argv[])
43    {
44        int error = 0;
45        if (InitCVIRTE (0, argv, 0) == 0)    /* Initialize CVI libraries */
46            return -1;   /* out of memory */
47
48
49
50        if ((OGLPanelHandle = LoadPanel (0, "project9.uir", OGLPANEL)) < 0)
51            return -1;
52
53        // Convert to OpenGL control
54        OGLControl_ID = OGLConvertCtrl(OGLPanelHandle,OGLPANEL_OGL_CTRL);
55        if (OGLControl_ID<0)
56        {
57            OGLGetErrorString (OGLControl_ID, buf, 255);
58            MessagePopup("OGLConvertCtrl Error", buf);
59            goto Error;
60        }
61
62        DisplayPanel (OGLPanelHandle);
63        //Setup CVIOGL control
64        SetOGLAttributes();
65        RunUserInterface();
66
67        //Discard CVIOGL control
68        OGLDiscardCtrl(OGLPanelHandle,OGLControl_ID);
69       return 0;
70
71    Error :
72       //Discard CVIOGL control
73        OGLDiscardCtrl(OGLPanelHandle,OGLControl_ID);
74        DiscardPanel(OGLPanelHandle);
75       return error;
76    }//main
```

Figure 9–4
project9 Header and *main* Function Listing *(continued)*

Lines 26–40 list the help message that is displayed on a pop-up panel (Figure 9–3) when you click on the **HELP** command button (Figure 9–1). The help message explains how to manipulate the OpenGL plot using the keyboard and the mouse buttons.

At line 54 the picture control (**DATA PLOT**) is converted to an OpenGL control using the *CVI*'s OpenGL library function *OGLConvertCtrl*. Note that the names of all *CVI* OpenGL library functions start with the prefix "OGL." Henceforth, the term *library function* will refer to *CVI*'s OpenGL Library function. The first argument in the library function *OGLConvertCtrl* is the panel handle (`OGLPanelHandle`) and the second argument is the picture control (`OGLPANEL_OGL_CTRL`) that you want converted. This function returns the control identifier (`OGLControl_ID`) for this OpenGL control that will be used in successive function calls when referring to this control. The converted control identifier is the same as the control identifier of the picture control. If the converting operation is successful, the control identifier is always a positive value. A negative value indicates an error. If there is any callback function and callback data associated with the picture control, the converted OpenGL control inherits them also.

If the value of the control identifier is negative, indicating an error, it is handled between lines 55 and 60. The library function *OGLGetErrorString* takes the error code as one of its arguments and returns the equivalent string that is displayed in a pop-up message panel. It there is an error, the program jumps to the label `Error` at line 71, where the OGL control and the panel are discarded and the program terminates, returning an error. Note that the *OGLGetErrorString* function can return the error string for both the User Interface Library and the OpenGL instrument driver since the error codes for these libraries are identified uniquely. The error codes for the User Interface Library is in the range –1 through –999 and the error codes for the OpenGL instrument driver are in the range –1001 through –1030. The error returns for all the library functions should be checked. For keeping this project code simple, the errors are not checked for most of the functions in this project.

Line 64 sets the attributes of the OpenGL control in the user-defined function *SetOGLAttributes*. This function is discussed in the section *Setting OpenGL Attributes* below.

After you have exited the project, you call the *OGLDiscardCtrl* library function at line 68 to release the resources by removing the OpenGL control (`OGLControl_ID`) from the panel and memory and deleting any plots on the control.

Load Data File

When you select the data file name from the **Select File to Plot** ring control, the *LoadPlotCB* callback function is invoked whose listing is shown in Figure 9–5.

```
//Select the data file, set the attributes, and plot on the OGL control
int CVICALLBACK LoadPlotCB (int panel, int control, int event,
        void *callbackData, int eventData1, int eventData2)
{
      int Index;
      switch (event)
        {
        case EVENT_COMMIT:
                //Get filename from ring control
                GetCtrlIndex (OGLPanelHandle, OGLPANEL_LOAD, &Index);
                GetLabelFromIndex (OGLPanelHandle,OGLPANEL_LOAD , Index,
                                                            FileName );
                //Plot OGL using selected file data
                PlotOGLData();
                break;
        }
      return 0;
} //LoadPlotCB
```

Figure 9–5
LoadPlotCB Source Listing

This callback function obtains the name of the data file selected from the ring control and plots the data contained in this file by calling the user-defined function *PlotOGLData*.

Setting OpenGL Attributes

Before you can plot on the OpenGL control, you need to set up the various attributes that will control the appearance of your plot. In the *SetOGLAttributes* function the OpenGL control background color, lighting attributes, axes labels, and axes grid attributes are set. The listing for the *SetOGLAttributes* function is shown in Figure 9–6.

You specify these attributes using the OpenGL library function *OGLSetCtrlAttribute.* This function sets the attribute selected on the OpenGL control specified. The first argument of *OGLSetCtrlAttribute* is the panel handle; the second argument is the OpenGL control identifier whose attribute you want to set. Here the attributes of the control identifier `OGLControl_ID` will be customized. Recall that this is the picture control that you created on the GUI and converted to an OpenGL control in the *main* function. The control attribute whose value needs to be set is the third argument of this function and will be explained as we walk through the source code. The value for this attribute is the last argument of the *OGLSetCtrlAttribute* function and dependent on the particular attributes.

```
//Set OGL attributes
static int SetOGLAttributes(void)
{
    //Enable Properties Pop-up panel
     OGLSetCtrlAttribute (OGLPanelHandle, OGLControl_ID,
                                            OGLATTR_ENABLE_PROPERTY_POPUP, 1);
    //Setup the OGL control background color
    OGLSetCtrlAttribute (OGLPanelHandle, OGLControl_ID, OGLATTR_BGCOLOR,
                                                            OGLVAL_DK_GRAY);


    //Setup Lighting Attributes
     OGLSetCtrlAttribute(OGLPanelHandle,OGLControl_ID,OGLATTR_LIGHTING_ENABLE,1);
     OGLSetCtrlAttribute(OGLPanelHandle,OGLControl_ID,OGLATTR_LIGHT_SELECT, 1);
     OGLSetCtrlAttribute(OGLPanelHandle,OGLControl_ID,OGLATTR_LIGHT_ENABLE, 1);
     OGLSetCtrlAttribute(OGLPanelHandle,OGLControl_ID,OGLATTR_LIGHT_DISTANCE, 3.0);
     OGLSetCtrlAttribute(OGLPanelHandle,OGLControl_ID,OGLATTR_VIEW_DISTANCE,2.5);
     OGLSetCtrlAttribute (OGLPanelHandle, OGLControl_ID, OGLATTR_PROJECTION_TYPE,
                                                          OGLVAL_PERSPECTIVE);


     //Setup Axis Labels
     OGLSetCtrlAttribute(OGLPanelHandle,OGLControl_ID,OGLATTR_XNAME_VISIBLE,1);
     OGLSetCtrlAttribute(OGLPanelHandle,OGLControl_ID,OGLATTR_YNAME_VISIBLE,1);
     OGLSetCtrlAttribute(OGLPanelHandle,OGLControl_ID,OGLATTR_ZNAME_VISIBLE,1);
     OGLSetCtrlAttribute(OGLPanelHandle,OGLControl_ID,OGLATTR_XNAME,"X Axis");
     OGLSetCtrlAttribute(OGLPanelHandle,OGLControl_ID,OGLATTR_YNAME,"Y Axis");
     OGLSetCtrlAttribute(OGLPanelHandle,OGLControl_ID,OGLATTR_ZNAME,"Z Axis");
     OGLSetCtrlAttribute(OGLPanelHandle,OGLControl_ID,OGLATTR_XLABEL_VISIBLE,1);
     OGLSetCtrlAttribute(OGLPanelHandle,OGLControl_ID,OGLATTR_YLABEL_VISIBLE,1);
     OGLSetCtrlAttribute(OGLPanelHandle,OGLControl_ID,OGLATTR_ZLABEL_VISIBLE,1);


     //Setup grid attributes
     OGLSetCtrlAttribute (OGLPanelHandle, OGLControl_ID, OGLATTR_XY_GRID_COLOR,
                                                            OGLVAL_LT_GRAY);

     OGLSetCtrlAttribute (OGLPanelHandle, OGLControl_ID,
                                                 OGLATTR_XY_GRID_VISIBLE, 1);


     OGLSetCtrlAttribute (OGLPanelHandle, OGLControl_ID,
                                      OGLATTR_YZ_GRID_COLOR, OGLVAL_LT_GRAY);
     OGLSetCtrlAttribute (OGLPanelHandle, OGLControl_ID,
                                      OGLATTR_YZ_GRID_VISIBLE, 1);


     OGLSetCtrlAttribute (OGLPanelHandle, OGLControl_ID,
                                     OGLATTR_XZ_GRID_COLOR, OGLVAL_LT_GRAY);
     OGLSetCtrlAttribute (OGLPanelHandle, OGLControl_ID,
                                     OGLATTR_XZ_GRID_VISIBLE, 1);



     return 0;
} //SetOGLAttributes
```

Figure 9–6
SetOGLAttributes Source Listing

When you right-click on the OpenGL control, a **Properties** pop-up panel is displayed (Figure 9–11) from where you can set the various OpenGL control properties. You can set the various properties of the OpenGL control at run-time also. This properties panel is active by default. By setting the value to `0` for the attribute `OGLATTR_ENABLE_PROPERTY_POPUP` in the function *OGLSetCtrlAttribute,* you can disable this pop-up panel. At line 5 it is shown enabled. The use of this function is redundant here but is included only if you would like to disable the **Properties** pop-up panel. You can also invoke the **Properties** panel by calling the *OGLPropertiesPopup* function. This function takes the panel handle as its first argument and the OpenGL control identifier as the second argument. At line 8 the OpenGL control background color is set to dark gray by specifying the `OGLATTR_BGCOLOR` attribute.

The lighting attributes for the OpenGL control are set at lines 12–18 and determine the lighting for the scene. With OpenGL, the lighting must be enabled or disabled explicitly. This is accomplished at line 12, where setting a 1 in the fourth argument of *OGLSetCtrlAttribute* enables the attribute `OGLATTR_LIGHTING_ENABLE`. This is the first lighting attribute that needs to be enabled in order to use the other lighting attributes. If this attribute is disabled, a global white ambient light illuminates the entire scene uniformly and will make the object appear two-dimensional. No lighting calculations concerning normals (the direction of light source), number of light sources, distance from the object, material properties of object, and lighting model will be performed. Note that for the OpenGL lighting model, white light consists of equal amounts of red, green, and blue light (RGB).

The `OGLATTR_LIGHT_SELECT` attribute in the *OGLSetCtrlAttribute* function at line 13 specifies which of the available lights to select for the individual attributes to be set. Here light `1` is selected, as shown in the fourth argument of *OGLSetCtrlAttribute*. You have a choice to select from lights 1 through 4. You turn on this light by setting the `OGLATTR_LIGHT_ENABLE` to a value `1` in the fourth argument of this function at line 14.

The `OGLATTR_LIGHT_DISTANCE` attribute at line 15 specifies the distance of the light selected from the center of the object. Here the distance is set to `3.0`. This value is based on the coordinate system of the object plotted.

The `OGLATTR_VIEW_DISTANCE` attribute at line 16 specifies the distance between the view position and the view center. The view distance here is `2.5` and is based on the coordinate system defined by the plot area. This attribute can be used to zoom into or out of the scene.

In the next line, the attribute `OGLATTR_PROJECTION_TYPE` determines the type of projection to display the object. There are two types of projections: orthographic and perspective. In an *orthographic projection* the size of the viewing volume does not change from one end of the scene to the other end. The actual size of the objects and angles between them is maintained. In a *perspective projection*, the view of a scene is what is more visible to a human eye (or camera) in real life, where closer objects appear to be larger than those farther away. At line 17 the projection is set to perspective projection using the attribute value as `OGLVAL_PERSPECTIVE`.

The attributes `OGLATTR_XNAME_VISIBLE`, `OGLATTR_YNAME_VISIBLE` and `OGLATTR_ZNAME_VISIBLE` at lines 21–23 specify whether the axes names are displayed. Lines 24–26 display the text names for the X-axis, Y-axis, and Z-axis as will appear on the OpenGL control when the data is plotted. The attributes specified for this are `OGLATTR_XNAME`, `OGLATTR_YNAME`, and `OGLATTR_ZNAME`, respectively. The attributes `OGLATTR_XLABEL_VISIBLE`, `OGLATTR_YLABEL_VISIBLE`, and `OGLATTR_ZLABEL_VISIBLE` at lines 27–29 make the names visible for the X-axis, Y-axis, and Z-axis, respectively.

At lines 33–36 the attribute `OGLATTR_XY_GRID_COLOR` for the XY plane grid color is set to light gray and displayed using the attribute `OGLATTR_XY_GRID_VISIBLE`. If you wish, you can create a color of your choice by using the *MakeColor* library function. Similarly, at lines 38–46 you are setting the grid color attribute for the YZ and XZ planes and making these grids visible.

Plotting Data

After you have set the OpenGL attributes, you need to plot the data using the *PlotOGLData* function whose listing is shown in Figure 9–7. Before you plot the data, you need to set up the various coordinates and plotting attributes. The starting point for the Z-axis plot area's coordinated system is set to `0.0` at line 10 using the attribute `OGLATTR_PLOTAREA_ZSTART`. The plot area coordinate system defines the space where plotting occurs.

At lines 12 and 13 the Z-axis size for the plot area's coordinate system is specified using the attribute `OGLATTR_PLOTAREA_ZSIZE`. You have to experiment with this value to proportionate your plot to lie within the OpenGL control. Too big a value will make the top part of the Z-axis scale lie outside the control region, and too small a value will not give you sufficient resolution on the Z-axis.

At lines 14 and 15 if you set the attribute `OGLATTR_VIEW_AUTO_DISTANCE` to a value `1`, it specifies that the view distance must be set automatically, ignoring the value of `OGLATTR_VIEW_DISTANCE` (line 16). Otherwise, setting the value to `0` signifies not to set the view distance automatically, but instead, to use the value indicated for this attribute (`OGLATTR_VIEW_DISTANCE`), which in this case is `2.5`. `OGLATTR_VIEW_DISTANCE` was explained above in the section *Setting OpenGL Attributes*.

```
//Plot OGL using data from the selected data file
static int PlotOGLData(void)
{

    int numPlots, i;
    int plotHandle;
    double X_Gain, Y_Gain, X_Offset,Y_Offset;

    // Setup control
    OGLSetCtrlAttribute (OGLPanelHandle, OGLControl_ID,OGLATTR_PLOTAREA_ZSTART, 0.0);

    OGLSetCtrlAttribute (OGLPanelHandle,
                                    OGLControl_ID,OGLATTR_PLOTAREA_ZSIZE,0.75);
    OGLSetCtrlAttribute (OGLPanelHandle,
                            OGLControl_ID,OGLATTR_VIEW_AUTO_DISTANCE, 0);
    OGLSetCtrlAttribute (OGLPanelHandle, OGLControl_ID,OGLATTR_VIEW_DISTANCE, 2.5);
    OGLGetCtrlAttribute (OGLPanelHandle,
                                OGLControl_ID,OGLATTR_NUM_PLOTHANDLES,&numPlots);

    //Delete any existing plots
    for (i=0;i<numPlots;i++)
    {
        OGLGetCtrlAttribute(OGLPanelHandle,OGLControl_ID,
                                OGLATTR_FIRST_PLOTHANDLE, &plotHandle);
        OGLDeletePlot(OGLPanelHandle,OGLControl_ID,plotHandle,0);
    }


     //Read data from the file into an array
      FileToArray (FileName, ReadArray, VAL_DOUBLE, POINTS_X * POINTS_Y , 1,
                                VAL_GROUPS_TOGETHER,
                                      VAL_GROUPS_AS_COLUMNS, VAL_ASCII);


    //Define the color map
    CreateColorMap (VAL_DK_GRAY, VAL_CYAN, VAL_MAGENTA, VAL_BLUE, VAL_RED);

    // Plot the color scale next to 3D plot to show this color map
    if (CreateColorScale() != 0)
    {
        MessagePopup ("Initialization Error", "Not able to plot on Color Map scale");
        return -1;
    }
     //Specify the gains and offsets
      X_Gain = PI/(POINTS_X-1);
      Y_Gain = PI/(POINTS_Y-1);
      X_Offset =0;
      Y_Offset =0;

    //Plot the OGL data
    plotHandle = OGLPlot3DUniform (OGLPanelHandle, OGLControl_ID,
                            ReadArray, POINTS_X, POINTS_Y,
                            OGLVAL_DOUBLE, X_Gain, X_Offset,  Y_Gain, Y_Offset);
```

Figure 9–7
PlotOGLData Source Listing *(continued)*

```
54       //Set color scheme of the plot
55       OGLSetPlotColorScheme (OGLPanelHandle, OGLControl_ID, plotHandle,
56                              OGLVAL_COLORMAP, MapColor, 5, VAL_YELLOW, 1, NULL, 0, 0);
57
58       //Setup plot attributes
59       OGLSetPlotAttribute(OGLPanelHandle,OGLControl_ID,plotHandle,
60                                           OGLATTR_SURFACE_STYLE,OGLVAL_SMOOTH);
61       OGLSetPlotAttribute(OGLPanelHandle,OGLControl_ID,plotHandle,
62                                           OGLATTR_SURFACE_SPECULAR_FACTOR,1.0);
63       OGLSetPlotAttribute(OGLPanelHandle,OGLControl_ID,plotHandle,
64                                           OGLATTR_SURFACE_SHININESS,50);
65
66       OGLSetPlotAttribute(OGLPanelHandle,OGLControl_ID,plotHandle,
67                                             OGLATTR_WIRE_STYLE, OGLVAL_SOLID);
68       OGLSetPlotAttribute(OGLPanelHandle,OGLControl_ID,plotHandle,
69                                             OGLATTR_WIRE_COLOR, OGLVAL_BLACK);
70
71       //Display plotHandle
72       OGLRefreshGraph(OGLPanelHandle, OGLControl_ID);
73
74       return 0;
75    }//PlotOGLData
```

Figure 9–7
PlotOGLData Source Listing *(continued)*

The library function *OGLGetCtrlAttribute* obtains the number of available plot handles in the variable `numPlots` using the attribute `OGLATTR_NUM_PLOTHANDLES` at lines 17 and 18. Between lines 21 and 26, if `numPlots` is greater than zero, it is used to delete any plots on the OpenGL control before plotting the new data.

The attribute `OGLATTR_FIRST_PLOTHANDLE` called in the function *OGLGetCtrlAttribute* at lines 23 and 24 obtains the first plot handle from the list of handles. If there are no plots in the control, the value of this attribute is zero. To delete the plot from the control, the library function *OGLDeletePlot* at line 25 deletes the plot using the plot handle obtained at lines 23 and 24.

The *FileToArray* library function at lines 30–32 opens the data file that you selected (Figure 9–2) and reads the data into an array `ReadArray`. You saw the *FileToArray* function used in *Chapter 4, Table Control,* and therefore it is not explained here.

The `ReadArray` data is an array consisting of the value plotted on the Z-axis at various locations of the XY grid. This is a two-dimensional array with the dimensions specified as

```
ReadArray[POINTS_Y][POINTS_X]
```

where POINTS_X is the size of the row dimension of the array, and POINTS_Y is the size of the column dimension of the array. These are specified in the arguments of the *OGLPlot3DUniform* library function, which is called at line 51 to create a uniform three-dimensional plot of the data on the OpenGL control.

It is more meaningful to display the plotted values in color in a way that each color represents a certain magnitude value of data. This is achieved by calling the user-defined function *CreateColorMap* at line 36, which assigns the color to the data plotted. We will see how this is accomplished in the section *Creating a Color Map* below. To associate the color to the magnitude of the data value, it is necessary to show the color representation of the data plotted by means of a color mapping scale. This is displayed as an intensity plot on the graph control labeled **Color Map** (Figure 9–1).

The X_Gain, Y_Gain arguments and X_Offset, Y_Offset arguments used in the *OGLPlot3DUniform* library function at lines 51–53 are used to specify the gain and offset applied to the x and y index values of ReadArray. The default values of the gain are 1.0, and for the offsets are 0. If the gains and offsets are used, the value of ReadArray[y][x] is the z-axis value, located on the graph at

```
[x*X_Gain + X_Offset, y*Y_Gain + Y_Offset]
```

Before you can plot data, you need to set up the color scheme for your data plot. The *OGLSetPlotColorScheme* library function at lines 55 and 56 determines the *color scheme* used and the color assigned to the data values in the array MapColor. The values for this array are set in the function *CreateColorMap,* as explained below in the section *Creating a Color Map*. You have a choice of the following color schemes:

- **None.** A solid color plot is displayed using the default color.
- **Shaded.** The color of the plot is interpolated between black and the default color.
- **Gray Scale.** The color of the plot is interpolated between black and white.
- **Color Spectrum.** The color of the plot uses seven predetermined colors distributed evenly over the range of the data values.
- **Color Map.** The color values are determined from the user-defined color map.
- **Color Array.** Colors values are determined from the user-defined color array.

In this project **Color Map** is used for the color scheme. Note that instead of using the *OGLSetPlotColorScheme* function, you can also assign the color scheme in the library function *OGLSetPlotAttribute* using the `OGLATTR_COLOR_SCHEME` attribute and selecting one of the color schemes above.

Lines 59–69 set up the various plot attributes. These attributes decide the appearance of your plotted data. You can set the point attributes, wire attributes, and surface attributes. Let us digress for a moment and explain these plot attributes.

The point attributes can specify the size of the point used for plotting, whether the points are smooth or not, and the RGB color of the point.

In OpenGL all the objects in a scene (or plot) are made of wires. Each line of wire corresponds to an edge of a primitive (typically, a side of a polygon). *Wireframe* is an object that consists of line segments only. The wire attributes specify the *wireframe* model features, such as the style of wire used (solid, dash, dot, dash-dot, and dash-dot-dot), the wire used to draw: smooth or otherwise, and the RGB color of the wire.

The surface attributes are used to specify the surface style. You can choose from the following surface styles: none, in which no surface is displayed; smooth; or flat. You can also set the RGB color of the surface, the surface's opaqueness or transparency, the *specular reflectance* of the material, and the surface shininess. *Specular reflectance* is the effect the material has on the reflected light, which is also dependent on the position of viewer. Shiny materials such as glass or high-gloss finishes have high specular reflectance, while objects like cork or carpets have low specular reflectance.

Examining the source code again, the surface style attribute, `OGLATTR_SURFACE_STYLE`, at lines 59 and 60 is set to `OGLVAL_SMOOTH` for a smooth surface plot. At lines 61 and 62, the specular reflectance factor is set using the attribute `OGLATTR_SURFACE_SPECULAR_FACTOR`. The range of specular reflectance factor is between `0.0` to `1.0`. The surface shininess is specified at lines 63 and 64 by assigning a value for the attribute `OGLATTR_SURFACE_SHININESS`. The range of shininess is between 0 (no shine) to `OGL_MAX_SURFACE_SHININESS` (maximum shininess has a value of 128). For an object to appear shiny, the `OGLATTR_SURFACE_SPECULAR_FACTOR` must be a nonzero value. You have to experiment with the attributes `OGLATTR_SURFACE_SPECULAR_FACTOR` and `OGLATTR_SURFACE_SHININESS` to obtain the desired shininess for your object.

The wire style and color is specified at lines 66–69 using the attributes `OGLATTR_WIRE_STYLE` and `OGLATTR_WIRE_COLOR`, respectively. The wire styles were discussed above and you can choose from among the following values: OGLVAL_NONE, OGLVAL_SOLID, OGLVAL_DASH, OGLVAL_DOT, OGLVAL_DASH_DOT, and OGLVAL_DASH_DOT_DOT. The wire color is set to black here. You can also use any color by setting the RGB values in the library function *MakeColor*, as indicated above.

The settings and changes only take effect when the library function *OGLRefreshGraph* is called at line 72 to refresh an OpenGL control. Note that to refresh the control, this function must be called in the same thread in which the control was created; otherwise you will get OGLErrorInvalidThread error.

Creating a Color Map

The *CreateColorMap* function is called in the *PlotOGLData* function to map colors to the magnitude of data values. The source listing for the *CreateColorMap* function is shown in Figure 9–8. Five color values are passed as arguments in the *CreateColorMap* function that are assigned to the array MapColor. You can specify the colors in the argument of the *CreateColorMap* function to indicate the colors for the different range of data values. For example, the MapColor array takes the color passed in the argument lowColor and assigns it to the range -10.0 to -5.0. The MapColor array is used in the function *OGLSetPlotColorScheme* at lines 55 and 56 in the *PlotOGLData* function (Figure 9–7) to map color to the magnitude of data values.

```
//Create a color map for data
void CreateColorMap (int lowColor, int MediumLow, int medColor,int MediumHigh, int highColor)
{
    MapColor[0].dataValue.valDouble = -10.0;
    MapColor[0].color = lowColor;
    MapColor[1].dataValue.valDouble = -5.0;
    MapColor[1].color = MediumLow;
    MapColor[2].dataValue.valDouble = 0.0;
    MapColor[2].color = medColor;
    MapColor[3].dataValue.valDouble = 5.0;
    MapColor[3].color = MediumHigh;
    MapColor[4].dataValue.valDouble = 10.0;
    MapColor[4].color = highColor;
} //CreateColorMap
```

Figure 9–8
CreateColorMap Source Listing

Creating a Color Scale

It is useful to show the representation of colors assigned to the data plot on the GUI. For this the **Map Color** control is created to act as a legend for the colors showing the data range. The *CreateColorScale* function is called from the *PlotOGLData* function above for this purpose. The source listing for *CreateColorScale* is shown in Figure 9–9.

The *CreateColorScale* function sets up the array to plot the range of data using the *PlotIntensity* library function. Some of the features of the *PlotIntensity* function are highlighted here.

The *PlotIntensity* function plots data whose colors correspond to the magnitude of data values in a two-dimensional array, `ColorMapArray`. This

```
//Display color map on the Color Map scale
int CreateColorScale (void)
{

    double ColorMapArray[5][2];
    int NumberOfXpoints, NumberOfYpoints, NumberOfColors,
        InterpolateColors,InterpolatePixels ;

    ColorMapArray[0][0] = -10.0;
    ColorMapArray[0][1] = -10.0;
    ColorMapArray[1][0] = -5.0;
    ColorMapArray[1][1] = -5.0;
    ColorMapArray[2][0] =  0.0;
    ColorMapArray[2][1] =  0.0;
    ColorMapArray[3][0] =  5.0;
    ColorMapArray[3][1] =  5.0;
    ColorMapArray[4][0] =  10.0;
    ColorMapArray[4][1] =  10.0;

    // Plot this array on the color scale graph control
    NumberOfXpoints = 2;
    NumberOfYpoints = 5;
    NumberOfColors =   5;
    InterpolateColors =   0;
    InterpolatePixels =    0;

    if (PlotIntensity (OGLPanelHandle, OGLPANEL_COLORMAP, ColorMapArray,
              NumberOfXpoints, NumberOfYpoints, VAL_DOUBLE, MapColor, VAL_YELLOW,
                                NumberOfColors,  InterpolateColors, InterpolatePixels) <= 0)
        return -1;
    return 0;
}//CreateColorScale
```

Figure 9–9
CreateColorScale Source Listing

function uses the `MapColor` array created in the *CreateColorMap* function above as one of its arguments to map the colors to the data magnitude values. The argument `InterpolateColors` indicates how to assign colors to Z array data values that do not exactly match the data values in the `ColorMapArray`. If `InterpolateColors` is zero, the data value is assigned the color associated with the next-higher `MapColor` data value. If `InterpolateColors` is nonzero, the data value is assigned a color that is computed using a weighted mean of the colors associated with the `MapColor` data values above and below the `ColorMapArray` data value. Similarly, the argument `InterpolatePixels` indicates how the adjacent pixels assigned to the data values are colored. When `InterpolatePixels` is zero, an unassigned pixel is given the same color as the closest assigned pixel. When `InterpolatePixels` is nonzero, an unassigned pixel is given a data value using a weighted mean of the data values associated with the four closest assigned pixels and allocated the color from `MapColor` to represent this value.

Printing the OpenGL Plot

After you have manipulated the OpenGL plot to display the data to your liking, you may want to save it in a file on your disk or send it to the printer. To do so you click on the **PRINT** command button on the GUI (Figure 9–2). This will invoke the *PrintCB* callback function whose source listing is shown in Figure 9–10.

The *SetWaitCursor* function, at lines 9 and 21, is part of the User Interface Library that sets the state of the wait cursor. The wait cursor is active when you pass a value `1` in the function argument, and is inactive when a `0` is passed. While the wait cursor is active, all other cursor styles are overridden until the cursor is deactivated. Here you need to activate the wait cursor, as you will copy the scaled bitmap image of the panel using the library function *OGLCopyScaledCtrlBitmap* (line 10) to the destination specified. Using the *OGLCopyScaledCtrlBitmap* function, you not only have a choice to copy the output to a file or a printer but can copy the image to a control on a different panel. The *OGLCopyScaledCtrlBitmap* function requires some explanation. Its function prototype is given below, and its arguments explained.

```
int OGLCopyScaledCtrlBitmap( int PanelHandle,
int OGL_Control_ID, int Maximum_Size,
int New_Width, int New_Height,
int Destination_Panel_Handle,
int Destination_Control_Id);
```

```
//Invoked from the PRINT command button
int CVICALLBACK PrintCB (int panel, int control, int event,
        void *callbackData, int eventData1, int eventData2)
{
    int error = OGLNoError, response;
    char PrintFileName[MAX_PATHNAME_LEN], buf[256];
    switch (event) {
        case EVENT_COMMIT:
            SetWaitCursor (1);
            OGLCopyScaledCtrlBitmap (OGLPanelHandle, OGLControl_ID, 0, -1, -1, -1, -1);
             response = ConfirmPopup ("PRINT LOCATION",
                                                  "Do you want to print to a file");
             if (response ==0) //Send output to printer
                           PrintPanel (OGLPanelHandle, "", 1, VAL_FULL_PANEL, 1);
            else //Output to File
            {
                   PromptPopup ("PRINT FILE NAME", "Enter Print File Name",
                                         PrintFileName, MAX_PATHNAME_LEN-1);
                   PrintPanel (OGLPanelHandle, PrintFileName, 0, VAL_FULL_PANEL, 1);
            }
            SetWaitCursor (0);
            break;
    }
    if (error != OGLNoError)
    {
       OGLGetErrorString (error, buf, 255);
        MessagePopup ("OGL Error", buf);
    }
    return error;
}//PrintCB
```

Figure 9–10
PrintCB Source Listing

In the *OGLCopyScaledCtrlBitmap* function, the first argument is the panel handle and the second argument is the OpenGL control identifier. The `Maximum_Size` argument allows you to specify whether the bitmap is to be scaled to the maximum possible resolution. Be aware that doing so will use a large amount of memory and will override the `New_Width` and `New_Height` values specified in the next two arguments. You can specify a `New_Width` and `New_Height` for the image copied by entering new values in these arguments. If you want to keep the same values as the displayed image, use `-1` for these arguments. The `Destination_Panel_Handle` is the handle of the panel on which the `Destination_Control_Id` control resides where you want to copy the bitmap image. Use a `-1` for `Destination_Panel_Handle` if the control lies on the same panel as the OpenGL control. Use a `-1` for `Destination_Control_Id` if the control is the parent control from which the OpenGL control is created. In such a case, be sure that the `Destination_Panel_Handle` is also set to `-1`.

Lines 11–20 determine the location of the copied image (file or printer). If you select a file, a pop-up panel will ask you to enter the file name, and the bitmap image is saved to this file in your default project folder. Otherwise, the Windows **Print** dialog window appears where you can select the printer and the printer options to send the image to the printer.

The User Interface Library function *PrintPanel* at lines 14 and 19 is used to print the panel selected. At line 14, the second argument in this function is an empty string indicating that no destination file is specified for the output, and therefore the output is sent to the printer. The *PrintPanel* function at line 19 sends the output to the file name specified in the second argument.

OpenGL Properties Panel

You should be aware of the **Properties** panel (Figure 9–11) that pops up when you right-click on the OpenGL control during program execution. By default the **Properties** panel is active when you create an OpenGL control. You can disable this panel as explained in the section *Setting OpenGL Attributes* above. This pop-up panel is built into the OpenGL control and allows you to change the attributes of the OpenGL control at run-time. At run-time you can use the **Properties** panel to override the OpenGL attributes that you had set through the software. Other attributes that were not specified in the software can also be changed using the **Properties** panel. Click on the tabs on **Properties** panel to change the settings on the various dialog windows that appear. The plotted data display is changed after you click on the **Apply** command button.

Summary

In this chapter you were given a taste of the concepts and terms used in OpenGL. The purpose of this chapter was to show how you could use OpenGL from *CVI*. You learned how to use the various functions in *CVI*'s OpenGL Library to control and plot data. The attributes of the library functions used in the project were explained, although there are many attributes that could not be covered here. To gain an understanding of the attributes available for a function, you need to bring up the function panel and look at a description and the values available for that attribute.

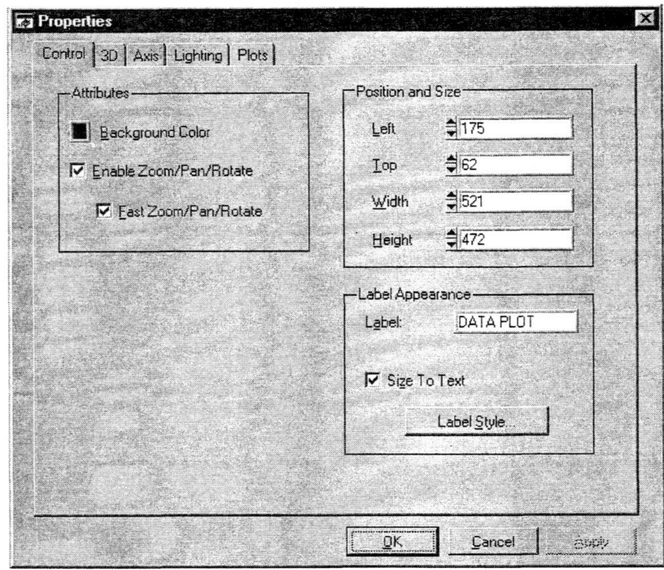

Figure 9–11
OpenGL Control **Properties** Panel

Library Function Prototypes and Definitions

This section lists alphabetically the *CVI* library functions that were introduced in this chapter.

OGLConvertCtrl Function

The *OGLConvertCtrl* function converts an existing picture control into an OpenGL control and returns the identifier for the converted control. Its prototype is shown below and its arguments explained in Table 9–1.

```
int OGLControlID = OGLConvertCtrl (int panelHandle,
                                   int Picture_to_Convert);
```

Table 9–1 OGLConvertCtrl *Function*

Input/ Output	Name	Type	Description
Input	*panelHandle*	integer	panel handle loaded in memory
	Picture_to_Convert	integer	control identifier of picture control to convert to OpenGL control
Output	`OGLControlID`	integer	identifier of the converted control; negative value indicates an error

OGLDeletePlot Function

The *OGLDeletePlot* function deletes the specified plot. Its prototype is shown below and its arguments explained in Table 9–2.

```
int status = OGLDeletePlot (int panelHandle,
                            int OGLControl_ID, int Plot_Handle,
                                                          int Refresh);
```

Table 9–2 OGLDeletePlot *Function*

Input/ Output	Name	Type	Description
Input	*panelHandle*	integer	panel handle loaded in memory
	OGLControl_ID	integer	identifier to reference OpenGL control
	Plot_Handle	integer	plot handle to delete returned by library functions *OGLPlot3DScatter* or *OGLPlot3DUniform*
	Refresh	integer	refresh control after plot is deleted; options are: `1` Yes `0` No
Output	*status*	integer	error status; 0 represents operation was successful; negative value indicates an error

OGLDiscardCtrl Function

The *OGLDiscardCtrl* function removes an OpenGL control from the panel and memory. The plots on the control are also deleted. Its prototype is shown below and its arguments explained in Table 9–3.

```
int status = OGLDiscardCtrl (int panelHandle,
                                    int OGLControl_ID);
```

OGLGetCtrlAttribute Function

The *OGLGetCtrlAttribute* function obtains the value of an OpenGL control attribute for the control specified. Its prototype is shown below and its arguments explained in Table 9–4.

```
int status = OGLGetCtrlAttribute (int panelHandle,
                int OGLControl_ID, int ControlAttribute,
                                 void *Attribute_Value);
```

OGLGetErrorString Function

The *OGLGetErrorString* function retrieves a text string from the error codes of the OpenGL instrument driver and the User Interface Library functions. Its prototype is shown below and its arguments explained in Table 9–5.

```
int status = OGLGetErrorString (int Error_Code,
                        char Buffer[], int Buffer_Length);
```

Table 9–3 OGLDiscardCtrl *Function*

Input/ Output	Name	Type	Description
Input	*panelHandle*	integer	panel handle loaded in memory
	OGLControl_ID	integer	identifier to reference OpenGL control
Output	*status*	integer	error status; 0 represents operation was successful; negative value represents error

Table 9–4 OGLGetCtrlAttribute *Function*

Input/ Output	Name	Type	Description
Input	*panelHandle*	integer	panel handle loaded in memory
	OGLControl_ID	integer	identifier to reference OpenGL control
	Control Attribute	integer	attribute whose value is required
Output	*Attribute_Value*	void*	value returned for the desired attribute
	status	integer	error status; 0 represents operation was successful; negative value represents error

Table 9–5 OGLGetErrorString *Function*

Input/ Output	Name	Type	Description
Input	*Error_Code*	integer	error code for which to retrieve the string text
	Buffer_Length	integer	length of the buffer containing the text
Output	*Buffer*	char[ ]	buffer to hold the returned error code text
	status	integer	error status; 0 represents operation was successful; negative value represents error

OGLPlot3DUniform Function

The *OGLPlot3DUniform* function creates a three-dimensional plot on the OpenGL control specified. Its prototype is shown below and its arguments explained in Table 9–6.

```
int status = OGLPlot3DUniform (int panelHandle,
             int OGLControl_ID, void *Array_of_Z_Values,
             int Number_of_X_points,
             int Number_of_Y_points,
             int Data_Type, double X_Gain,
             double X_Offset,
             double Y_Gain, double Y_Offset);
```

Table 9–6 OGLPlot3DUniform *Function*

Input/ Output	Name	Type	Description
Input	*panelHandle*	integer	panel handle loaded in memory
	OGLControl_ID	integer	identifier of OpenGL control on which the plot is drawn
	Array_of_Z_Values	void*	array of z values at various locations of XY grid
	Number_of_X_points	integer	size of row dimension of *Array_of_Z_Values*
	Number_of_Y_points	integer	size of column dimension of *Array_of_Z_Values*
	Data_Type	integer	data type of *Array_of_Z_Values* array; allowable data types are: ■ character ■ short integer ■ integer ■ float ■ double-precision ■ unsigned short integer ■ unsigned integer ■ unsigned character
	X_Gain	double	gain to be applied to the x location of *Array_of_Z_Values* index values; this is explained in the section *Plotting Data*
	X_Offset	double	offset to be applied to the x location of *Array_of_Z_Values* index values; this is explained in the section *Plotting Data*
	Y_Gain	double	gain to be applied to the y location of *Array_of_Z_Values* index values; this is explained in the section *Plotting Data*
	Y_Offset	double	offset to be applied to the y location of *Array_of_Z_Values* index values; this is explained in the section *Plotting Data*

(continued)

Table 9–6 OGLPlot3DUniform *Function (continued)*

Input/ Output	Name	Type	Description
Output	*status*	integer	error status; 0 represents operation was successful; negative value represents error

OGLPropertiesPopup Function

The *OGLPropertiesPopup* function displays the OpenGL properties panel and allows you to modify the control and plot attributes. Its prototype is shown below and its arguments explained in Table 9–7.

```
int status = OGLPropertiesPopup (int panelHandle,
                                 int OGLControl_ID);
```

OGLRefreshGraph Function

The *OGLRefreshGraph* function refreshes an OpenGL control. Its prototype is shown below and its arguments explained in Table 9–8.

```
int status = OGLRefreshGraph (int panelHandle,
                              int OGLControl_ID);
```

Table 9–7 OGLPropertiesPopup *Function*

Input/ Output	Name	Type	Description
Input	*panelHandle*	integer	panel handle loaded in memory
	OGLControl_ID	integer	identifier of OpenGL control to modify
Output	*status*	integer	error status; 0 represents operation was successful; negative value represents error

Table 9–8 OGLRefreshGraph *Function*

Input/ Output	Name	Type	Description
Input	*panelHandle*	integer	panel handle loaded in memory
	OGLControl_ID	integer	identifier of OpenGL control to refresh
Output	*status*	integer	error status; 0 represents operation was successful; negative value represents error

OGLSetCtrlAttribute Function

The *OGLSetCtrlAttribute* function sets the value of an OpenGL control attribute for the control specified. Its prototype is shown below and its arguments explained in Table 9–9.

```
int status = OGLSetCtrlAttribute (int panelHandle,
                                  int OGLControl_ID, int ControlAttribute,
                                                     void *Attribute_Value);
```

Table 9–9 OGLSetCtrlAttribute *Function*

Input/ Output	Name	Type	Description
Input	*panelHandle*	integer	panel handle loaded in memory
	OGLControl_ID	integer	identifier to reference OpenGL control
	Control Attribute	integer	attribute whose value is to be set
	Attribute_Value	void*	value to be set for the attribute desired
Output	*status*	integer	error status; 0 represents operation was successful; negative value represents error

OGLSetPlotAttribute Function

The *OGLSetPlotAttribute* function sets the value of an OpenGL plot attribute for the control specified. Its prototype is shown below and its arguments explained in Table 9–10.

```
int status = OGLSetPlotAttribute (int panelHandle,
                      int OGLControl_ID, int Plot_Handle,
                         int Plot_Attribute, void Attribute_Value);
```

OGLSetPlotColorScheme Function

The *OGLSetPlotColorScheme* function sets the color scheme to be used when the plot is drawn on an OpenGL control. This function also sets the data that may be required for each of those color schemes. Its prototype is shown below and its arguments explained in Table 9–11.

```
int status = OGLSetPlotColorScheme (int panelHandle,
                    int OGLControl_ID, int Plot_Handle,
                    int Color_Scheme,
                    ColorMapEntry *Color_Map_Entries,
                    int Number_of_Colors,
                    int High_Color, int Interpolate_Colors?,
                    int *Color_Array_Entries,
                    int X_Size, int Y_Size);
```

PlotIntensity Function

The *PlotIntensity* function plots a solid rectangle on a graph control. The plot consists of pixels whose colors correspond to the magnitude of the data values in a two-dimensional array. The coordinates of the plot are represented by the locations of the data values in the array. Its prototype is shown below and its arguments explained in Table 9–12.

```
int plotHandle = PlotIntensity (int panelHandle,
                 int Control_ID, void *ZArray,
                 int numberOfXpoints, int numberOfYpoints,
                 int zDataType, ColorMapEntry ColorMapArray[],
                 int hiColor, int numberOfColors,
                 int interpolateColors,
                 int interpolatePixels);
```

Table 9–10 OGLSetPlotAttribute *Function*

Input/ Output	Name	Type	Description
Input	*panelHandle*	integer	panel handle loaded in memory
	OGLControl_ID	integer	identifier of OpenGL control on which the plot is drawn
	Plot_Handle	integer	OpenGL plot handle whose attribute is to be set
	Plot_Attribute	integer	attribute whose value is to be set
	Attribute_Value	any type	value to set for the plot attribute
Output	*status*	integer	error status; 0 represents operation was successful; negative value represents error

Table 9–11 OGLSetPlotColorScheme *Function*

Input/ Output	Name	Type	Description
Input	*panelHandle*	integer	panel handle loaded in memory
	OGLControl_ID	integer	identifier of OpenGL control on which the plot is drawn
	Plot_Handle	integer	OpenGL plot handle whose attribute is to be set
	Color_Scheme	integer	color scheme used for the plot; select from the following available choices: ■ **None** ■ **Shaded** ■ **Gray Scale** ■ **Color Spectrum** ■ **Color** ■ **Color Array** For an explanation of the color schemes, see the section *Plotting Data*

(continued)

Table 9–11 OGLSetPlotColorScheme *Function (continued)*

Input/ Output	Name	Type	Description
	Color_Map_Entries	Color Map-Entry*	array of ColorMapEntry structure defining how the data values are translated to color values
	Number_of_ Colors	integer	number of entries in *Color_Map_Entries* array
	High_Color	integer	color to be used for data that is above the highest data value in the *Color_Map_ Entries* array
	Interpolate_ Colors?	integer	choice of how to assign colors for data points that do not exactly match the values in *Color_Map_Entries* array; for an explanation, see the section *Creating a Color Scale*
	Color_Array_Entries	integer*	array of colors for each data point in the plot; specified when the *Color_Scheme* is **Color Array,** otherwise use NULL
	X-Size	integer	X size of *Color_Array_Entries*
	Y_Size	integer	y size of *Color_Array_Entries*
Output	*status*	integer	error status; 0 represents operation was successful; negative value represents error

Table 9–12 PlotIntensity *Function*

Input/ Output	Name	Type	Description
Input	*panelHandle*	integer	panel handle loaded in memory
	Control_ID	integer	identifier of control on which to plot the data
	ZArray	void*	two-dimensional array of data values to convert to colors

(continued)

Table 9–12 PlotIntensity *Function (continued)*

Input/ Output	Name	Type	Description
	numberOfXpoints	integer	number of points to display along the x-axis in each row
	numberOfYpoints	integer	number of points to display along the y-axis in each column
	zDataType	integer	data type of ZArray; allowable data types are: ■ double ■ float ■ integer ■ unsigned integer ■ short integer ■ unsigned short integer ■ character ■ unsigned character
	ColorMapArray	Color-Map Entry	array of `ColorMapEntry` structure defining how the data values in *ZArray* are translated to color values
	hiColor	integer	color used for data that is above the highest data value in *ColorMapArray*
	numberOf Colors	integer	number of entries in *ColorMapArray*
	interpolate Colors	integer	how to assign colors for data points that do not exactly match the values in *ColorMapArray*; for an explanation, see the section *Creating a Color Scale*
	interpolate Pixels	integer	indicates how the adjacent pixels assigned to the data values are colored; for an explanation, see the section *Creating a Color Scale*
Output	*plotHandle*	integer	handle for the plot

PrintPanel Function

The *PrintPanel* function prints the panel selected to the file or printer specified. Its prototype is shown below and its arguments explained in Table 9–13.

```
int printStatus = PrintPanel (int panelHandle,
                              char fileName[], int scaling,
                              int scope, int confirmDialogBox);
```

Table 9–13 PrintPanel *Function*

Input/ Output	Name	Type	Description
Input	*panel Handle*	integer	panel handle loaded in memory
	fileName	char[ ]	name of output file created in the current folder if complete pathname is not specified; leave blank to send output to the printer
	scaling	integer	selects scaling mode for printing; available choices are: 1 expand object to the size of an entire page 0 print the object at same relative size on paper as on the screen
	scope	integer	selects portion of panel to print; avail able choices are: ■ VAL_VISIBLE_AREA. Print area visible on screen including Menu bars, scroll bars, and a frame. ■ VAL_FULL_PANEL. Print entire panel without menu bars, scroll bars, or frames.
	ConfirmDialogBox	integer	if enabled, shows the dialog box before printing to confirm the selected print attributes
Output	*printStatus*	integer	status of print operation; error status; 0 represents operation was successful; negative value represents error

SetWaitCursor Function

The *SetWaitCursor* function specifies the state of the wait cursor. When enabled, all other cursor styles are overridden to display the wait cursor. Its prototype is shown below and its arguments explained in Table 9–14.

```
int status = SetWaitCursor (int waitCursorState);
```

Table 9–14 SetWaitCursor *Function*

Input/ Output	Name	Type	Description
Input	*waitCursorState*	integer	state of wait cursor; choices are: 1 wait cursor active 0 wait cursor inactive
Output	*status*	integer	error status; 0 represents operation was successful; negative value represents error

BIBLIOGRAPHY

Deitel, H. M., and Deitel, P. J., *C How to Program*. Upper Saddle River, NJ: Prentice Hall, 1994.

Eckel, B., *Thinking in C++*. Upper Saddle River, NJ: Prentice Hall, 2000.

Edward, H., *Building an Interactive Web Page with DataSocket*. National Instruments Corporation (Application Note 127).

Hewlett-Packard Company. *HP 3631A Triple Output DC Power Supply, User's Guide*. Dallas, TX: Hewlett-Packard, April 1998 [Part Number E3631-90002].

Khalid, S. F., *LabWindows/CVI Programming for Beginners*. Upper Saddle River, NJ: Prentice Hall, 2000.

National Instruments Corporation. *DAQ Hardware Overview Guide*. Austin, TX: National Instruments, January 2000 [Part Number 370097A-01].

National Instruments Corporation. *Getting Started with LabWindows/CVI*. Austin, TX: National Instruments, 1999 [Part Number 320680E-01].

National Instruments Corporation. *Hot Technologies for Measurements*. Austin, TX: National Instruments, 2000 [Part Number 350656A-01]

National Instruments Corporation. *Integrating the Internet into Your Measurement System, DataSocket Technical Overview*. Austin, TX: National Instruments, 1998.

National Instruments Corporation. *LabWindows/CVI Advanced Analysis Library Reference Manual*. Austin, TX: National Instruments, 1998 [Part Number 320686D-01].

National Instruments Corporation. *LabWindows/CVI Instrument Driver Developer's Guide*. Austin, TX: National Instruments, 1999 [Part Number 320684E-01].

National Instruments Corporation. *LabWindows/CVI Programmer's Reference Manual*. Austin, TX: National Instruments, 1999 [Part Number 320685E-01].

National Instruments Corporation. *LabWindows/CVI User Interface Reference Manual*. Austin, TX: National Instruments, 1996 [Part Number 320683C-01].

National Instruments Corporation. *LabWindows/CVI User Manual*. Austin, TX: National Instruments, 1999 [Part Number 320681E-01].

National Instruments Corporation. *LabWindows/CVI VXI Course Manual*. Austin, TX: National Instruments, January 1999 [Part Number 320945D-01].

National Instruments Corporation. *NI-DAQ User Manual for PC Compatibles*. Austin, TX: National Instruments, January 2000 [Part Number 3216644F-01].

National Instruments Corporation. *NI-VISA Programmer Reference Manual*. Austin, TX: National Instruments, March 2000 [Part Number 370132A-01].

National Instruments Corporation. *NI-VISA User Manual*. Austin, TX: National Instruments, June 1998 [Part Number 321074D-01].

National Instruments Corporation. *PCI E Series User Manual*. Austin, TX: National Instruments, January 2000 [Part Number 370097A-01].

Press, W., Teukolosky, S., et al. *Numerical Recipes in C: The Art of Scientific Computing*, New York: Cambridge University Press, 1994.

Segal, M., and Akeley, K. *The Design of the OpenGL Graphics Interface*. Mountain View, CA: Silicon Graphics Computer Systems, 1994.

Wolfe, R. *Short Tutorial on VXI/MXI*. Austin, TX: National Instruments Corporation, April 1996 (Application Note 030).

Woo, M., Neider, J., et al. *OpenGL Programming Guide*. Reading, MA: Addison Wesley Longman, 2000.

INDEX

*IDN?, 204, 339, 342
*RST, 204, 340

A

A/D, 223
Accessing the DataSocket Server, 99–104
Accessory tab, 233
ActiveX, 122
  container
    DataSocket, 98
  control
    DataSocket, 98
ADC, 222
Adding High-Level Instrument Driver Functions, 359–60
address bus
  decoding, 181
  purpose of, 181
  widths, 181
address space
  A16 space, 181
  A24 space, 181
  A32 space, 181
AdvProject folder, 2
Agilent Technologies, 335
alias channel name
  creating, 239
aliasing, 225
Am913, 281
Am9513, 268, 273
  counter size, 267, 274
Am9513 MIO
  counter numbers, 267
Am9513A System Timing Controller, 264
American Standard Code for Information Interchange. *See* ASCII
analog input
  configure, 239
Analog Input Functions, 248–59
analog input signals, 222

Analog Input tab, 232
Analog Input/Output Parameters, 225–27
analog *multiplexer*, 223
analog output
  configure, 239
Analog Output channels
  polarity
    Bipolar, 232
    Bipolar External Reference, 232
    Unipolar, 232
    Unipolar External Reference, 232
Analog Output Functions, 259–61
Analog Output tab, 232
analog-to-digital converter, 222. *See* ADC
Analyzing the Reader Code, 90–97
Analyzing the Source Code-Project1-1, 3–22
Analyzing the Writer Code, 80–89
ANSI C, 94
AOClearWaveforms, 260, 261, 275
AOGenerateWaveforms, 260, 261, 275, 276
AOUpdateChannel, 259, 275
AOUpdateChannels, 259, 276
API. *See* application programming interface
APPL?, 208
application programming interface, 77
ARB, 366
ASCII, 361
AssertSysReset, 190, 210
ATTR_CELL_TYPE, 129
ATTR_COLUMN_WIDTH, 132
ATTR_ENABLE_COLUMN_SIZING, 124
ATTR_ENABLE_ROW_SIZING, 124
ATTR_ROW_HEIGHT, 128
ATTR_SIZE_MODE, 124, 127, 128, 130, 132
Attribute Editor, 346, 350, 351, 353, 354, 355
    creating new group label in list box, 353
  adding new attributes to list box, 353
  dialog window, 355
  indenting attributes in list box, 354
  launching, 351
  moving attributes in list box, 354
  used for, 351
Auto Size (when loaded)
  Size/Scroll Option
    check box, 133

B

backplane, 177
Base I/O Address, 247
Basics of Programming with VISA, 191–97
Begin Normal Operation, 190
Bipolar, 232
Bipolar External Reference, 232
Browsing the Table Control Dialog Window, 125–36

C

cage, 177
callback function
  purpose, 333
calling convention
  for CVI's libraries and DLL functions, 317
capture waveform, 254
Change Input Control Type
  selection box, 329
Channel List String
  dialog box, 338, 339
Chassis Address, 237
Chassis ID, 237
class drivers
  purpose, 333
class template, 346
  definition, 336
  General Purpose, 337
  used with instruments, 337

ClipboardGetTableVals, 150, 151
ClipboardPutTableVals, 150, 151
code width, 227
  equation, 227
CodeBuilder
  not used, 2
communication
  GPIB, 332
  serial interface, 332
  VXI, 332
configuration address space, 181
configuration registers, 181
configuration window
  registers, 181
Configuring SCXI, 237
ContinuousPulseGenConfig, 269, 270, 272, 278
ContinuousUpdate, 47
count limit action
  COUNT_CONTINUOUSLY, 267
  COUNT_UNTIL_TC, 267
Count Register, 264, 266, 268, 273
counter, 263
  Count Register, 264
  GATE input, 264
  input/output, 263
  OUT output, 264
  pulse state, 264
  SOURCE input, 264
  terminal count, 264
  toggle state, 264
Counter Applications, 265–74
  used for, 265
counter event, 263
Counter Fundamentals, 263–65
Counter I/O tab, 235
Counter/Timer
  purpose, 223
CounterEventOrTimeConfig, 266, 268, 270, 272, 274, 278
CounterMeasureFrequency, 273, 281
CounterRead, 266, 268, 273, 281
counters/timers, 222
CounterStart, 266, 267, 268, 273, 281
CounterStop, 266, 268, 281
crate, 177
CreateMetaFont, 61
Creating A Function Panel, 308–19, 321
Creating a Function Tree, 304–8
Creating an Instrument Driver, 334–47
Creating Graph Legends, 46–50
Creating Instrument Driver Documentation, 360–62
Creating the Instrument Driver text file, 360–62
Creating the Instrument Driver Windows Help, 362
CVI's OpenGL Library functions
  referred to as, 372
cviogl.fp, 367

D

D/A, 222, 223
DAC, 226, 259
  purpose, 223
DAQ, 176, 219, 220
  library functions, 219
  meaning, 220
  PCMCIA card, 220
DAQ board
  consists of, 222
  generate analog output signals, 222
  MIO, 222
  read analog input signals, 222
  used in various configurations, 220
DAQ board parameters
  Base I/O Address, 246
  Direct Memory Address, 246
  Interrupt Level, 246
  setup, 246

DAQ Channel Wizard, 219, 229, 237, 238, 255, 259
DAQ Designer, 227
DAQ Designer Tool, 219
DAQ drivers, 228
DAQ Library Functions
  types of, 248
DAQ system
  consists of, 220
DAQ Troubleshooting Wizard, 230
DAQ-STC, 265, 267, 268, 273, 281, 294. *See also* Data Acquisition System Timing Controller
  counter numbers, 267
  counter size, 267, 274
Data Acquisition. *See* DAQ
  fundamental steps, 248
Data Acquisition Board Architecture, 222–23
Data Acquisition Overview, 220–22
Data Acquisition System Timing Controller, 265
data bus
  width, 181
Data Neighborhood, 238
Data Neighborhood folder, 229, 237
DataSocket
  access method, 77
  communicates with, 77
  data socket transfer protocol, 77
  data transfer rate, 78
  file scheme, 77
  file size limitations, 78
  file transfer protocol, 77
  Overview, 76–77
  publisher, 77
  streaming data, 76
  subscriber, 77
  technology, 75
  what is, 76
DataSocket Applications, 97–99
DataSocket Data file format, 77, 89
DataSocket Data Files, 77–78
DataSocket object
  configurations, 83
    AutoUpdate mode, 84
    non-AutoUpdate mode, 84
DataSocket Server, 77, 78, 79, 82, 89
  communication, 77
DataSocket Server Manager
  configuring, 99
  path, 99
  Permissions Groups, 101
  Predefined Data Items, 102
  Server Settings, 100
  settings, 78
  what is, 99
  window, 99
DataSocket Server Manager Configurations, 99–104
DDE, 76
Default Resource Manager, 200
Default Setup Command
  dialog box, 338
DelayedPulseGenConfig, 270, 271, 284
DeleteGraphPlot, 21, 24, 41
DeleteTableRows, 138, 152
Deleting High-Level Instrument Driver Functions, 359
Device Class. *See* VXI Device Classes
device number, 229
Devices and Interfaces folder, 229, 230
differential, 225
DIG_Line_Config, 261, 284
DIG_Prt_Config, 261, 284
digital applications
  categories of, 261
Digital I/O, 261
  configure, 239
digital I/O port
  configuration, 261
Digital I/O tab, 235
Digital Input/Output Functions, 261–63

digital input/output ports, 222
digital lines, 261
digital signals, 222
digital-to-analog, 222
digital-to-analog converter. *See* DAC
Direct Memory Access, 247
DiscardPanel, 4, 25, 53
DisplayPanel, 4
DMA. *See* Direct Memory Address
DMM, 333
driver function
  names, 338
driver name, 335
DS_ControlLocalServer, 82, 105
DS_DiscardObjHandle, 87
DS_GetAttrHandle, 95
DS_GetAttrValue, 95, 105
DS_GetDataType, 108
DS_GetDataValue, 94, 109
DS_GetLastMessage, 85, 94, 111
DS_GetLibraryErrorString, 83
DS_GetStatus, 91, 112
DS_Open, 83, 84, 85, 87, 89, 91, 95, 112, 113
DS_SetAttribute, 87
DS_SetAttrValue, 87, 114
DS_SetDataValue, 87, 116
DS_Update, 87, 96, 116
duty cycle
  definition, 268

E

E1441A Arbitrary Waveform Generator, 197
E3631A.doc, 361
E3631A.hlp, 362
E3631A_error_message
  listing, 357
Easy I/O for DAQ library functions
  analog input functions, 248
  analog output functions, 259
  digital input/output functions, 262
  support
    Counter Timers, 248
    Digital I/O, 248
Easy I/O for DAQ library functions, 248
Edit Attribute
  dialog window, 353, 354
Edit Cell
  command button, 134
Edit Cell Group
  command button, 134
Edit Column
  command button, 130
  dialog window, 130
Edit Default Cell Values
  command button, 130, 132
  dialog boxes, 132
  dialog window, 130
Edit Graph dialog box
  Callack Function, 35
  Constant Name, 35
  Control Settings
    Control Mode, 37
    Copy Original Data, 37
    Data Mode
      Discard, 37
      Retain, 37
    Data Mode, 37
    Edit Axis Settings
      Auto Divisions, 38
      Auto Scale, 39
      Axis Name, 38
      dialog box, 38
      Display Format, 39
      Divisions, 38
      Eng. Units, 39
      Gain, 38
      Log Scale, 39
      Loose Fit, 39

Edit Graph dialog box, *(continued)*
Loose Fit Units, 39
Mark Origin, 40
Maximum, 38
Minimum, 38
Offset, 38
Padding, 39
Precisions, 38
Reverse Axis, 39
Show Grid, 39
Show Lables, 39
Use Label Strings, 39
Use Label Strings command button, 39
Enable Zooming, 37
Right Y-axis, 38
Smooth Update
advantage, 37
disadvantage, 37
X-axis, 38
Y-axis, 38
Control Settings, 37
Cursors..., 35
Edit Cursors
Color, 35
Cross Hair Style, 35
Cursor Number, 35
Enabled, 35
Mode, 35
Free Form, 35
Snap to Point, 35
Number of Cursors, 35
Point Style, 35
Use Right Y Axis, 35
Edit Group
dialog window, 353, 354
Edit Row
command button, 127
Edit State, 123
Edit Table
dialog window, 125
Editing High-Level Instrument Driver Functions, 356–58
encrypted hyper text transfer protocol, 77
Error Query
dialog window, 342
Error Query Response Contents
pull-down box, 342
Ethernet, 78, 221
Event Counting and Timing, 266–68
Expected ID Query Response
dialog box, 339

F

file transfer protocol, 77
FileToArray, 141, 153, 378
fillmode, 257, 258
Grouped by Channel, 258
Grouped by Channels, 258
Grouped by Scan, 258
FillTableCellRange, 141, 153
FindResourcesCB, 200
firewall, 97
Format Choices
dialog box, 340, 342
Format String
dialog box, 340, 342
indicator box, 340
framebuffer, 366
frequency
counters used to measure, 273
Frequency Measurement, 273–74
ftp, 77
function name
prefix, 305
function panel, 332
change control
Edit>>Change Control Type, 329
changing control type, 329

create control
Create>> Return Value, 313
Create>>Binary..., 324
Create>>Global Variable..., 327
Create>>Message..., 328
Create>>Numeric..., 322
Create>>Output..., 311
Create>>Ring..., 326
Create>>Slide..., 323
definition, 302
dialog box
Create Binary Control, 324, 326
Create Function Panel Window Node, 307
Create Input Control, 310
Create Message Control, 328
Create Output Control, 311
Create Return Value Control, 313
Create Ring Control, 326
Edit Create Global Variable Control, 327
Edit Slide Control, 323, 324
execute code
Code>>Run Function Panel, 303
inserting
Code>>Insert Function Call, 303
save function panel
Save .FP File, 313
Save .FP File As..., 307
Select Function Panel, 303
selecting, 303
used in instrument drivers, 332
view
View>>Recall Function Panel, 304
function panel controls, 321–30
binary control, 321, 324
global control, 321, 327
input control, 309
message control, 321, 328
numeric control, 321
output control, 309
return value control, 309
ring control, 321, 326
slide control, 321, 323
Function Panel Editor, 352
Function Panel Window, 307
function tree, 303
classes, 303
Create>> Instrument, 305
Create>>Classs, 305
Create>>Function Panel Window..., 307
creating
New>>Function Tree (*.fp)., 304
dialog box
Create Class Node, 305
Function Tree Editor, 304, 305, 317, 352
Function/Arbitrary Waveform Generator, 197, 333

G

gain, 227, 379
gate mode, 267
options, 274
COUNT_WHILE_GATE_HIGH, 267
COUNT_WHILE_GATE_LOW, 267
START_COUNTING_ON_FALLING_EDGE, 268
START_COUNTING_ON_RISING_EDGE, 268
UNGATED_SOFTWARE_START, 267
General Command Strings
dialog window, 338
General Information
dialog window, 337, 338
generated code box, 302
Generated Driver Files Review, 347
generated files
.sub file, 350
action/status functions, 349
close function, 349

generated files, *(continued)*
  configuration functions, 348
  function panel, 347
  include file, 350
  intialization functions, 347
  measure output functions, 349
  source file, 349
  utility functions, 349
geometric primitives, 366
GetActiveTableCell, 146, 156, 166
GetAttribute, 350
GetAxisScalingMode, 41, 62
  axisScaling attribute
    VAL_AUTOSCALE, 63
    VAL_LOCK, 63
    VAL_MANUAL, 63
GetBitmapFromFile, 136, 156
GetCtrlAttribute, 19, 20
GetCtrlIndex, 14, 42
GetCtrlVal, 19
GetGraphCursor, 42, 56, 63
GetGraphCursorIndex, 42, 64
GetLabelFromIndex, 14
GetNumTableRows, 146, 156
GetSystemTime, 87
GetTableCellFromVal, 157
GetTableCellFromValue, 149
GetTableCellVal, 146, 159
GLU, 366
GLUT, 366, 367
GPIB, 176, 191
GPIB Address
  control box, 343
Graph Attributes and Cursors, 34
graph control, 258
Graph controls
  cursor controls, 33
  legends, 33
  panning, 33
  plotting objects, 33
    arc, 56
    line, 56
    oval, 58
    rectangle, 58
    text, 52
  zooming, 33
Graph cursors
  number of, 35
  viewing plotted data, 35
graph legend
  purpose, 46
group label
  selecting, 354
GroupByChannel, 259, 286
GUI
  programmatically creating, 1

H

Hardware Configurations, 246–48
Height
  dialog box, 127
  row, 127
help text
  adding to function panel, 316
Hewlett-Packard, 335
HideBuiltInCtrlMenuItem, 136, 159
HLA. *See* VISA High-Level Access
How Does DataSocket Communicate?, 76–77
HTML. *See* Hyper Text Markup Language
http. *See* hyper text transfer protocol
https. *See* encrypted hyper text transfer protocol
Hyper Text Markup Language, 98
hyper text transfer protocol, 77

I

I/O Interface
list box, 336
types, 336
Icon Editor
utility, 136
icons
Create Binary..., 324
Create Global Variable..., 327
Create Input..., 310
Create Message..., 328
Create Numeric..., 322
Create Output..., 311
Create Return Value, 313
Create Ring..., 326
Create Slide..., 323
Insert Function Call, 303
Run Function Panel, 303
ID Query Command
dialog box, 339
ID/Logical Address, 184
IEEE 488.2 commands, 204
input range, 226
Insert Row Above
command button, 127
Insert Row Below
command button, 127
InsertListItem, 9
InsertTableColumns, 159
InsertTableRows, 138, 150, 159
InstallCtrlCallback, 9
Installing and Setting Up the DAQ Board, 228–38
InstallPopup, 56
INSTR, 200
INSTR Resource
purpose, 194
instrument
error queue, 342
firmware revision, 342
identification string, 339
reset, 340
self-test, 340
Instrument Descriptor, 347
format of, 192, 193
syntax, 192
instrument driver
adding functions, 359
attribute data types
hidden
ViAddr, 346
ViBoolean, 346
ViInt32, 346
ViReal64, 346
ViSession, 346
ViString, 346
attribute definition, 333
creating
text file, 360
Windows Help, 362
creation steps, 334
definition, 332
deleting functions, 359
files created, 345
files generated, 337
inherent attribute, 355
interfaces, 332
prefix, 337
selecting attribute, 354
instrument driver architecture
IVI, 332
VXIplug&play, 331, 332
Instrument Driver Development Wizard, 334, 337, 350, 356
creating driver
based on existing driver, 336
new instrument driver, 336
files created
instrument driver .sub file, 335

Instrument Driver Development Wizard,
(*continued*)
instrument driver include file, 335
instrument driver source file, 335
instrument function panel file, 334
launching, 335
using the, 334
Instrument Driver Introduction, 332–34
instrument driver *session*, 347
instrument driver source code
modifying, 356
instrument prefix, 335
dialog box, 336, 337, 338
instrument's identification string
entering, 340
Instrument>>Load, 308
instrumentation amplifier, 223
Interactive Execution Window, 303
Interface Type, 192
Interrupt Level, 247
Introduction to OpenGL, 366–67
IVI, 237, 331
architecture, 332
enhancements, 332
functions implemented, 345
skeleton driver files, 337
IVI drivers
communication
using attributes, 332
IVI engine, 332, 333, 347, 349, 351
IVI features
Instrument Interchangeability, 333
Instrument Simulation, 333
Instrument state-caching, 333
multithread safety, 334
IVI Foundation, 332, 333
IVI standard
required functions
*Prefix*_ClearErrorInfo, 346
*Prefix*_GetAttribute<type>, 346
*Prefix*_GetErrorInfo, 346
*Prefix*_GetNextCoercionRecord, 346
*Prefix*_InitWithOptions, 346
*Prefix*_InvalidateAllAttributes, 346
*Prefix*_IviClose, 346
*Prefix*_IviInit, 346
*Prefix*_LockSession, 346
*Prefix*_ReadInstrData, 346
*Prefix*_SetAttribute<type>, 346
*Prefix*_UnlockSession, 346
*Prefix*_WriteInstrData, 346
Ivi_AddAttribute, 354
Ivi_CheckAttribute, 349
Ivi_GetAttribute, 349
Ivi_LockSession, 334
Ivi_SetAttribute, 349
Ivi_UnlockSession, 334
IW. *See* Interactive Execution Window

L

LAN. *See* local area network
Legend Control
Function Panel, 46
instrument driver
legend.fp, 46
library functions, 46
new in *CVI* 5.5, 46
LGCreateLegendControl, 48, 64
attribute
relativePosition, 65
LGInsertLegendItemForPlot, 49, 50, 66
LGSetLegendCtrlAttribute, 48, 50, 67
attributes, 67
Library Functions
AOClearWaveforms, 275
AOGenerateWaveforms, 276
AOUpdateChannel, 275
AOUpdateChannels, 276

AssertSysReset, 210
ClipboardGetTableVals, 151
ClipboardPutTableVals, 151
ContinuousPulseGenConfig, 278
CounterEventOrTimeConfig, 278
CounterMeasureFrequency, 281
CounterRead, 281
CounterStart, 281
CounterStop, 281
CreateMetaFont, 61
DelayedPulseGenConfig, 284
DeleteGraphPlot, 24
DeleteTableRows, 152
DIG_Line_Config, 284
DIG_Prt_Config, 284
DiscardPanel, 25
DS_ControlLocalServer, 105
DS_GetAttrValue, 105
DS_GetDataType, 108
DS_GetDataValue, 109
DS_GetLastMessage, 111
DS_GetLibraryErrorString, 111
DS_GetStatus, 112
DS_Open, 113
DS_SetAttrValue, 114
DS_SetDataValue, 116
DS_Update, 116
FileToArray, 153
FillTableCellRange, 153
GetActiveTableCell, 156
GetAxisScalingMode, 62
GetBitmapFromFile, 156
GetGraphCursor, 63
GetGraphCursorIndex, 64
GetNumTableRows, 156
GetTableCellFromVal, 157
GetTableCellVal, 159
GroupByChannel, 286
HideBuiltInCtrlMenuItem, 159
InsertTableColumns, 159
InsertTableRows, 159
LGCreateLegendControl, 64
LGInsertLegendItemForPlot, 66
LGSetLegendCtrlAttribute, 67
MakeDir, 118
MakePoint, 161
MakeRect, 163
NewCtrl, 26
NewCtrlMenuItem, 164
NewPanel, 26
nidaqAICreateTask, 288
nidaqAIDestroyTask, 288
nidaqAIRead, 290
nidaqAIScanOp, 290
nidaqAISinglePointOp, 291
nidaqAISingleScanOp, 292
nidaqAIStart, 292
nidaqAIStop, 293
nidaqGetErrorString, 294
OGLConvertCtrl, 386
OGLDeletePlot, 387
OGLDiscardCtrl, 388
OGLGetCtrlAttribute, 388
OGLGetErrorString, 388
OGLPlot3DUniform, 389
OGLPropertiesPopup, 391
OGLRefreshGraph, 391
OGLSetCtrlAttribute, 392
OGLSetPlotAttribute, 393
OGLSetPlotColorScheme, 393
PlotArc, 67
PlotIntensity, 393
PlotLine, 70
PlotOval, 70
PlotRectangle, 71
PlotText, 72
PlotY, 28
PrintPanel, 397

Library Functions, *(continued)*
- *PulseWidthOrPeriodMeasConfig*, 294
- *ReadFromDigitalLine*, 296
- *ReadFromDigitalPort*, 297
- *RefreshGraph*, 30
- *SetActiveTableCell*, 166
- *SetAxisScalingMode*, 73
- *SetBreakOnLibraryErrors*, 118
- *SetDir*, 120
- *SetPanelAttribute*, 30
- *SetTableCellAttribute*, 166
- *SetTableCellRangeAttribute*, 168
- *SetTableCellRangeVals*, 169
- *SetTableCellVal*, 166
- *SetTableColumnAttribute*, 169
- *SetTableRowAttribute*, 170
- *SetWaitCursor*, 398
- *ShowBuiltInCtrlMenuItem*, 170
- *SortTableCells*, 172
- *viAssertUtilSignal*, 210
- *viClose*, 211
- *viFindNext*, 212
- *viFindRsrc*, 212
- *viIn16*, 212
- *viOpen*, 212
- *viOpenDefaultRM*, 215
- *viOut16*, 215
- *viRead*, 215
- *viSetAttribute*, 216
- *viStatusDesc*, 217
- *viWrite*, 218
- *WriteToDigitalLine*, 298
- *WriteToDigitalPort*, 299

LLA. *See* VISA Low-Level Access
local area network, 75
Logical Address
- represented as, 181

M

mainframe, 177
MakeColor, 376, 381
MakeDir, 88, 118
MakePoint, 141, 142, 161
MakeRect, 142, 163, 164
Manual Zooming and Panning, 43–46
Manuals
- *LabWindows/CVI Instrument Driver Developers Guide*, 333, 351, 353, 354, 363
- *LabWindows/CVI User Interface Reference*, 20
- *NI-VISA Programmer's Reference Manual*, 194

marker cursor, 35
MAX. *See* Measurement &Automation Explorer
maximum terminal count value
- setting, 267

Measurement & Automation Explorer, 229, 238
Measurement & Automation folders
- Data Neighborhood, 229
- Devices and Interfaces, 229
- IVI, 229
- Scales, 229
- Software, 229

MEMACC Resource
- purpose, 194

Message-Based device
- capabilities
  - Asynchronous Communication Capability, 188
  - Instrument Capability, 188
  - Master Capability, 188
  - *Programmable Handlers*, 188
  - *Programmable Interrupters*, 188
- communication registers
  - Data High register, 189
  - Data Low register, 189
  - Protocol/Signal register, 189

Response register, 189
metafonts, 50
available choices, 52
creating, 59
Min Num Lines Visible
control box, 130
MIO, 261, 265
Analog Input Circuitry, 222
analog multiplexer, 222
instrumentation amplifier, 222
sample and hold circuitry, 222
block diagram, 222
MIO DAQ
components of, 222
Mode, 129, 232
Moving Around in the Table Control, 123–24
multifunction input/output. *See* MIO
multi-rate scanning, 248
MXIbus, 179

N

NEGATIVE_POLARITY, 270
NewCtrl, 9, 10, 26
controlLeft, 26
controlTop, 26
NewCtrlMenuItem, 164
NewPanel, 8, 15
attributes
panelHorizontalCoord, 27
panelVerticalCoord, 27
NI-DAQ syntax, 253, 259
structure, 253
nidaqAICreateTask, 248, 249, 252, 254, 255, 288
task types, 254
nidaqAIDestroyTask, 255, 288
nidaqAIRead, 248, 255, 290
nidaqAIScanOp, 256, 257, 259, 290
nidaqAISinglePointOp, 256, 291
nidaqAISingleScanOp, 256, 292
nidaqAIStart, 248, 255, 292
nidaqAIStop, 249, 255, 293
nidaqDestroyTask, 249
nidaqGetErrorString, 254, 294
nidaqRead, 255
non-instrument driver, 332
nonreferenced single-ended, 225
normal vector, 366
normals, 375
NRSE. *See* nonreferenced single-ended

O

Object Linking and Embedding, 77
offset, 379
what is, 181
OGL_MAX_SURFACE_SHININESS, 380
OGLATTR_BGCOLOR, 375
OGLATTR_COLOR_SCHEME, 380
OGLATTR_ENABLE_PROPERTY_POPUP, 375
OGLATTR_FIRST_PLOTHANDLE, 378
OGLATTR_LIGHT_DISTANCE, 375
OGLATTR_LIGHT_ENABLE, 375
OGLATTR_LIGHT_SELECT, 375
OGLATTR_LIGHTING_ENABLE, 375
OGLATTR_NUM_PLOTHANDLES, 378
OGLATTR_PLOTAREA_ZSIZE, 376
OGLATTR_PLOTAREA_ZSTART, 376
OGLATTR_PROJECTION_TYPE, 375
OGLATTR_SURFACE_SHININESS, 380
OGLATTR_SURFACE_SPECULAR_FACTOR, 380
OGLATTR_SURFACE_STYLE, 380
OGLATTR_VIEW_AUTO_DISTANCE, 376
OGLATTR_VIEW_DISTANCE, 375, 376

OGLATTR_WIRE_COLOR, 380
OGLATTR_WIRE_STYLE, 380
OGLATTR_XLABEL_VISIBLE, 376
OGLATTR_XNAME, 376
OGLATTR_XNAME_VISIBLE, 376
OGLATTR_XY_GRID_COLOR, 376
OGLATTR_XY_GRID_VISIBLE, 376
OGLATTR_YLABEL_VISIBLE, 376
OGLATTR_YNAME, 376
OGLATTR_YNAME_VISIBLE, 376
OGLATTR_ZLABEL_VISIBLE, 376
OGLATTR_ZNAME, 376
OGLATTR_ZNAME_VISIBLE, 376
OGLConvertCtrl, 372, 386
OGLCopyScaledCtrlBitmap, 383, 384
OGLDeletePlot, 378, 387
OGLDiscardCtrl, 372, 388
OGLErrorInvalidThread, 381
OGLGetCtrlAttribute, 378, 388
OGLGetErrorString, 372, 388
OGLPANEL_OGL_CTRL, 372
OGLPlot3DUniform, 379, 389
OGLPropertiesPopup, 375, 391
OGLRefreshGraph, 381, 391
OGLSetCtrlAttribute, 373, 375, 392
OGLSetPlotAttribute, 380, 393
OGLSetPlotColorScheme, 379, 380, 381, 393
OGLVAL_PERSPECTIVE, 376
OLE. *See* Object Linking and Embedding
OLE for Process Control, 77
opaqueness, 380
OPC. See OLE for Process Control
OPC Servers, 84
OPC tab, 233
OpenGL
  ARB Consortium, 366
  color scheme, 379
  definition, 366
  distance from object, 375
  header file
    cviogl.h, 370
  instrument driver, 367
    CVI's OpenGL, 367
    cviogl.fp, 367
    error codes, 372
    OpenGL 3D Plotting Control, 367
  lighting attributes, 375
  manipulating plotted data, 368
    pan, 368
    rotate, 368
    zoom, 368
  material properties, 375
  number of light sources, 375
  platforms, 366
  plot attributes, 380
  point style attributes, 380
  projections
    orthographic, 375
    perspective, 376
  Properties panel, 375, 385
  surface style attributes, 380
  Utility Kit. *See* GLUT
  Utility Library. *See* GLU
  wire style attributes, 380
  wire styles, 381
OpenGL Architecture Review Board. *See* ARB
OpenGL instrument driver, 370
OpenGL Plot
  pan plotted data, 367
  rotate plotted data, 367
  zoom plotted data, 367
OpenGL Project, 367–85
Operate Mode
  icon, 133
Options >> Create Object File..., 320

Options>>Generate Function Prototypes, 317
oscilloscope, 333

P

pan
  meaning, 34
parallel, 221
PCMCIA DAQ card, 220
PC-TIO-10
  counter numbers, 267
pixel, 366
PlotArc, 57, 67
PlotData, 41
PlotIntensity, 382, 393
PlotLine, 56
PlotOval, 58, 70
PlotRectangle, 58, 71
PlotStripChart, 255
PlotText, 59, 72
Plotting Geometric Patterns on Graph Control, 50–59
PlotUniformCB, 15, 22
PlotY, 22, 41
Point
  structure, 141
point by point, 254
Polarity, 232
Polarity/Range, 232
port
  digital, 261
POSITIVE_POLARITY, 270
power supply, 333, 335, 337, 339, 352
Prefix, 346
Prefix.doc, 361
Prefix.hlp, 362
Prefix_CheckAttribute, 349
*Prefix*_ClearErrorInfo, 346
*Prefix*_close, 345
*Prefix*_error_message, 345
*Prefix*_error_query, 345
*Prefix*_GetAttribute, 346, 349
*Prefix*_GetAttribute<type>, 346
*Prefix*_GetErrorInfo, 346
*Prefix*_GetNextCoercionRecord, 346
*Prefix*_init, 345, 347
Prefix_InitAttributes, 350
*Prefix*_InitWithOptions, 346, 347
  options
    range checking, 347
    simulation, 347
    state-caching, 347
    status checking, 347
*Prefix*_InvalidateAllAttributes, 346
*Prefix*_IviClose, 346
*Prefix*_IviInit, 346
*Prefix*_LockSession, 334, 346
*Prefix*_ReadInstrData, 346
*Prefix*_reset, 345
*Prefix*_revision_query, 345
*Prefix*_self_test, 345
*Prefix*_SetAttribute, 346, 349
*Prefix*_SetAttribute<type>, 346
*Prefix*_UnlockSession, 334, 346
*Prefix*_WriteInstrData, 346
PrintPanel, 385, 397
Project window, 335
Project window list, 2
project1-1
  main, 3
  project1-1.c, 2
project5
  code for
    Finding System Resources, 200
    header and the main function, 198
  Configuring the function generator code, 205
  Setting up Communication code, 202

Projects
MyFunctions.prj, 304, 308, 319
project1-1.prj, 2
project2-1.prj, 34
project2-2.prj, 46, 332
project2-3.prj, 50
project3Receive.prj, 90
project3Send.prj, 78, 90
project4-1.prj, 134
project5.prj, 197
project6-1.prj, 249, 256
project6-2.prj, 256
project9.prj, 367
protocol, 77
Pulse Generation, 268–72
Pulse Measurement, 272–73
pulse period
measure, 272
pulse polarity
definition, 269
pulse train, 268
pulse width
measure, 272
pulse width/period
options, 272
type of, 272
PulseWidthOrPeriodMeasConfig, 272, 273, 294
PXI, 191

Q

Quick Edit Window, 133

R

range tables, 333
dialog window, 354
Range, Gain and Code Width, 226–27
ReadFromDigitalLine, 262, 296
ReadFromDigitalPort, 262, 297
Real-Time System Integration. *See* RTSI
Rect
structure, 163
referenced single-ended, 225
RefreshGraph, 21, 30
rendering, 366
Reset Command
dialog box, 340
Reset Delay(s)
control box, 343
Resistance Temperature Detector. *See* RTD
Resizing Rows and Columns, 124
Resman, 189, 190, 192
caveats, 190
resolution, 225, 226
Resource Descriptor
combo box, 343
resource name, 347
RS-485, 221
RSE. *See* referenced single-ended
RTD, 224
RTSI
use, 223
RunUserInterface, 82

S

sample and hold circuitry, 223
sampling rate, 225
scale types
linear, 241
polynomial, 241
table scale, 241
scaling factor, 241

SCPI, 208
  commands syntax, 338
  format, 338
SCXI
  chassis, 221
  modules, 221
SCXI chassis folder, 237
Select an Instrument Driver
  dialog window, 336
Selection State, 123
Self Test
  dialog window, 340
Self Test Delay(s)
  control box, 343
Self-Test Command
  dialog box, 340
Self-Test Response Contents
  control box, 340
serial interface, 221
SetActiveTableCell, 146
SetAttribute, 350
SetAxisScalingMode, 42, 73
SetBreakOnLibraryErrors, 89, 118
SetCtrlAttribute, 9, 10, 22, 48, 66, 124, 125, 126
SetCtrlVal, 42
SetDir, 88, 120
SetPanelAttribute, 8, 10, 30
SetTableCellAttribute, 128, 142, 146, 166
SetTableCellRangeAttribute, 128, 150, 168
SetTableCellRangeVals, 141, 169
SetTableCellVal, 141, 166
SetTableColumnAttribute, 124, 125, 138, 169
SetTableRowAttribute, 124, 125, 126, 127, 170
settling time, 226
  meaning, 226
SetWaitCursor, 383, 398
shininess, 380
ShowBuiltInCtrlMenuItem, 170
Signal Conditioning, 220, 223–24
  Amplification, 224
  Filtering, 224
  Isolation, 224
  Linearization, 224
  Transducer Excitation, 224
  where used, 223
    amplify signals, 224
    different inputs/outputs, 223
    filter unwanted noise, 224
    isolation, 224
    multiplex channels, 223
    noisy environment, 223
Signal Conditioning eXtensions for Instrumentation. *See* SCXI
Signal-to-Noise ratio, 224
Silicon Graphics, 366
simulation, 333
SinePattern, 20, 34, 41, 47, 49
Size Mode, 128, 132
  dialog box, 127, 128, 132
  options, 127, 131
Size/Scroll Options
  command button, 132
  dialog window, 132
slew rate, 226
Slot Zero Controller, 177
Software folder, 230
SortTableCells, 145, 146, 172
Source Editor, 1, 2, 352
specular reflectance, 380
specular reflectance factor, 380
Standard Commands for Programmable Instrumentation. See SCPI
Standard Operations
  dialog window, 339
state caching, 333
STC. *See* Am9513A System Timing Controller
strip chart, 258
surface
  opaqueness, 380
  shininess, 380

surface, *(continued)*
  specular reflectance, 380
  transparency, 380
switch, 333
Sysreset line, 191
SYST:ERR?, 342

T

Table Column Controls
  Insert Row Above, 130
  Insert Row Below, 130
table control
  Cell Appearance
    background color, 129
    grid lines, 129
    horizontal grid, 129
    vertical grid, 129
  cell mode, 129
  cell types
    numeric cells, 122, 123
    picture cells, 122, 123
    string cells, 122, 123
  Column group, 130
  creating, 125
  default attributes, 125
  default state
    Selection State, 123
  Edit Default Cell Values
    dialog window, 128
  Edit Row dialog window, 127
    Height dialog box, 127
    Size Mode, 127
  manipulating
    cells, 122
    columns, 122
    rows, 122
  navigation, 123
  non-existent prior to CVI 5.5, 121
  Numeric Attributes group, 129
    Show Inc/Dec Arrows, 129
  resizing
    column width, 124
    row height, 124
    using mouse, 124
  Row group, 127
  search data, 122
  sort data, 122
  states
    Edit State, 123
    Selection State, 123
  String Attributes group, 129
    Wrap Mode, 129
      options, 129
  Type of data, 128
Table Control Basics, 123–24
Table Control Events, 125
Table Control Overview, 122
Table Control Project, 134–35
Table Control States, 123
Table Mode, 125
  Column mode, 125
  default setting, 125
  dialog box, 125
  Grid mode, 125
  inheritance, 125
  Row mode, 125
Target Directory
  dialog box, 338
task identifier, 248, 254
TCP, 76
TCP/IP, 76
test
  DAQ channels, 243
  dialog window, 343
Test Results
  window, 343

Testing the Function Panel Functions, 319–21
Testing the Instrument Driver, 362–63
texel, 366
texture coordinates, 366
texture element. *See texel*
texture map, 366
The DAQ Designer Tool, 227–28
timebase, 273
timebase source
  units of, 271
TIO ASCI, 265
transducer/sensor, 220
Transmission Control Protocol. *See* TCP
transparency, 380
Triple Output DC Power Supply, 335, 356
TST?, 340
Type, 128
Type of Instrument
  list box, 337

U

Uniform, 15
Unipolar, 232
Unipolar External Reference, 232
UpdateDSCallback, 81, 84, 85
URL, 76, 77
URL scheme, 77
USB, 221
User Interface Editor, 1, 8, 125, 367
User Interface Library
  error codes, 372
userint.h, 164
Using DAQ Library Functions, 248–74
Using the Attribute Editor, 347–50
Using The DAQ Channel Wizard, 238–46, 253
Using the System Clipboard, 124

V

VAL_CELL_NUMERIC, 129
VAL_CELL_PICTURE, 129
VAL_CELL_STRING, 129
VAL_SIZE_TO_CELL_IMAGE, 128
VAL_SIZE_TO_CELL_IMAGE_AND_TEXT, 128
VAL_SIZE_TO_CELL_TEXT, 128, 130
VAL_TABLE_COLUMN_RANGE, 164
VAL_TABLE_ENTIRE_RANGE, 164
VAL_TABLE_ROW_RANGE, 164
VAL_USE_EXPLICIT_SIZE, 124, 128, 132
vertex, 366
viAssertUtilSignal, 190, 210
viClose, 192, 197, 208, 211
viFindNext, 200, 212
viFindRsrc, 200, 212
viIn, 205
viIn16, 196, 212
viOpen
  Access Mode argument, 194
  DefaultRM argument, 192
  Instrument_Descriptor argument, 192
  InstrumentHandle argument, 194
  Timeout argument, 194
viOpen, 192, 194, 195, 196, 197, 202, 204, 212
viOpenDefaultRM, 192, 197, 200, 202, 215
viOut, 205
viOut16, 196, 215
  generic form, 196
viOut32, 196
viOut8, 196
viRead
  generic form, 195
viRead, 204, 205, 215
VISA, 175, 177, 191, 332, 340, 342
  basic steps of communicating, 191

VISA, *(continued)*
communicates with
VXI, GPIB, Serial interfaces, 191
data types, 191
assigned to instrument driver attributes, 346
ViSession, 200
ViStatus, 198
functions, 191
high level API, 191
High-Level Access, 195
High-Level Access/Low_Level Access
differences between, 195
interface independent, 191
Low-Level Access, 195
object-oriented language, 191
operations, 191
platform independent, 191
portability, 191
Resource, 191
session
meaning of, 192
standard, 177
status codes, 200
size, 200
VISA Default Resource Manager, 191, 192
different from VXI Resource Manager, 192
VISA Resource
INSTR, 194
viSetAttribute, 203, 216
viStatusDesc, 204, 217
viWrite
generic form, 195
viWrite, 195, 196, 204, 218
VME, 175
VMEbus, 176, 178
VPP-4.3, 177
VXI, 175, 191
A Short History, 176–77
address space
decoding, 183
Address Space and Configuration Registers, 181–86
Chassis, Modules and Connectors, 177–78
Communicating with Message-Based Devices, 187–89
Controlling the VXI System, 178–80
Device Class
decoding, 186
Extended, 183, 187
Memory, 183, 187
Message-Based, 183, 187
Message-Based Devices. *See* also VXI - Communicating with Message-Based Devices
Register-Based, 183, 187
Device Classes, 186–87
Introduction to VXI, 176
Resource Manager, 183, 186, 188, 189, 190
allocation of interrupt lines, 190
functions, 190
location, 189
power cycle, 190
sends word serial command, 190
sets up Commander/Servant hierarchy, 190
Resource Manager interaction with
ID/Logical Address Register, 190
MODID lines, 190
Status/Control Register, 190
VXI *chassis*, 177
VXI configuration registers, 181
A24/A32 Offset Register, 185
Device Type Register
Device Type, 184
memory computations, 184
Model Code, 184

Required Memory/Model Code, 183
ID/Logical Address Register, 183, 184, 185, 186
Address Space
decoding, 183
Address Space, 183
Device Class, 183
Device Type, 184
Logical Address, 183, 184
Manufacturer ID Code, 183
operation mode, 181
purpose, 183
Status/Control Register
A24/A32 Active, 185
A24/A32 Enable, 185
MODID, 185
Pass, 185
Ready, 185
Sysfail Inhibit, 185
Status/Control Register, 185
VXI Resource Manager, 192
VXI system
controlling using
High-speed MXIbus link, 179
VXI mainframe linked through GPIB, 179
VXI-based embedded computer, 178
VXIbus, 176, 181
base level commands, 188
board connectors, 178
board connectors
J (plug), 178
P (plug), 178
board width, 178
boards, 177
cards, 177
commander
definition, 187
Commander/Servant hierarchy, 187
hardware, 176
multimaster backplane, 187
number of slots, 178
servant devices, 188
what is, 176
VXIbus modules, 177
VXIbus protocol
extended longword serial, 188
longword serial, 188
word serial, 188
VXI-MXI extender, 179
VXIplug&play
functions implemented, 345
model, 332
skeleton driver files, 337
VXI*plug&play* Consortium, 176
VXIplug&play standard
required functions
*Prefix*_close, 345
*Prefix*_error_message, 345
*Prefix*_error_query, 345
*Prefix*_init, 345
*Prefix*_reset, 345
*Prefix*_revision_query, 345
*Prefix*_self_test, 345

W

What is a Function Panel?, 302–4
WhiteNoise, 14
Width, 132
dialog box, 132
wireframe, 380
WriteToDigitalLine, 262, 263, 298
WriteToDigitalPort, 262, 263, 299

Z

zoom
meaning, 34

The Author

Shahid F. Khalid has more than 30 years of experience in software engineering. He is presently a Senior Engineering Specialist at Boeing, Canoga Park, California, where he has worked for the past five years. He is presently working with Boeing Commercial Avionics Systems (BCAS), Seattle, Washington, on the design and implementation of the Next Generation Automatic Test Equipment software using *LabWindows/CVI*. In the past, at Boeing, Shahid has designed and written software in *LabWindows/CVI* for: testing the Delta IV rocket engine controller; creating applications for laser beam alignment, test equipment diagnostics software for VXI platforms; and data base application; and has worked on company's proprietary projects.

Mr. Khalid is the author of the book *LabWindows/CVI Programming for Beginners*, which is the first book written on *LabWindows/CVI*.

Shahid has worked as a software contractor in aerospace and defense with companies like GTE Government Systems and Loral Electro-Optical Systems. He has worked at Logicon Incorporated and at Jet Propulsion Laboratories in Pasadena, California. He has been a Unit Head of a software development team at Singer-Kearfott in New Jersey, where he was involved with the development of test software for the inertial navigation guidance for the US Navy Mark VI program. Shahid has worked as an analog circuit design engineer and computer programmer at various companies in Kansas, Tennessee, and New Jersey.

Shahid has a Bachelor's degree in Physics and Mathematics, and a Master's degree in Mathematics from Panjab University, Pakistan. He also has a Master's Degree in Electronic Engineering from Kansas University, Kansas, and a Master's Degree in Computer Science from Stevens Institute of Technology, New Jersey.

He is married and is the father of two daughters and a son. After living in Pakistan, Kansas, Tennessee, and New Jersey, he now lives in Agoura Hills, California. When he is not writing books, he likes to travel to different countries, swim, and read books on computers and programming languages.

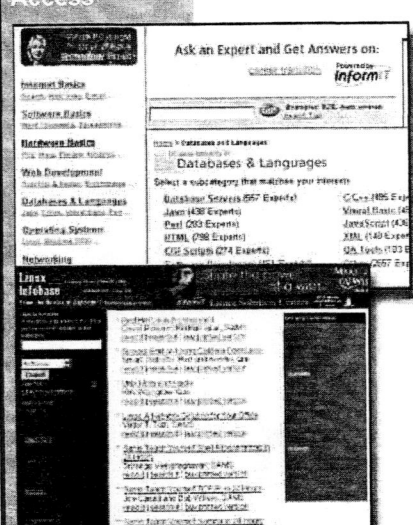

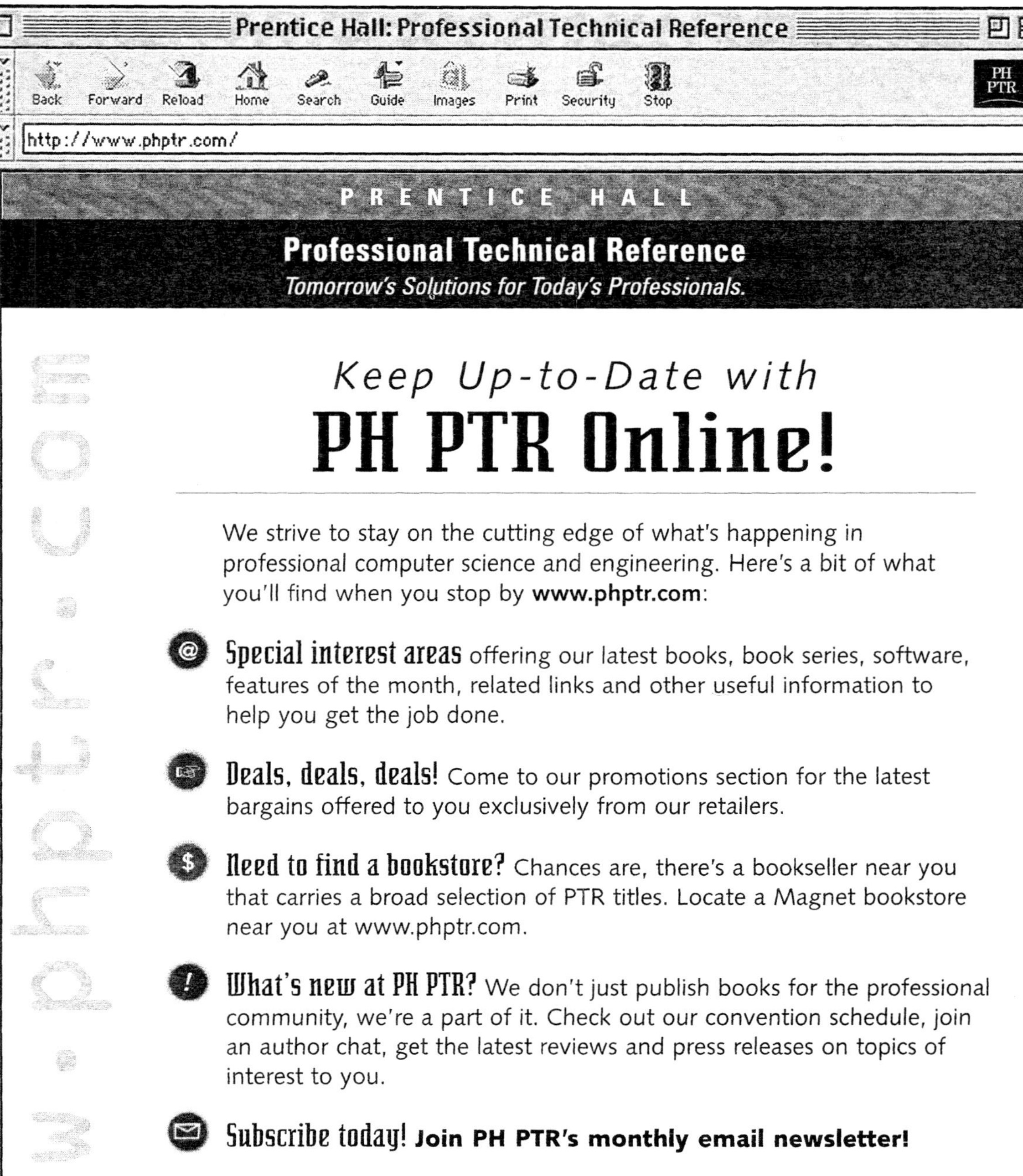
Prentice Hall: Professional Technical Reference
Back Forward Reload Home Search Guide Images Print Security Stop
PH PTR
http://www.phptr.com/
PRENTICE HALL
Professional Technical Reference
Tomorrow's Solutions for Today's Professionals.
www.phptr.com
Keep Up-to-Date with
PH PTR Online!
We strive to stay on the cutting edge of what's happening in professional computer science and engineering. Here's a bit of what you'll find when you stop by www.phptr.com:
Special interest areas offering our latest books, book series, software, features of the month, related links and other useful information to help you get the job done.
Deals, deals, deals! Come to our promotions section for the latest bargains offered to you exclusively from our retailers.
Need to find a bookstore? Chances are, there's a bookseller near you that carries a broad selection of PTR titles. Locate a Magnet bookstore near you at www.phptr.com.
What's new at PH PTR? We don't just publish books for the professional community, we're a part of it. Check out our convention schedule, join an author chat, get the latest reviews and press releases on topics of interest to you.
Subscribe today! Join PH PTR's monthly email newsletter!
Want to be kept up-to-date on your area of interest? Choose a targeted category on our website, and we'll keep you informed of the latest PH PTR products, author events, reviews and conferences in your interest area.
Visit our mailroom to subscribe today! http://www.phptr.com/mail_lists

LICENSE AGREEMENT AND LIMITED WARRANTY

READ THE FOLLOWING TERMS AND CONDITIONS CAREFULLY BEFORE OPENING THIS SOFTWARE PACKAGE. THIS LEGAL DOCUMENT IS AN AGREEMENT BETWEEN YOU AND PRENTICE-HALL, INC. (THE "COMPANY"). BY OPENING THIS SEALED SOFTWARE PACKAGE, YOU ARE AGREEING TO BE BOUND BY THESE TERMS AND CONDITIONS. IF YOU DO NOT AGREE WITH THESE TERMS AND CONDITIONS, DO NOT OPEN THE SOFTWARE PACKAGE. PROMPTLY RETURN THE UNOPENED SOFTWARE PACKAGE AND ALL ACCOMPANYING ITEMS TO THE PLACE YOU OBTAINED THEM FOR A FULL REFUND OF ANY SUMS YOU HAVE PAID.

1. **GRANT OF LICENSE:** In consideration of your payment of the license fee, which is part of the price you paid for this product, and your agreement to abide by the terms and conditions of this Agreement, the Company grants to you a nonexclusive right to use and display the copy of the enclosed software program (hereinafter the "software") on a single computer (i.e., with a single CPU) at a single location so long as you comply with the terms of this Agreement. The Company reserves all rights not expressly granted to you under this Agreement.

2. **OWNERSHIP OF SOFTWARE:** You own only the magnetic or physical media (the enclosed software) on which the software is recorded or fixed, but the Company retains all the rights, title, and ownership to the software recorded on the original software copy(ies) and all subsequent copies of the software, regardless of the form or media on which the original or other copies may exist. This license is not a sale of the original software or any copy to you.

3. **COPY RESTRICTIONS:** This software and the accompanying printed materials and user manual (the "Documentation") are the subject of copyright. You may not copy the Documentation or the software, except that you may make a single copy of the software for backup or archival purposes only. You may be held legally responsible for any copying or copyright infringement which is caused or encouraged by your failure to abide by the terms of this restriction.

4. **USE RESTRICTIONS:** You may not network the software or otherwise use it on more than one computer or computer terminal at the same time. You may physically transfer the software from one computer to another provided that the software is used on only one computer at a time. You may not distribute copies of the software or Documentation to others. You may not reverse engineer, disassemble, decompile, modify, adapt, translate, or create derivative works based on the software or the Documentation without the prior written consent of the Company.

5. **TRANSFER RESTRICTIONS:** The enclosed software is licensed only to you and may not be transferred to any one else without the prior written consent of the Company. Any unauthorized transfer of the software shall result in the immediate termination of this Agreement.

6. **TERMINATION:** This license is effective until terminated. This license will terminate automatically without notice from the Company and become null and void if you fail to comply with any provisions or limitations of this license. Upon termination, you shall destroy the Documentation and all copies of the software. All provisions of this Agreement as to warranties, limitation of liability, remedies or damages, and our ownership rights shall survive termination.

7. **MISCELLANEOUS:** This Agreement shall be construed in accordance with the laws of the United States of America and the State of New York and shall benefit the Company, its affiliates, and assignees.

8. **LIMITED WARRANTY AND DISCLAIMER OF WARRANTY:** The Company warrants that the software, when properly used in accordance with the Documentation, will operate in substantial conformity with the description of the software set forth in the Documentation. The Company does not warrant that the software will meet your requirements or that the operation of the software will be uninterrupted or error-free. The Company warrants that the media on which the software is delivered shall be free from defects in materials and workmanship under normal use

for a period of thirty (30) days from the date of your purchase. Your only remedy and the Company's only obligation under these limited warranties is, at the Company's option, return of the warranted item for a refund of any amounts paid by you or replacement of the item. Any replacement of software or media under the warranties shall not extend the original warranty period. The limited warranty set forth above shall not apply to any software which the Company determines in good faith has been subject to misuse, neglect, improper installation, repair, alteration, or damage by you. EXCEPT FOR THE EXPRESSED WARRANTIES SET FORTH ABOVE, THE COMPANY DISCLAIMS ALL WARRANTIES, EXPRESS OR IMPLIED, INCLUDING WITHOUT LIMITATION, THE IMPLIED WARRANTIES OF MERCHANTABILITY AND FITNESS FOR A PARTICULAR PURPOSE. EXCEPT FOR THE EXPRESS WARRANTY SET FORTH ABOVE, THE COMPANY DOES NOT WARRANT, GUARANTEE, OR MAKE ANY REPRESENTATION REGARDING THE USE OR THE RESULTS OF THE USE OF THE SOFTWARE IN TERMS OF ITS CORRECTNESS, ACCURACY, RELIABILITY, CURRENTNESS, OR OTHERWISE.

IN NO EVENT, SHALL THE COMPANY OR ITS EMPLOYEES, AGENTS, SUPPLIERS, OR CONTRACTORS BE LIABLE FOR ANY INCIDENTAL, INDIRECT, SPECIAL, OR CONSEQUENTIAL DAMAGES ARISING OUT OF OR IN CONNECTION WITH THE LICENSE GRANTED UNDER THIS AGREEMENT, OR FOR LOSS OF USE, LOSS OF DATA, LOSS OF INCOME OR PROFIT, OR OTHER LOSSES, SUSTAINED AS A RESULT OF INJURY TO ANY PERSON, OR LOSS OF OR DAMAGE TO PROPERTY, OR CLAIMS OF THIRD PARTIES, EVEN IF THE COMPANY OR AN AUTHORIZED REPRESENTATIVE OF THE COMPANY HAS BEEN ADVISED OF THE POSSIBILITY OF SUCH DAMAGES. IN NO EVENT SHALL LIABILITY OF THE COMPANY FOR DAMAGES WITH RESPECT TO THE SOFTWARE EXCEED THE AMOUNTS ACTUALLY PAID BY YOU, IF ANY, FOR THE SOFTWARE.

SOME JURISDICTIONS DO NOT ALLOW THE LIMITATION OF IMPLIED WARRANTIES OR LIABILITY FOR INCIDENTAL, INDIRECT, SPECIAL, OR CONSEQUENTIAL DAMAGES, SO THE ABOVE LIMITATIONS MAY NOT ALWAYS APPLY. THE WARRANTIES IN THIS AGREEMENT GIVE YOU SPECIFIC LEGAL RIGHTS AND YOU MAY ALSO HAVE OTHER RIGHTS WHICH VARY IN ACCORDANCE WITH LOCAL LAW.

ACKNOWLEDGMENT

YOU ACKNOWLEDGE THAT YOU HAVE READ THIS AGREEMENT, UNDERSTAND IT, AND AGREE TO BE BOUND BY ITS TERMS AND CONDITIONS. YOU ALSO AGREE THAT THIS AGREEMENT IS THE COMPLETE AND EXCLUSIVE STATEMENT OF THE AGREEMENT BETWEEN YOU AND THE COMPANY AND SUPERSEDES ALL PROPOSALS OR PRIOR AGREEMENTS, ORAL, OR WRITTEN, AND ANY OTHER COMMUNICATIONS BETWEEN YOU AND THE COMPANY OR ANY REPRESENTATIVE OF THE COMPANY RELATING TO THE SUBJECT MATTER OF THIS AGREEMENT.

Should you have any questions concerning this Agreement or if you wish to contact the Company for any reason, please contact in writing at the address below.

Robin Short
Prentice Hall PTR
One Lake Street
Upper Saddle River, New Jersey 07458

About the CD

The CD-ROM included with *Advanced Topics in LabWindows/CVI* contains the following:

The accompanying CD-ROM includes a trial version of LabWindows/CVI 6.0 plus all the project files discussed in this book. All the projects on the CD have been run and tested thoroughly and neither the author nor Prentice Hall is liable for any errors or consequential damages occurring from the use of the software.

The "AdvProjects" folder must be copied to the hard disk and the Read-only and the Archive attributes must be removed from all the files before running the projects. These projects have been created and run using the full version of LabWindows/CVI 5.5 and should have no problems running with the full version of LabWindows/CVI 6.0. However, due to the disabling of some of the menu commands in the evaluation version of LabWindows/CVI 6.0 you may not be able to run some of the projects in their entirety.

HiQ is not functional on the CD. If you are interested in HiQ, please contact National Instruments at www.ni.com.

The CD-ROM can be used on Microsoft Windows® 95/98/NT/2000®

License Agreement

Use of the software accompanying *Advanced Topics in LabWindows/CVI* is subject to the terms of the License Agreement and Limited Warranty, found on the previous two pages.

Technical Support

Prentice Hall does not offer technical support for any of the programs on the CD-ROM. However, if the CD-ROM is damaged, you may obtain a replacement copy by sending an email that describes the problem to: disc_exchange@prenhall.com.